Potato in progress: science meets practi

Organized by:

Agricultural Promotion Projects (APP) Emmeloord
P.O. Box 822
3700 AV Zeist
The Netherlands
tel: +31 30 6933 489
fax: +31 30 6974 517
www.potato2005.com
www.potatoreporteronline.com

Potato in progress

science meets practice

edited by:
A.J. Haverkort
P.C. Struik

Wageningen Academic
Publishers

Subject headings:
Marketing
Agronomy
Crop protection

ISBN 9076998841

First published, 2005

Wageningen Academic Publishers
The Netherlands, 2005

National Recommendation Committee Potato 2005

- Drs. R.P.J. Bol, Director, Department of Trade and Industry, Ministry of Agriculture, Nature and Food Quality
- M.J.E.M. Jager, Queens Commissioner Province of Flevoland
- Mr. W.L.F.C. ridder van Rappard, Mayor of Municipality of Noordoostpolder
- Th.A.M. Meijer, Chairman Product Boards Arable Products
- G.J. Doornbos, President of the Dutch Organisation for Agriculture and Horticulture (LTO Nederland)
- M.J. Varekamp, Chairman of Dutch Potato Organisation (NAO)

Advisory committee Potato 2005 congress

- A. Andreatta, HZPC representative, Brazil
- Dr. H. Böhm, Europlant Pflanzenzucht GmbH, Germany
- Y. Cohen, Mehadrin-Tnuport, Israel
- Mr. R. van Diepen, Dutch Potato Organisation (NAO), the Netherlands
- I. Dykes, British Potato Council (BPC), UK
- R.H. Gall, New Zealand Vegetable & Potato Growers Federation (Vegfed), New Zealand
- J.E. Godfrey, R.J. & A.E. Godfrey, United Kingdom
- J.H.J. Haarhuis, Agricultural Promotion Projects, the Netherlands
- Ir. J. Hak, Hak & Partners b.v / GMV (FME CWM)., the Netherlands
- Dr. M. Huarte, National Potato Program Coordinator, INTA Balcarce, Argentina
- F.J. Lawrence, Potatoes South Africa, South Africa
- Ir. P. Oosterveld, (NAK) Dutch General Inspection Service for agricultural seeds and seed potatoes, the Netherlands
- H.W. Platt, Agriculture & Agri-Food Canada, Canada
- Ir. G.H. Siebenga, ABN AMRO, the Netherlands
- Dr. H.G. Zandstra, International Potato Center (CIP), Peru

Scientific committee Potato 2005 congress

- Prof. Dr. A.F. Abou-Hadid, Central Laboratory for Agricultural Climate, Egypt
- Dr. P. Anderson, International Potato Center (CIP), Peru
- Dr. B.V. Anisimov, Ministry of Agriculture of Russian Federation, Russia
- Prof. Dr. J.F. Guenthner, Idaho Center for Potato Research & Education, University of Idaho, USA
- Prof. Dr. A. Hanafi, Institut Agronomique et Veterinaire Hassan II, Morocco
- Dr. A.J. Haverkort, Plant Research International, the Netherlands
- Ing. M. Martin, Arvalis - Institut du Végétal / ITPT, France
- Dr. ir. I. Mastenbroek, (NAK) Dutch General Inspection Service for agricultural seeds and seed potatoes, the Netherlands
- Prof. Dr. Ir. P.C. Struik, Wageningen University, the Netherlands
- Dr. Ir. A. Veerman, Applied Plant Research (PPO), the Netherlands

Agriculture Promotion Projects (APP) - Board of directors

- Johan H.J. Haarhuis, Chairman, the Netherlands
- Herman P. Fousert, Treasurer, the Netherlands
- Jan van der Endt, Secretary, the Netherlands
- Arie H. Kauffman, Honorary Chairman, the Netherlands

Table of contents

Preface

When the organizers of the Potato 2005 congress discussed with the scientific committee on the topic of the congress we decided that we would not focus on one single scientific discipline. The audience and contributors represented a wide diversity of interests ranging from breeding to consumer behaviour. Each stakeholder in the potato industry is interested to know what science has to offer, and scientists are interested in knowing what drives the industry. The congress offered an excellent platform to present highlights in science and practice and also hosted a meeting of the Global Initiative on Late Blight (GILB).

The 40 papers presented in the seven sessions of this book reflect the rapid developments in the potato industry. The nutritional value of the tuber, volatile consumer moods and saturating and upcoming markets are discussed under the headings of health and consumer behaviour, and trade. In Western Europe and the United States markets of traditional potato products have reached a maximum and consumers look for alternatives taking into account health and diversity in food products to choose from. In upcoming markets, especially in Asia, potato is rapidly becoming part of daily meals and snacks. As a consequence areas and yields increase and efficient supply chains develop everywhere in the world.

The need to optimize the efficient use of resources to increase sustainability is reflected in the sections dealing with breeding and seed production, decision support systems, and production and storage. Consumers and processors increasingly demand the raw material to be produced in a sustainable way. The trend is that - like food safety - companies will bring this in the pre-competitive domain: sustainably produced beyond doubt.

The most important potato disease - late blight - is dealt with in the proceedings of the GIBL workshop. This disease is costing the industry worldwide over € 5 billion and continues to be a serious threat. Many disciplines such as genomics, population dynamics and chemistry cooperate intimately in joint research programmes to combat the disease. Papers on other diseases and pests are grouped in a section on crop protection.

Potato is progressing globally - with increasing impact on food supply and added value - providing many opportunities for science to meet practice.

A.J. Haverkort
P.C. Struik
(editors)

Health and consumer behaviour

Nutritionally relevant aspects of potatoes and potato constituents

H.J. Buckenhüskes
German Agricultural Society, DLG-Competence Centre Agriculture and Food Business, Eschborner Landstraße 122, D-60489 Frankfurt, Germany

Abstract

Parallel to the increasing standard of living in the western countries the consumption of potatoes decreased steadily. The reasons for that are multifarious; however, one of the major arguments should be the prejudice that potatoes are causing overweight. Therefore it is the objective of this paper to discuss some nutritionally relevant aspects of potatoes and potato constituents. Special attention will be given to the biological value of potato protein, the question of carbohydrate content and the glycaemic index, biological active substances as well as the occurrence of acrylamide in some potato products.

Keywords: Potato, protein, biological value, glycaemic index, biological active substances, acrylamide

Introduction

The potato crop is indigenous in various parts of South America and it was introduced into Europe through two independent ways: The Spaniards are believed to have first brought it from Peru to Spain between 1540 and 1560. On the other hand it was brought from Venezuela to England by the slave-trader John Hawkins in 1565 (Bücher, 1975). Initially, the crop was grown as an attraction in the pleasure-grounds of princes and kings. Moreover the potato was used as a medicinal plant and therefore grown by pharmacists or physicians, in Spain in particular. Up to the 20th century potatoes and sauerkraut represented the most important sources available protecting man - especially seamen - from scurvy, a fact which later on was learned to be caused by its high content of vitamin C.

However, the large-scale cultivation of the crop began only in the beginning of the 19th century and today potatoes belong to the most important foods of the western civilisation. From a nutritional point of view this advancement was due to the overall nutritive value of the potatoes, which at first is caused by their high content of carbohydrates. In 1979 Teuteberg propounded the hypothesis that the population explosion in some European countries during the late 18th and the early 19th century perhaps was rather caused by the fast increasing consumption of potatoes than by the decline of epidemics or the growing standard of living.

Parallel to the increasing standard of living the consumption of potatoes decreased steadily, especially since the end of the Second World War. The reasons for that are multifarious; however, one of the major arguments should be the prejudice that potatoes are causing overweight and obesity. Therefore it is the objective of this presentation to discuss some nutritionally relevant aspects of potatoes and potato constituents.

Major composition of potatoes

The chemical composition of potatoes - that is its content of nutrients - depends on multifarious factors like variety, the kind of soil on which they were grown, agricultural conditions including fertilization, stage of maturity, storing conditions etc. Therefore it is only feasible to give an idea about the major nutritionally relevant components as it is done in Table 1. In order to enable the nutritional significance of the given data the table is supplemented by recommendations for the daily intake of some nutrients which were published by a joint committee of the Austrian, German and Swiss Societies of nutrition in 2000 (DACH, 2000).

Table 1. Average composition of potatoes and its possible contribution to the recommended nutrient supply of persons between 25 and 51 years of age.

	Raw potato[1]	Recommended intake per day[2]		% of daily intake delivered by 150 g potatoes	
	per 100 g	men	women	men	women
Energy [kcal]	70	2,900	2,300	3.6%	4.6%
Protein [g]	2.0	59	47	5.1%	6.4%
Fat [g]	0.1				
Carbohydrates [g]	14.8				
Water [g]	77.8	2,600	2,250		
Minerals [g]	1.0				
Dietary fibre [g]	2.1	30	30	10.5%	10.5%
Vitamin B_1 [mg]	0.11	1.2	1.0	13.8%	16.5%
Vitamin B_2 [mg]	0.05	1.4	1.2	5.4%	6.3%
Vitamin B_6 [mg]	0.31	1.5	1.2	30.7%	38.3%
Folic acid [mg]	0.02	0.4	0.4	7.5%	7.5%
Pantothenic acid [mg]	0.40	6	6	10%	10%
Vitamin C [mg]	17	100	100	25.5%	25.5%
Vitamin E [mg]	0.05	14	12	0.5%	0.6%
Calcium[mg]	6.4	1,000	1,000	1%	1%
Iron [mg]	0.43	10	15	6.5%	4.3%
Magnesium [mg]	21	350	300	9.0%	10.5%
Manganese [mg]	0.15				
Phosphorous [mg]	50	700	700	10.7%	10.7%
Potassium [mg]	418				
Sodium [mg]	2.7				
Zinc [mg]	0.34	10	7	5.2%	7.4%

[1] Souci *et al.*, 2000.

[2] DACH, 2000.

The data in Table 1 characterise the potato as a low energy and low sodium but highly nutritious food. In fact the potato is a rich source of the vitamins C, B_1, B_2, B_6, and folic acid, the minerals potassium and magnesium, and last but not least of dietary fibre.

The amount of nutrients as well as the energy content of the wide range of potato products depends on a large scale on the kind of product; this means the technological processes and especially the kind of heating applied (see Table 2).

Table 2. Nutrients of differently prepared potatoes (per 100 g edible product; Biesalski, 2005).

	Cooked peeled potatoes	Cooked unpeeled potatoes	Potato dumpling	Potato chips	Creamed potatoes	Fried potatoes	French fried potatoes
Energy [kcal]	68	57	108	535	78	117	124
Protein [g]	2.0	1.6	3.1	5.5	2.1	1.7	2.3
Fat [g]	0.1	0.1	0.6	39.4	2.5	6.7	5.1
Carbohydrates [g]	14.2	11.8	21.8	40.6	11.4	12.1	16.6
Water [g]	79.8	63.3	69.7	7.9	80.7	75.7	71.1
Dietary fibre [g]	2.3	1.8	2.9	3.0	1.6	2.0	2.5
Vitamin B_1 [mg]	0.08	0.07	0.09	0.22	0.08	0.07	0.10
Vitamin B_2 [mg]	0.04	0.03	0.05	0.10	0.06	0.04	0.05
Vitamin B_6 [mg]	0.22	0.21	0.26	0.89	0.21	0.19	0.31
Folic acid [µg]	15	22	17	20	17	14	25
Vitamin C [mg]	12	11	12	8	11	10	17
Vitamin E [mg]	0.1	0.1	0.2	6.1	0.1	1.4	3.2

Potato protein

As indicated in Table 1 potatoes consist only two percent of protein, however, protein of very high biological quality. The quality of a protein depends on its amount and proportion of essential amino acids. This means that a protein that is rich in all of the amino acids essential for the human body would score higher on the scale of biological quality than a protein deficient in one or more essential amino acids. One possibility to compare the biological quality of proteins is the so called biological value. The biological value specifies how much gram of the body nitrogen can be renewed or formed by 100 g of absorbed food nitrogen. With other words the biological value is based on the nitrogen actually retained in the body in relation to the total nitrogen absorbed by the body after digestion (Elmadfa and Leitzmann, 2004):

$$\text{Biological value of a protein} = \frac{\text{Retained nitrogen}}{\text{Absorbed nitrogen}} \times 100$$

If the protein provides all the needs for protein synthesis at a rate equal to protein turnover, then the biological value will be 100. The standard is whole chicken egg protein. Since the protein synthesis of our body depends on our age, the biological value of protein is calculated with reference to growth in the young.

From Table 3 it can be seen, that the potato protein is of very high biological value which of course additionally can be improved by combination with some specific other food proteins. This fact should be of interest under circumstances which may stride along with protein deficiencies, e.g. illness or strict vegetarian diet where people eat no animal foods whatever.

Table 3. Biological value of proteins from different sources (Kasper, 2004).

Source	Biological value
Whole egg and potato (35%/65%)	137
Whole egg and milk (71%/29%)	122
Whole egg and wheat (86%/32%)	118
Whole chicken egg	100
Potatoes	90 - 100
Cow milk	84 - 88
Beef	83 - 92
Edamer cheese	85
Soybean	84
Rice	83
Fish	83
Rye flour	76 - 83
Maize	72 - 76
Beans	73
Wheat flour	59

Glycaemic index

With an average amount of 15% carbohydrates - which for its part mainly consists of starch - potatoes belong to the carbohydrate rich victuals. As mentioned above, this fact was the basis of its today's importance. On the other hand the same fact is the background for the prejudice that potatoes are causing overweight and obesity.

Background for this discussion is the fact that carbohydrate containing meals produce an increase in blood glucose, which for its part depends on several factors such as sex, the consistency of food, the kind of carbohydrates, accompanied fat and protein, the presence of alcohol, etc. (Elmadfa and Leitzmann, 2004). One possible indicator of the relative glycaemic response to dietary carbohydrates is the glycaemic index, which first was described in the 1980s and which today is discussed very vehemently.

To determine the glycaemic index (GI) of a victual, probationers are given on the one day a portion of the respective food that provides 50 grams of carbohydrate and on another day a control food - white bread or pure glucose - that provides the same amount of carbohydrate. Blood samples from the candidates are taken prior to eating and at regular intervals after eating over the next two hours. The determined changes in blood glucose over the time are plotted as a curve. The glycaemic index is calculated as the area under the glucose curve after the respective food is eaten, divided by the corresponding area after the control food is eaten. The quotient is multiplied by 100 to represent a percentage of the control food. For example, a glycaemic index of 65% means that the blood sugar response of the investigated food is 65% of the response of 50 g glucose or white bread.

The consumption of high-glycaemic index foods results in higher and more rapid increases in blood glucose levels than the consumption of low-glycaemic index foods. From a physiological point of view a rapid increase in blood glucose leads to an increased secretion of insulin by the pancreas and an inhibition of glucagon release, which is the antagonist of insulin. The physiological processes induced by this situation may cause a sharp decrease in blood glucose levels (hypoglycaemia) followed by the feeling of hunger and the need of fresh and extended food intake.

However, these relations are not only a question of the glycaemic index but just so a question of the amount of food eaten. In order to take both aspects into consideration, another concept was established the so called glycaemic load (GL). The glycaemic load of a food is calculated by multiplying the glycaemic index by the amount of carbohydrate in grams provided by a food and dividing the total by 100. This means that each unit of the glycaemic load represents the equivalent blood glucose-raising effect of 1 gram of pure glucose or white bread. The practical relevance can be seen with the following example: The carbohydrate in watermelon has a high GI. But since there isn't a lot of it, the watermelon's glycaemic load is relatively low.

From a practical point of view foods are divided into three groups: A GI of more than 70 is said to be high, a GI of 55 to 70 is medium, and a GI of less than 55 is low. Otherwise a GL of 20 or more is high, a GL of 11 to 19 inclusive is medium, and a GL of 10 or less is low.

The glycaemic index and the glycaemic load of some potato products are summarized in Table 4. It can be seen that the glycaemic index as well as the glycaemic load are depending on the kind of product and the mean of production and/or preparing.

Table 4. Glycaemic index and glycaemic load of some potato products (Foster-Powell et al.*, 2002).*

Product	Glycaemic index	Glycaemic load
Glucose	100	10
Creamed potatoes	85	17
Baked potatoes	85	26
French fries	75	22
Cooked peeled potatoes	50	14

Bioactive substances

Besides the traditional nutrients water, carbohydrates including dietary fibre, protein, fat, vitamins, minerals, and trace elements, the plants are provided with numerous substances which are produced by their secondary metabolism and which are designated as secondary plant products or phytochemicals. The total number of such phytochemicals is supposed to lie between 60,000 - 100,000. According to Ames (1983) a well balanced human diet will contain around 1.5 g of such substances per day. In nature these phytochemicals fulfil biological functions like antagonistic activities against insects or microorganisms, growth regulators or attractants. That some of these substances possess considerable effects on the health and wellbeing of man - that means that they are biologically active - belongs to the primary experiences of the mankind. Epidemiological studies have shown that the consumption of fruits and vegetables and therefore the intake of increased amounts of phytochemicals is associated with reduced risk of some chronic diseases.

Although the research on the effects of biologically active substances was intensified during the last decades it is no question that there are more open questions than answered ones. Table 5 presents an overview over the most important classes of bioactive substances as well as their supposed effects on the human body. Two of these groups are present in potatoes in considerable amounts: carotinoides and polyphenols.

The total amount of carotinoides of potatoes depends at first on the variety and further on factors like the degree of maturity at the time of harvesting or the time and conditions of

Table 5. Bioactive substances (plus dietary fibre) and their supposed effects (Watzl and Leitzmann, 1999).

Bioactive substances	Indications to the following effects									
	A	B	C	D	E	F	G	H	K	L
Carotinoides	X		X		X			X		
Glucosinolates	X	X						X		
Monoterpenoids	X	X								
Phytic acid	X		X		X				X	
Phytoestrogens	X		X							
Phytosterols	X							X		
Polyphenols	X	X	X	X	X	X	X		X	
Protease inhibitors	X		X							
Saponins	X	X			X			X		
Sulfides	X	X	X	X	X	X	X	X		X
Dietary fibre	X				X			X	X	X

A = anticarcinogenic
B = antimicrobial
C = antioxidative
D = antithrombic
E = immunomodulatory
F = anti-flammatory
G = regulating blood pressure
H = cholesterol-lowering
K = regulating blood glucose level
L = digestion-supporting

storing. There is a general correlation that the yellow fleshy varieties contain more carotinoides than the white fleshy. Related to the fresh weight the total carotinoid content may range from 38 to 500 microgram per 100 g of potatoes. The main carotinoides to be found are violaxanthin, lutein, zeaxanthin, antheraxanthin and neoxanthin, whereas ß-carotene is one of the minor components (Adler, 1971; Tevini 1984; Breithaupt and Bamedi, 2002).

Chu *et al.* (2002) investigated the antioxidant and antiproliferative activities on common vegetables and determined in potatoes a total phenolic content of 38 mg/ 100 g. In this case the phenolic content was expressed as milligrams of gallic acid equivalent per 100 g of fresh weight of the edible part of potatoes. Sixty percent of the total amount was found as free phenolics whereas 40% were found in a bound form. The bound phenolics can survive upper gastrointestinal digestion and may ultimately be broken down in the colon by fermentation by the microflora of the large intestine.

In the investigations of Chu *et al.* (2002) potato extracts showed antioxidative activities equivalent to 83 mg of vitamin C /100 g. Approximately 13% of the total antioxidant activity should be contributed by the natural vitamin C content itself, whereas the remaining activity results at first from the phenolic substances and the carotinoides.

Last but not least, Chu *et al.* (2002) have found a slight antiproliferative activity of potato extracts on the growth of $HepG_2$ human liver cancer cells in vitro, however, to get clear evidence further investigations are needed.

Acrylamide

In April 2002 the National Food Administration of Sweden had for the first time drawn the attention to the occurrence of acrylamide in food, a substance which is suspected of being carcinogenic. Up to that time acrylamide was known as a synthetic substance that is used commercially in the production of polyacrylamide or which occurs in tobacco smoke. Within a short time it was clear, that the acrylamide was formed in the foodstuffs during production or preparation processes. By analysing some food which was prepared by frying, roasting or baking acrylamide contents between 100 and more than 2,000 µg/kg were found (Haase *et al.*, 2003).

The mechanism of the formation has not yet been fully explained. However, there is some evidence that the formation is influenced by:

- the occurrence of low molecular proteinous substances, especially asparagine;
- the availability of low molecular carbohydrates, especially reducing sugars like glucose or fructose;
- the temperature;
- a low moisture content (low water activity).

Influenced by the combination of these factors acrylamide should be produced by the mechanisms of the so called Maillard reaction (Figure 1). The Maillard reaction is one of the most important reactions in the formation of colour and flavour of food.

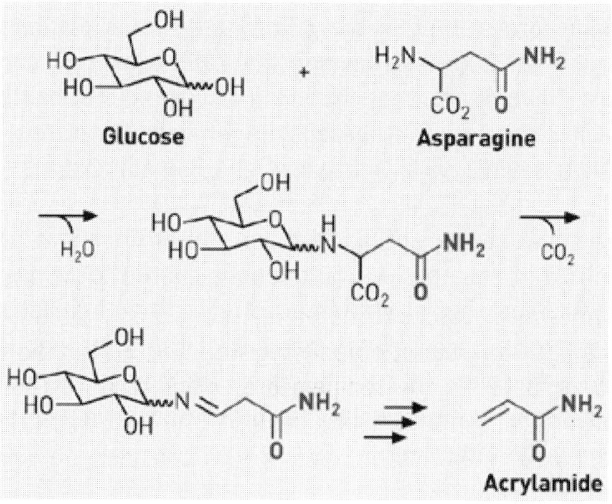

Figure 1. Formation of acrylamide from glucose and asparagine (according to Mottram et al.*, 2002 and Stadler* et al.*, 2002; picture: pubs.acs.org/cen/topstory/ 8040/8040notw2.html).*

Investigations carried out during the last two years have shown that the formation of acrylamide in deep-fat fried potato products is influenced by the raw material (potato variety, growing place) as well as by technological parameters (time, temperature). Concerning the raw material the content of reducing sugars is highly correlated with the degree of acrylamide formation (Figure 2).

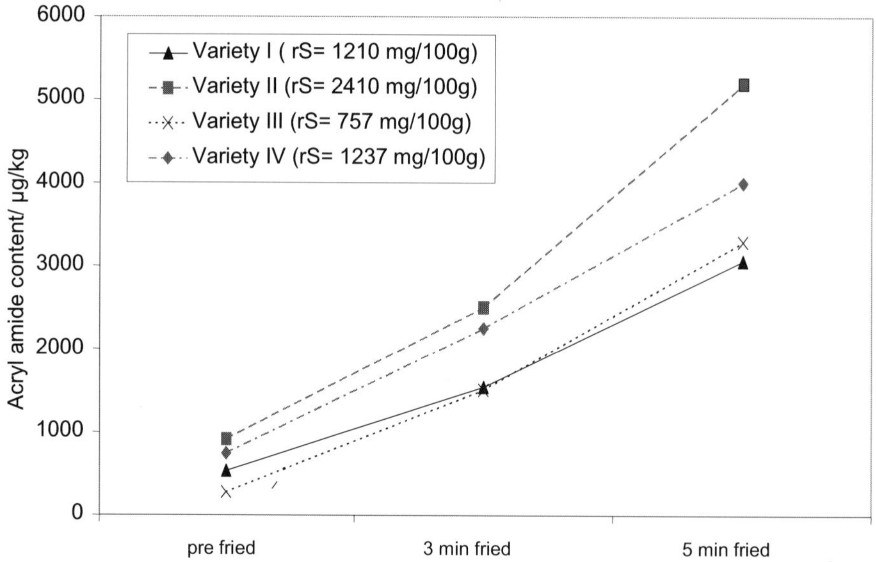

Figure 2. Influence of the content of reducing sugars (rS) of potatoes of unknown variety on the formation of acrylamide after different times of frying (E.H. Reimerdes; K. Franke, personal communication, 20.04.2005)

With regard to this relation it has to be mentioned that the carbohydrates in the potatoes are subjected to a dynamic balance between starch and sugars:

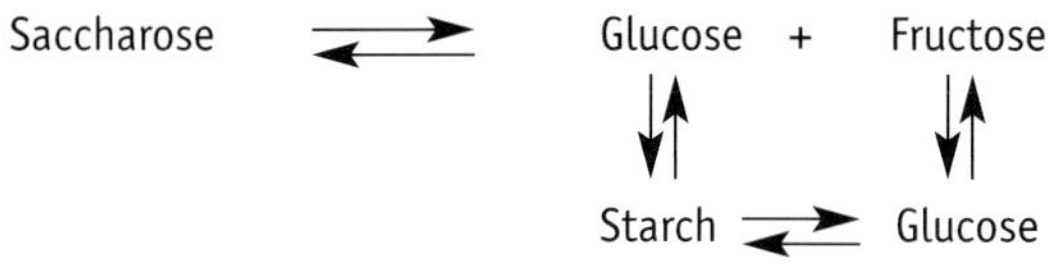

In case of fully ripe harvested potatoes this balance is dependent on the storage temperature. At temperatures below 10 °C the amount of reducing sugars as well as the amount of saccharose increases as more and as faster as the temperature comes to 0 °C (Adler, 1971). Since the formation of acrylamide occurs not only during industrial processing but even during preparing potatoes in the households it seems to be necessary to consider the complete line of potato storage, handling and distribution.

Conclusion

Despite the old prejudice that potatoes are causing overweight and despite the recently raised discussions about low carb food as well as the occurrence of acrylamide in potato products it can be stated that potatoes have been in the past and that they will be in the future a valuable part of a modern, light and healthy nutrition.

References

Adler, G. (1971). Kartoffeln und Kartoffelerzeugnisse (Potatoes and potato products). Paul Parey, Berlin and Hamburg, Germany, 208 pp.

Ames, B.N. (1983). Dietary Carcinogens and Anticarcinogens. Science 221, 1256-1262.

Biesalski, H.-K. (2005). Die Kartoffel - kalorienarmer Nährstofflieferant mit wertvollen Inhaltsstoffen (The potato - low energy nutrient supplier with valuable components). LCI - Moderne Ernährung Heute, April, 1-6.

Breithaupt, D.E. and A. Bamedi (2002). Carotenoids and carotenoid esters in potatoes (Solanum tuberosum L.): New insights into an ancient vegetable. J. Agric. Food Chem. 50, 7175-7181.

Brücher, H. (1975). Domestikation und Migration von Solanum tuberosum L. (Domestication and migration of Solanum tuberosum L.) Kulturpflanze, Band 23, 11-74.

Chu, Y.-F., J. Sun, X. Wu and R.H. Liu (2002). Antioxidant and Antiproliferative Activities of Common Vegetables. J. Agric. Food Chem. 50, 6910-6916.

DACH: Deutsche Gesellschaft für Ernährung, Österreichische Gesellschaft für Ernährung, Schweizerische Gesellschaft für Ernährungsforschung, Schweizerische Vereinigung für Ernährungsforschung (Ed.) (2000). Referenzwerte für die Nährstoffzufuhr (Reference values for nutrient supply). 1st edition, Umschau Braus, Frankfurt, Germany, 240 pp.

Elmadfa, I. and C. Leitzmann (2004). Ernährung des Menschen (Human nutrition). 4th edition. Eugen Ulmer, Stuttgart, Germany, 660 pp.

Haase, N.U., B. Matthäus and K. Vosmann (2003). Minimierungsansätze zur Acrylamid-Bildung in pflanzlichen Lebensmitteln - aufgezeigt am Beispiel von Kartoffelchips (Minimalization concepts for the formation of acryl amide in vegetable foodstuffs - demonstrated an the example of potato chips). Deutsche Lebensmittel-Rundschau 99, 87-90.

Kasper, H. (2004). Ernährungsmedizin und Diätetik (Nutrition medicine and dietetics). 10th edition. Elsevier, München, Germany, 634 pp.

Mottram, D.S., B.L. Wedzicha and A.T. Dodson (2002). Acrylamide is formed in the Maillard reaction. Nature 419, 448-449.

Souci, S.W., W. Fachmann and H. Kraut (2000). Food Composition and Nutrition Tables. 6th edition. Medapharm Scientific Publishers, Stuttgart, 1182 pp.

Stadler, R.H., I. Blank, N. Varga, F. Robert, J. Hau, P.A. Guy, M.-C. Robert and S. Riediker (2002). Acrylamide from Maillard reaction products. Nature 419, 449-450.

Teuteberg, H.J. (1979). Die Rolle von Brot und Kartoffeln in der historischen Entwicklung der Nahrungsgewohnheiten (The role of bread and potatoes in the historical development of nutritional habits). Ernährungs-Umschau 26, 149-154.

Tevini, M., W. Iwanzik and G. Schönecker (1984). Analyse, Vorkommen und Verhalten von Carotinoiden in Kartoffeln und Kartoffelprodukten (Analysis, occurrence and behaviour of carotinoides in potatoes and potato products). In: Forschungskreis der Ernährungsindustrie e.V. Yearbook 1984., Hannover, Germany, 213 pp.

Watzl, B. and C. Leitzmann (1999). Bioaktive Substanzen in Lebensmitteln (Bioactive substances in food). 2nd edition. Hippokrates, Stuttgart, Germany, 254 pp.

Curriculum vitae – Herbert J. Buckenhüskes

Herbert J. Buckenhüskes studied Food Technology at Hohenheim University at Stuttgart, Germany, from which he received the diploma degree in 1979. In 1980 he joined the Research Staff of the Institute of Food Technology, Department for Fruit- and Vegetable Technology of Hohenheim University, from which, in 1984, he received a PhD degree in General Natural Sciences and in 1990 the habilitation in Food Technology. In 1989 he joined the Gewürzmüller company at Stuttgart, Germany, at first he held the position as marketing manager, followed by manager for product development and product management and then as a research coordinator. Since 2004 he is working as an independent consultant for Food Science and since 2005 he is Head of Food Technology of the German Agricultural Society (DLG) at Frankfurt, Germany. Since 1998 Dr. Buckenhüskes is an Extraordinary professor at the Hohenheim University and from 1999-2001 he held the same position at the Suez Canal University of Ismailia, Egypt. Prof. Buckenhüskes was President (2000-2004) and currently serves as Vice-President of the German Association of Food Technologists (GDL). Since 2001 he is chairman of the working group "Ethics in Food Technology" within the GDL.

Company profile – The German Agricultural Society (DLG)

The German Agricultural Society (DLG) was founded in 1885 by Max Eyth. With more than 16,000 members it is one of the leading organisations in the German agricultural and food area. It is a politically and economically independent specialist organisation which is open for anybody to join. The DLG's mandate and understanding of its work is to promote technical and scientific progress. With its activities and initiatives the DLG sets standards and provides impetus for progress. It conducts knowledge sharing world-wide with leading international practitioners and other specialist organisations. The DLG promotes the product quality of foods on the basis of impartial and binding quality standards. Its quality understanding is based on current scientific findings and self-developed, recognised methods. With its international competitions the DLG is Europe's leader in quality assessment.

The potential of potatoes for attractive convenience food: focus on product quality and nutritional value

J. van Gijssel
Agrotechnology and Food Innovations B.V., Wageningen UR, BU Food Quality, P.O.Box 17, 6700 AA Wageningen, The Netherlands

Abstract

Consumer food choices are driven by convenience, health and wellness, pleasure and value.

Health and wellness is related to low calorie, low acrylamide, low Glycaemic Index and high Satiety Index. However, most consumers dislike to be confronted with negative health issues and prefer positive attributes such as convenience, pleasure and value. Convenience asks for ready to heat and ready to eat products with a fresh image and a long shelf-life. Mild preservation techniques and hurdle technology offer great opportunities for convenient processed potato products.

Pleasure and value is very important and the main trigger for success. It is a real challenge to combine attributes such as excellent taste and texture with health issues and convenience.

Keywords: convenience, processing, acrylamide, glycaemic index, satiety

Introduction

Product innovation and product acceptance require an integrated approach. Consumer preferences and perceptions will form the basis for desired technological developments.
The following seven future consumer types are described by Meulenberg (Linnemann *et al.*, 1999):

- Environment-conscious consumer prefers unprocessed foods (fresh) or from short production chains, foods from organic farming, focusses on technological efficiency.
- Nature and animal-loving consumer is interested in methods of primary production, concerned about genetic modification, considers animal welfare an important issue, focusses on ethical efficiency of production systems.
- Health-conscious consumer prefers fresh products that support health trends, e.g. low-calorie, low-fat, rich in vitamins and minerals, and all other sorts of foods with alleged health-protecting or health-promoting properties.
- Convenience consumer goes for snacks, fast food, take-out meals, ready-to-eat meals, foods that are easy to prepare and restaurant food.
- Hedonic consumer prefers (exotic) specialities, delicacies, foods with added value, food as entertainment and pleasant pastime, restaurant food, foods of high sensory quality.
- Price-conscious consumer prefers homemade meals, with ingredients of a favourable price/quality ratio (e.g. products from large-scale production, or alternative, cheaper raw materials).

- Variety-seeking consumer seeks diversity in raw materials, ingredients and fabricated foods for homemade meals, as well as diversity in the type of meal (from elaborate homemade meals to convenient dining out).

Overall, consumer food choices are driven by convenience, health and wellness, pleasure and value.

A wide variety of potatoes are available for production of ready to heat and ready to eat products.

All products require high quality standards, such as fresh taste, excellent texture, good colour, vitamin content and food safety. The consumer trends require flexibility of the food industry by which a large product variation can be offered.

The major challenges for the food industry are to reach 1) both a fresh image and a long shelf-life, 2) both a high quality flavour and texture and convenience, and 3) both food safety and health and nutritional value.

Fresh image and a long shelf-life

A fresh image can be obtained by using mild preservation techniques for extending shelf-life. The challenge is to find techniques that do not require an intensive time and temperature treatment in order to extend the shelf-life. A negative aspect of using severe conditions is that vitamins will be inactivated and the texture and colour of the product will change. New processing techniques are based on high pressure and electromagnetic treatment to inactivate microorganisms. A hurdle treatment is required in order to obtain a high product quality with a fresh image. A multidisciplinair approach combines hurdles from the raw material, process treatment, packaging and storage conditions. It is the combination that leads to shelf-life.

A high quality flavour and texture and convenience

Convenience, a ready to heat or ready to eat product requires processing at the food industry. The more processed the product, the lesser the preparation time for the consumer. However, the quality of the product will diminish. A ready meal that consists of potato, vegetable and meat will get a heat treatment based on an average. If the meal has multiple compounds the taste and texture will be equalized after severe treatment and storage.

Blanching and mild preservation techniques might help in retaining the fresh image, texture and bite.

Food safety and health and nutritional value

Food safety means no pathogens, no spoilage microorganisms, no chemicals, no toxins and low acrylamide content. Optimisation of heat treatment and introduction of hurdles will diminish the growth of microorganisms. Acrylamide can be formed at browning during heating processes. It is a reaction of the amino acid asparagine with reducing sugars. The acrylamide content of a product can be reduced by raw material selection, optimising potato storage, use of mild heat treatment at processing and frying until a golden yellow colour desired by the consumer.

Figure 1 demonstrates that the acrylamide content in French fries increases at higher frying times. Frying at 180 °C shows that breakdown or binding occurs from 4 to 5 min frying.

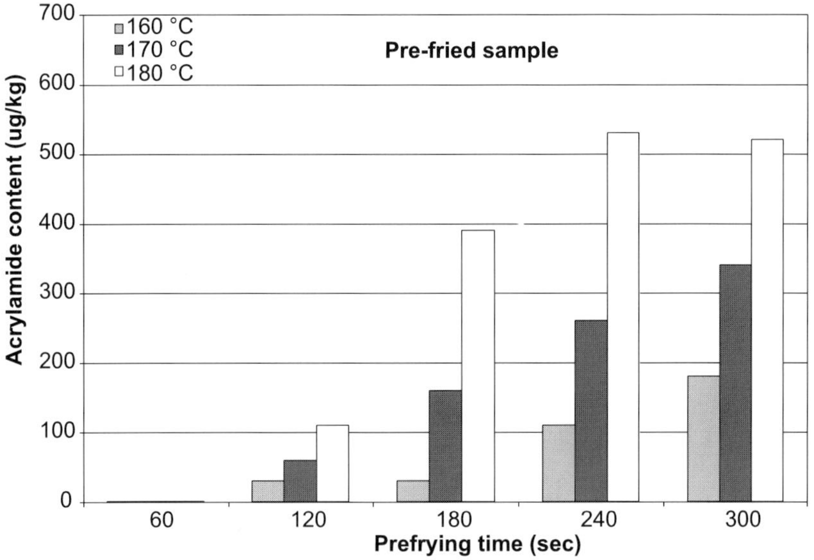

Figure 1. Effect of frying time and temperature on acrylamide content of pre-fried French fries. Reference for prefrying conditions is 1 min at 180 °C and for finish frying is 3 min at 175 °C (van Gijssel et al.*, 2003).*

Nutritional value of processed potatoes

Table 1 demonstrates the composition of processed potatoes such as French fries and boiled potatoes compared with the raw potato. During processing into French fries the moisture content will reduce in such a way that the carbohydrate content and protein content increase. The frying process adds fat to it, resulting in 11% fat in the final product. French fries have an unhealthy image of being rich in fat. This is untrue if it is compared with e.g. the 20 - 25% fat of a minced-meat ball of 100 g.

Health threats and opportunities

Important health threats are cardiovascular disease, tumors, obesity and diabetis. This has lead to the hype of low-carb diet in USA and UK. This type of diet is already on its return and will be followed by formulation of food with a low glycaemic index and information on satiety (Table 2).

Table 1. Nutritional value per 100 g of processed potatoes compared with raw potato (NEVO, 2001 and Schuten et al., *2004).*

	Raw potato (reference)	Boiled potato	French fries
Calories (kJ)	351	285	794
Total Fat (g)			11
Protein (g)	2	2	2.5
Total Carbohydrates (g)	19	15	20
Mono + di saccharides (g)	1	1	1
Polysaccharides (g)	18	14	19
Dietary fiber (g)	3.1	3.1	3.2
Sodium (mg)	2	2	160
Potassium (mg)	450	400	640
Calcium (mg)	6	8	15

Table 2. Acrylamide content, Glycaemic Index and Satiety Index of a few potato products.

	Acrylamide content (ppb) (NFA, Sweden, 2002)	Glycaemic Index (Foster-Powell *et al.*, 2002)	Satiety Index (Holt *et al.*, 1995)
White bread	50	100	100
Boiled potatoes	0	88	323
French fries	450	76	116
Potato crisps	1200	73	91

A low GI and a high satiety index are recommended in diminishing the calorie intake and blood sugar level in the battle against obesity and diabetes.

Factors that affect the GI of products are the amylose/amylopectin ratio, cooking and processing (retrogradation), fat and protein content, acids and mixed meals. This increases the complexity but offers possibilities to process the potato towards a desired application.

Conclusions

Potatoes are a valuable compound of our diet and are known to be an important carbohydrate source. For health reasons, it is required to have a low acrylamide content, a low calorie content, a high Satiety Index and a low Glycaemic Index. These aspects are driven from health issues and might give a preference to boiled potatoes. However, for pleasure and value consumers desire fried products and variety. For convenience it is desired to reach ready to heat and ready to eat and a fresh image. This requires mild preservation techniques in order to inactivate microorganism and hurdle technology to extend shelf life.

A multidisciplinairy approach from farm to fork is required to diminish the amount of acrylamide, to optimise the Glycaemic Index and Satiety Index and to maintain a high product quality.

Consumer behaviour is driven by convenience, health and wellness, pleasure and value. It is a real challenge to combine these trends towards attractive good tasting processed potato products that reach the targets for food safety and health.

References

Linnemann, A.R., G. Meerdink, M.T.G. Meulenberg and W.M.F. Jongen (1999). Consumer-oriented technology development. Trends in Food Science & Technology, 9, 409-414.

National Food Administration (2002). Acrylamide in Heat-Processed Foods. National Food Administration, Uppsala, Sweden [Unpublished] rev.26apr02 http://www.mindfully.org/Food/Acrylamide-Heat-Processed-Foods26apr02.htm

Foster-Powell, K, S.H.A. Holt and J.C. Brand-Miller (2002). International table of glycemic index and glycemic load values: 2002. American Journal of Clinical Nutrition, 76, 5-56.

Holt, S.H., J.C. Miller, P. Petocz and E. Farmakalidis (1995). A satiety index of common foods. European Journal of Clinical Nutrition, 49, 9, 675-690.

Gijssel, J. van, M. P. van Hoof, E.P.J. Boer, M.G. Sanders, C.L.J. Roelofsen, E. Slotboom, J.B. Weitkamp, D.J. Schaap, J.G. Slotboom, C. Zondervan, M.C. Spanjer, P.M. Rensen, M. Hiemstra, E.J.M. Konings and J.A. van Kooij (2003). Possibilities to reduce acrylamide content during processing of potatoes into chips and French fries. Summary report A&F no. B671, Agrotechnology and Food Innovations B.V., Wageningen UR, Wageningen, 9 pp.

Schuten, H.J., J. van Gijssel and E. Slotboom (2004). Effect of frying conditions on the fat content of French fries, A&F report 023, Agrotechnology and Food Innovations B.V., Wageningen UR, Wageningen, http://www.vwa.nl/download//rapporten/Voedselveiligheid/vwa_French_fries_report040129.pdf

NEVO tabel (2001), Nederlandse Voedingsmiddelentabel, Voedingscentrum, Den Haag, The Netherlands.

Curriculum vitae – J. van Gijssel

Janny van Gijssel studied Food Technology at Wageningen University, from which she received her MSc in 1988. In 1988, she started her professional career as food technologist at TNO Nutrition and Food Research in Zeist. From 1991 - 1995 she was in charge of EU FLAIR project CT 91-0050 'New technologies and raw materials for nutritious and attractive cereal products'. In 1996 she became Product Manager Convenience Food. In 1998, she became Application Specialist Snacks and Coated Foods at an international starch company AVEBE, Foxhol. In 2001 she made a new step and started as Programme Co-ordinator Potato, Vegetable and Fruit Processing at Agrotechnology and Food Innovations (successor of ATO) in Wageningen. One of the tasks she executes is theme leader of Work Package 2 Formation and Processing in EU HEATOX 'Heat-generated food toxicants, identification, characterisation and risk minimisation', which runs from November 2003 - 2006.

Company profile – Agrotechnology and Food Innovations, Wageningen UR

Agrotechnology & Food Innovations (A&F) is part of Wageningen University and Research Centre. A&F specialises in innovative, market-driven research for the food sector and agrotechnology. It is a modern R&D institute for innovative and leading technologies in the whole production chain. The strength of A&F is to offer concrete solutions.

Agrotechnology & Food Innovations has a great variety of research facilities and equipment, that range from labscale to semi-industrial (pilot) scale. As a result the institute is able to optimise new technologies and products, and to offer tailor-made solutions to industries.

Agrotechnology & Food Innovations key activities are:
- Quality in Chains;
- Biobased Products;
- Food Quality;
- Agrisystems and Environment.

Our expertise in all the separate links of the production and sale chains enables us to realise - together with our clients - creative and innovative based solutions based on scientific insights and technological know-how.

The contract partners of A&F can be found, among others, in the food industry, the chemical industry, the pharmaceutical industry, the starch and paper industry, the car industry, industries of building materials, paints, glues, inks and lubricants, retailers, auctions, transport companies, the European Commission and national authorities.

Adding value to potatoes by processing for the benefit of the consumer

M.J.H. Keijbets
Aviko BV, P.O. Box 8, 7220 AA Steenderen, the Netherlands

Abstract

An overview is given of developments in the potato processing industry in connection to consumer trends and market development. The demand of these developments on potato raw material and frying oil used is described. Innovation in processes and new product development are considered key drivers for the future of this industry.

Keywords: processing, French fries, consumer, market, innovation

Introduction

Potato processing into products like French fries, chips (crisps) and a multitude of potato specialties largely contributes to the way consumers nowadays eat and experience potatoes in the affluent Western societies and increasingly also worldwide. In the Western world 20-50% of the potatoes consumed in the daily diet is in a processed form. In this way potato processing adds substantial value to the potato chain.

In this paper an overview will be given of the developments in the (French fry) potato processing industry, the driving force behind these - the trends in the consumer market - and the demands on potato growing arising from these developments.

Processing capacity

In 2003 the total potato processing capacity and that for French fries and related products of seven EU countries (Belgium, France, Germany, Italy, the Netherlands, UK, Sweden) and the USA was almost equal (Table 1). These figures include also the processing into pre-peeled (mostly pre-cooked) potatoes in some countries. In 2004 13% of the Dutch potato processing capacity was used for pre-peeled potatoes (more than 400.000 tons!). In Canada about 2.500.000 million tons are processed into French fries. In Europe frozen potato products other than French fries have a growing market share in volume and even more in value because of

Table 1. Processing of raw potatoes 2003 (x 1000 tons).

	Total	French fries and related products
EU[1]	11.500	6.500
USA[2]	11.700	6.800

[1] UEITP (2004); [2] USDA (2004).

their higher added value. In the Dutch market for instance these account for about 30% of volume and 42% of value (Saulnier, 2002).

A threat to the potato processing industry is nowadays the worldwide overcapacity in a saturated (Western) market and the low profitability due to low margins on "bulk" French fries.

The rising demand for potato products in Central and Eastern Europe and in Asia will partly be filled by imported product but also by new production facilities to be erected in these areas.

Consumer trends

The major consumer trends in food consumption are (Sloan, 2003):
- convenience;
- pleasure, indulgence;
- health;
- sustainability (organic).

The importance of convenience and pleasure for the behaviour of the consumer in affluent Western type of societies is rather clear. Pre-fried French fries - but also pre-peeled potatoes- are typical convenient products that are chosen because of their good taste and texture (crispiness) characteristics. Microwavable French fries are a typical example of convenience and pleasure in one.

Health may become a key driver for the consumption of potato products for the coming years. Overweight and obesity - also rapidly growing among children - are becoming widespread among Western consumers and form a major concern for health authorities in the EU and the USA (van Kreijl *et al.*, 2004). Overweight is assumed to cause severe chronic health problems in the near future (diabetes, heart disease, joint degeneration). Potato products which are fried during the final preparation contain oil or fat and are widely considered to be unhealthy, at least have the image of unhealthy food because of their fat and energy (calorie) content. Salt reduction is another issue. A too high salt (sodium) consumption raises the risk for high blood pressure. Consumers will determine their food choices more and more by health issues: "What is good for my body and my wellbeing".

Sustainable production is becoming another key issue in the Western world. The consciousness that natural resources are limited and the prudent use of these is becoming more widespread. Modern intense potato growing practices as used for the industry in particular are considered to be non-sustainable because of their high usage of pesticides against *Phytophthora infestans*.

Market developments

Both in North America and the EU the consumption of deepfrozen potatoes no longer grows but stabilizes or even decreases. French fry consumption in Europe ranges from less than 2 kg in Central and Eastern Europe to 12-13 kg per caput per year in the UK and the Netherlands. A rising demand for potato products such as French fries is expected in Central and Eastern Europe and Asia.

In the Netherlands - a major potato processing country in the EU with a share of almost 30% - the same production and consumption trends are visible in deepfrozen French fries. However, a strong increase in production and consumption of chilled, pre-peeled, pre-cooked potatoes has emerged. It has to be seen if this trend of "fresh" pre-peeled potatoes will also develop in other (EU) countries. The convenient ("no potato peeling in the household anymore") and health drivers could favour this development.

Another driver will be the growing out-of-home eating market (food service). This growing market has specialized needs for potato products and meal solutions to which the potato processing industry has to respond.

In retail the purchasing power of the retailers is nowadays very concentrated and might even further concentrate. At the same time a heavy price war ánd the growth of hard and soft discounters in several countries put pressure on margins and on the position of the A-brands against private brands. All these developments will change the potato processing industry and urge them to invest even more in innovation than they do now.

Potato raw material

The potato raw material is essential for the processing industry, both in terms of quality and of price. New cultivar development remains crucial for their future development. In Europe nowadays a broad variety of cultivars is used for processing. The traditional Bintje on the mainland is more and more replaced by purposely bred French fry varieties such as Agria, Asterix, Felsina, Innovator, Lady Olympia, Ramos and Victoria. These new varieties have a better shape, more length, (some of them) less reducing sugar, are more resistant against diseases and drought and in general have a more consistent growing and yield pattern. Organic potato growing for the processing industry is limited up till now. Organic cultivars lack yield and processing quality leading to too high raw material costs and inferior product quality. The challenge for the future is to develop more sustainable cultivars, in particular cultivars (more) resistant against *Phytophthora* which require lower inputs without the use of genetic modification. Genomics programmes as in the Netherlands (Centre for BioSystems Genomics, 2003) are very promising in this respect. Public acceptance of genetic modification may develop in Europe in the coming decade and would be very helpful to reach the goal of resistance.

Breeding for low reducing sugar content is a long time goal for the processing industry. The discovery of acrylamide - a substance that is formed during the Maillard or browning reaction - in 2002 in finish fried French fries and in chips lays even more emphasis on this breeding goal. Acrylamide is known to cause cancer in animals and has been classified by the WHO's International Agency on Cancer (IARC) as possibly cancerogenic to humans (IARC, 1994). Acrylamide is formed during heating from reducing sugars and asparagine, an amino acid abundantly present in the potato tuber. It is clear now that the determining factor for acrylamide formation in potato products is the level of reducing sugars (Amrein *et al.*, 2003).

The emerging fresh pre-peeled potato market asks for more breeding effort towards cooking varieties for the industry. Here partly quite different breeding goals emerge such as small size, visual aspects after cooking and taste. A particular challenge is to obtain cultivars that have a sufficient yield for the farmer in the small grading sizes.

Frying oil

Palm oil still is the major frying oil used in the potato processing industry in Europe. Other oils are rapeseed und sunflower oil. Until recently palm oil was mainly used in the hardened form containing 20% trans fatty acids. Trans fatty acids are considered unhealthy because they behave as saturated fat in the human body (increase bad cholesterol and/or decrease the good one) or even worse. The Dutch industry decided in 2003 to use only trans fat free oil. The challenge now is to stimulate the end user, in particular in food service, to use trans fat free or low frying oils. Another challenge is to shift the composition of the frying oil in the direction of more unsaturated fat and less saturated fat than present in natural palm oil. Palm oil contains 49% saturated and 51% unsatured fatty acids, sunflower oil 12% saturated and 88% unsaturated fatty acids. This would mean less use of palm oil and more use of unsaturated oils in the industry.

Innovation in new processes and products

Process innovation is a continuous development in the processing industry. Machine suppliers innovate, develop better equipment and offer them to their customers: better and more intelligent peelers, graders, defects sorters, dryers, packaging machines etc. However, radical changes in potato processing technology are not visible yet. Many of these improvements are connected to a better quality or to a higher yield efficiency. In an ever-increasing competitive world and under increasing margin pressure a better cost efficiency is rather essential (Somsen, 2004). New consumer and market demands, however, have lead to the application of new technologies. An example is the coating technology now widely applied in industry. French fries and other potato products are coated with starch or flour batters after the dryer and before the fryer in a coating bath. In this way a very crispy crust is created with excellent holding functionality (Wu and Woerman, 1997). By adding herbs and spices to the batter new taste experiences have been developed for potato products. Low fat technologies are already available to some extent (fat reduction after frying) and will become more important in the near future (other heating technologies than frying). New packing materials and types are available as well.

New product development (NPD) is the key driver for absolute margin growth in the potato processing industry. French fries are rather "old" products now produced in very high volumes with low margins. In order to escape from this margin erosion the potato processing industry successfully invests in development and introduction of new products. As a result a broad range of "new" French fries (e.g. microwavable, crispy or spicy coated, fried in special oils like sunflower, canola and olive oil, skin on, special cuts etc.) has come into the market (Figure 1). All kinds of other potato products (e.g. wedges, hash browns, croquettes, potato dishes, kid products, filled products, organic fries, local specialties as Reibekuchen or pommes dauphines etc.) are available as well. This development will continue both to increase variety for the affluent consumer and to create new margin. An important driver for NPD in the coming years might be health: potato products with less fat, better fatty acid composition and less salt. At the same time the excellent nutritional profile of the potato, rich in vitamins and minerals, could be used more than is practised until now. Pre-cooked potatoes having no added fat are a good example of this category, but also fried and certainly oven finished pre-fried potato products perform better than generally assumed.

Figure 1. Some recent new potato products.

References

Amrein, T.M., S. Bachmann, A. Noti, M. Biedermann, M.F. Barbosa, S. Biedermann-Brem, K. Grob, A. Keiser, P. Realin, F. Escher, R. Amado (2003). Potential of acrylamide formation, sugars, and free asparagine in potatoes. Journal of agricultural and food chemistry, 51, 5556-5560.

Centre for BioSystems Genomics, Wageningen, the Netherlands (2003) www.cbsg.nl

IARC (1994). Acrylamide. IARC Monographs on the evaluation of carcinogenic risk of chemicals to humans, 60, IARC, Lyon, France, 389-433.

Saulnier, J.M. (2002). Just in time for new Euro rollout, prepare to pay more of them for fries (Dutch frozen potato products market). Quick Frozen Foods International, 43, 72-89.

Sloan, A.E. (2003). Top 10 trends to watch and work on: 2003. Food Technology, 57, 30-50.

Somsen, D. (2004). Production yield analysis in food processing. Applications in the French-fries and the poultry industries. Ph. D Thesis, Wageningen University, Wageningen, the Netherlands.

Union Européenne des Industries de Transformation de la Pommes de terre, Brussels, Belgium (2004). European Processing Industry - UEITP Statistics.

United States Department of Agriculture (2004). Potatoes 2003 Summary. September 2004. Agricultural Statistics Board, NASS, USDA

Van Kreijl, C.F., A.G.A.C. Knaap, M.C.M. Busch, A.H. Havelaar, P.G.N. Kramers, D. Kromhout, F.X.R. van Leeuwen, H.M.J.A. van Leent-Loenen, M.C. Ocké, H. Verkleij (Eds) (2004), Ons eten gemeten. Gezonde voeding en veilig voedsel in Nederland. RIVM-Rapport nr. 270555007, Bilthoven www.rivm.nl

Wu, Y and J.H. Woerman (1997). French fry formulations and method of making. US Patent 5.648.110.

Curriculum vitae – M.J.H. Keijbets

Martin J.H. Keijbets studied Food Technology at Wageningen University where he graduated in 1970. He joined the former Institute for Storage and Processing of Agricultural Produce in Wageningen (IBVL) in the same year to start a Ph.D. on potato cooking and cell wall characteristics which he received in 1974. From 1974 till 1988 he served IBVL as researcher in potato processing, department leader and deputy director. From 1988 until now Dr. Keijbets works at Aviko B.V. as head of the R&D department and from January 2005 as head of Innovation & Product Development.

Company profile – Aviko

Aviko was established in 1962 in Keppel in the Netherlands. The company moved to Steenderen and expanded in 43 years to become the no. 2 potato processor in Europe with a market share of about 20%. Aviko is a 100% daughter company of Royal Cosun. Aviko processes yearly with 1300 people 1.400.000 tons of potatoes in 7 processing facilities in the Netherlands, Germany and Poland. Aviko serves the retail, food service and industrial markets in Europe and outside with an extended range of frozen and chilled French fries, frozen potato specialties, chilled pre-cooked potatoes, pasteurised potatoes and dried potato flakes and granules.
www.aviko.com

Potato: a dull meal component?!

Wiro G.J. Sterk
Gfrost Diepvries BV, Nijverheidsweg 19, 4879 AP Etten-Leur, The Netherlands

Abstract

The consumer is very unpredictable; how can we understand his needs so that we can develop potato products, which will generate long term sales. The last decennia potato products have been under pressure in the traditional European potato countries (NL, BL, FR, G, UK) through pasta and rice products as part of the new exiting cuisines. The potato became a dull meal component, which was a necessity and fitting within the old eating culture, but was far from a great eating experience. How is the situation now? Are we meeting consumer demands and customer needs? How can we measure this, how can we steer R&D throughout the chain and how can we implement this effectively in order to achieve our goals?

The basic starting point is a system of Category Management. This system forces us to understand the consumer needs through clearly segment his specific needs, by shopping through the eyes of the consumer. This can be done by identifying trends, interviews with consumers and collecting shopper- and consumer data.

When the needs are identified, the answer of how we can play an active role in the needs of the consumer can be found. Never impose our needs to the consumer; this will not work!

Keywords: category management, trends, consumer needs, customer demands, shopper

Introduction

Category Management is a strong tool which delivers increased category returns by shaping the total category offer to reflect the needs of the consumer and shopper today - while anticipating those of tomorrow. It helps clients gain competitive category advantage; defines and segments the category in line with consumer usage and shopper behaviour and identifies the quantified category drivers and quick wins for the category, for retailers and suppliers. Importantly, it provides clarity of vision that, with some adaptation, can be successfully implemented in different channels, with suppliers driving the category agenda.

Insights into consumer and shopper behaviour identify the areas of maximum benefit for brand portfolio and NPD (New Product Development) action. These insights can then be used with retailers to ensure they will more actively support the development of your brands. For the first time, brand owners can not only explain the category profit which can be delivered by supporting branded initiatives, but, uniquely, also show the clear link between consumer brand communication, brand and NPD activities that will change consumer behaviour, as well as the changes required at the point of purchase in-store to influence shopper behaviour.

Trends

The first bases for Category Management is to understand consumer needs and customer demands through the current worldwide and local trends.

Worldwide trends:
- Getting connected.
- Solutions for individuals.
- Better health.
- Instant availability.
- Cruising the planet.
- Transformation of traditional structures.

Socio-demographic trends
- Increase of one and two person households.
- Strong education of number of members per household.
- Change of ethnical mix in European population.
- More professionally active women.
- Families are having children at an older age.
- Increase of elderly people with changed needs.

Distribution trends:
- Increase of private labels.
- Strong growth of soft discounters (wide range of products with core assortment at EDLP (Every Day Low Price).
- Introduction of new players in hard discount (reduced range - approx. 800 products - at EDLP).
- Internationalizing of buying activities.
- Reduction of product knowledge among buyers.
- Increase in food safety issues (Traceability).
- Increase of logistical issues (EDI).

Eating trends (Figure 1).

Consumer interviews

Through thorough interviews with consumer and shopper groups you can understand better their needs and wishes. As an example Farm Frites organised such a session. The main needs were in the area of spending better "quality time" with their children:
- Help to utilise efficiently the thrifty time: need of products with convenience.
- To please the child: products with Kid Appeal (fun value) and indulge value.
- To please parents and child together: products with affection value (sharing) and family concepts.
- Help to prevent confrontations between parents and child: problem solvers.
- To take over caring: high valuable products (quality, image, development stimulating products).

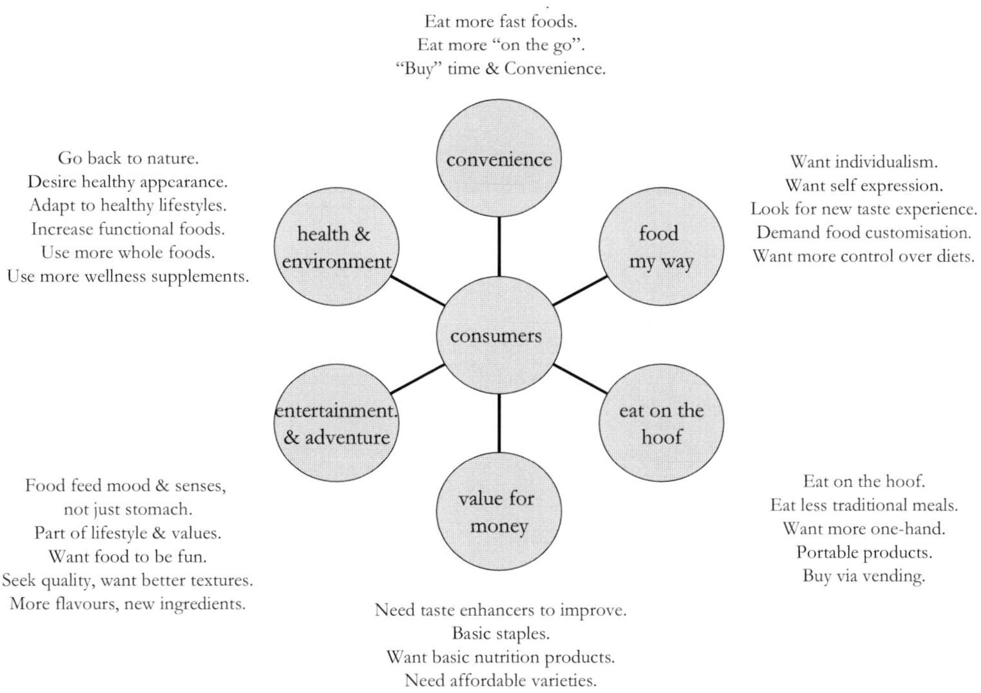

Figure 1. Eating trends in the developed world.

In the healthy section the Dutch mothers (are the shoppers) believe the following:
- Healthy eating is potatoes, vegetables and meat (last need is declining quickly).
- To cook as often as possible; only three times a week.
- Other meal components are sub-optimal (pasta, rice, pizza, deep frying etc.).
- Compensation theory: shortage of vegetables will be compensated with fruit, bread or dairy products.

Shopper and consumer data

General retail facts and figures (Nielsen data):
- The total annual turnover of supermarkets in Holland is € 24 milliard.
- Number of supermarkets in The Netherlands is 5200.
- An average supermarket has an annual turnover between €4 and €5 million.
- On average a supermarket has a turnover between €125,- and €150,- per square meter. At an EDLP (Every Day Low Price) this average is €500,-.
- A Dutch shopper goes on average 3.1 times per week to the supermarket and buys for € 20,- each time.
- Around 60% of the shoppers are going to the supermarket more than once a week.
- The average travel time between the house and the supermarket is approximately 8 minutes.
- A cashier in a supermarket scans for approximately € 30.000,- per week.

- The market share of Private Label in The Netherlands is 22% of the value share. For deep frozen potato products this is 45%, by non-food this is 12%.
- The average percentage of out of stock is around 3% for fast movers in the weekend and 5 to 6% for slow movers.
- Frozen potato products are developing poorly; the volume has decreased with 5% and the turnover with 6%. The category is loosing market share to Chilled Convenience.

Frozen Potato Target group (Table 1).

Table 1. Frozen potato products vs. family cycle (GFK Panel: market data research centre).

Targetgroep	Households without children Housewife <35 year	Households with children	Household without children Housewife 35-55 years	Household without children Housewife 55+ years
Frozen potato	26.9%	32.2%	23.3%	17.6%
Spending	13.3%	51.3%	15.6%	19.9%

The shoppers from households with children, who are spending 51.3% (Table 1) in the category, are looking for service products (GFK Panel):

- The mother is often the shopper.
- Influence of the children is important in buying behaviour.
- Searches for some extras on shopping floor.

Category management

Through above mentioned data it can be checked whether there are gaps in the current offering to the consumers in order to fulfil their needs. In this case we use the following segmentation / consumer decision tree (Figure 2).

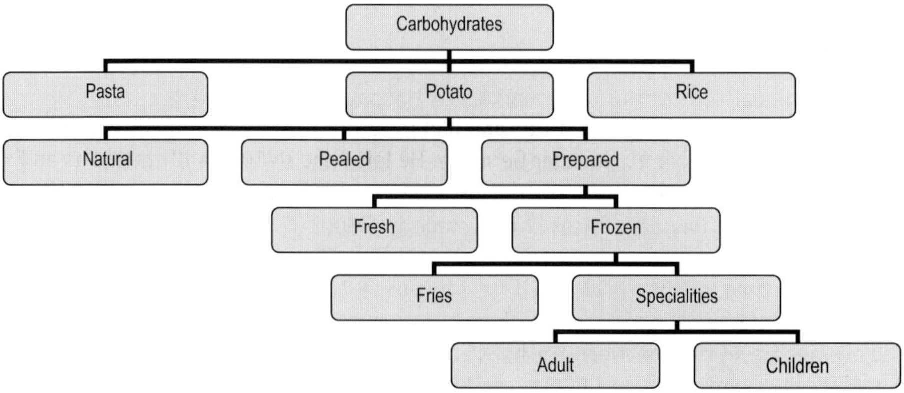

Figure 2. Consumer decision tree / segmentation.

In the consumer need segment children it has been identified that there were no products to fulfil their need. However according to all the data (trends, shopper research etc.) it was clearly identified that there is a big group of consumers searching for children needs, but the product must fulfil the following requirements / intrinsic values:

- convenience;
- health;
- fun value;
- indulgence;
- affection value;
- family concept;
- problem solvers;
- high valuable products.

This will then be passed on to idea generators sitting within the R&D and / or marketing department. Several ideas are matched with above criteria in order to see if the customer needs are met.

The result is: Shrek, a formed product with potato and peas in letter shapes. This meets all of the requirements / intrinsic values within the children segment.

Conclusion

Through clearly identifying the trends, consumer needs and consumer data, product development and product introductions will be successful. This will drive the business we are in forward, not by imposing our needs to the consumer; this will not work!

Use this method and a potato will never be a dull meal component again.

Curriculum vitae – Wiro G.J. Sterk

Wiro G.J. Sterk studied HTS Bedrijfskader in Eindhoven, The Netherlands. In 1987 his professional career started in his family company of processing frozen fish for all the major European Supermarkets and Foodservice. After Military Service he decided to develop his career outside the family company and worked for half a year in Toronto Canada (1988 / 1989) as a marketing and sales trainee for a European cheese and confectionary importer.

He could not get landed immigrant status and had to return to The Netherlands where he joined Frico Cheese as an Area Export Manager Germany. He left Frico in 1991 and started as an Export Manager for Boekos Meat Products in Boekel (The Netherlands), where he covered the marketing and sales throughout Europe. In 1993 he became European Trading Director within the Hazlewood Food Group to intensify sales from the UK to mainland Europe. He left the food Industry in 1998 for a short period to work as Director in an Marketing Agency supporting mainly the IT and Food industry. In 1999 he became Commercial Director at Vaco (Geest Plc) to improve the trading relationship with Ahold and support the ambition of Geest to grow in mainland Europe. Wiro Sterk became minor shareholder in 2001 of Gfrost in Etten Leur (The Netherlands) and acquired the company fully in 2004. He is also the founder and Chairman since 1998 of the "Ver(s)kenners", which is a group of decision makers within the retail, foodservice and food industry, who are developing the market in fresh and frozen food.

Company profile – Gfrost

Gfrost has been founded in 1978 and is a successful marketing, sales and distribution company within the frozen food industry. Its mission: "Gfrost Benelux discovers and brings the finest products from the world eating culture, so that our customers can enrich the eating moments of their consumers".

Gfrost works with partners from the food industry and through a full portfolio of support measurements it will develop niches to markets. Haägen Dazs, Van Dobben, Dr. Oetker have become now well established names in the Dutch market and we are momentarily developing the brand Farm Frites. The key drivers to their success are: continued innovation, knowledge, optimal sourcing, intensive distribution, pioneership and partnership. The driving force is: Bridging Demands For Foods!

Understanding consumer behaviour is not optional: do we change it or does it change us?

David F. Walker
Chairman, British Potato Council, 4300 Nash Court, Oxford Business Park South, Oxford, OX4 2RT, United Kingdom

Abstract

Understanding consumer behaviour is as fundamental to commercial organisations as a compass is to navigation. Without it the organisation might find its own way, but this will be by luck more than judgement. A good knowledge of the dynamics of its own demand enables a business to steer in a straighter line in the direction of sales. It reduces business risk by looking at demand from the consumer's perspective and establishing the reasons why they purchase. What is the need or want that is being fulfilled by purchasing our product or service? Does our product fulfil this need adequately? Could we do it better?

Tracking and reacting to the answers to these questions puts the business in the driving seat by enabling it to tailor products or services to meet consumer wants. But it is not easy. The organisations that do this well, the ones that are truly customer focused, are the most successful, but it takes brave decision making and the right culture within the organisation to achieve it, something not every organisation does or has.

In an industry such as ours, that historically comprises lots of small, independent and adversarial businesses, namely farms, ones that do not have the structure or wherewithal to do all of this themselves, I see it is a basic principle of the British Potato Council to fulfil this function for our levy payers.

Keywords: consumer behaviour, demand, added value, research, commodity

Introduction

Perhaps the essence of this paper is best summed up by the old, but nevertheless, valid marketing maxim - make what you can sell, don't try and sell what you can make. The meaning is simple, don't just produce things in the vague hope that you will be able to sell them but produce things that you know you can sell. In other words the starting point should be demand not supply. What do people want and can I satisfy that want?

I firmly believe that understanding consumer behaviour is a basic cost of doing business and I want to show in this paper, by giving some examples, how certain organisations have reacted to this task - some good and some catastrophic. Then I want to tell you more about how we at the BPC go about this task in the UK.

Some recent examples of companies who have failed to understand consumer behaviour

In picking some relevant examples I have to admit it was more a case of who to leave out as there have been some classic errors. One of the most famous instances of this in the UK has to be the Sinclair C5, a great product, fantastic invention but nobody wanted one. In more recent years there was the case of Ratners Jewellers, a successful mid-market, high street jeweller selling to the masses. However a simple, one sentence, faux pas by its eponymous Chairman, Gerald Ratner set off a downward spiral that the company was never to recover from. In a speech to the Institute of Directors, which the media instantly seized upon, Ratner had joked that his Ratners High street chain 'sold a pair of earrings for under a pound, which is cheaper than a prawn sandwich from Marks & Spencer, but probably wouldn't last as long'. The foot was then rammed further into his mouth when he followed up by saying a tacky sherry decanter was so cheap because it was 'total crap'. Needless to say, members of the public did not take kindly to being taken for fools, and in the furore that followed they stopped shopping at his stores. Mr Ratner left the company and the group was re-branded.

Atkins Nutritionals UK

A very recent and relevant case of a food company totally misunderstanding its target audience in the UK is Atkins Nutritionals, who went into administration in the UK in March, less than two years after it started. Atkins capitalised on the popularity of the diet in the US by bringing in its products such as milkshakes, cake mixes, cereal bars and chocolate bars to the UK market. However, over investment and changing consumer tastes led to the company's downfall. The company suffered for two main reasons. Firstly, the products it introduced were too far removed from UK tastes; secondly, the Atkins diet was based on a high-fat, high-protein intake which cultivated a bad image through the media. The diet, like many others, was fashion led and therefore short lived and perversely, the reason the brand was initially successful was partly due to it having the same name as the diet which also contributed to its sudden demise. As soon as the diet came under attack from the media and nutritionalists, Atkins products also suffered and were very soon de-listed.

Some instances of companies who have understood consumer behaviour - although in the first case late

McDonald's

Let's look first at McDonald's, one of food retailing's great success stories. A company that became an icon, taking the humble burger to the world and making eating out an everyday event through fast, filling, value for money meals.

For 30 or more years the company could do no wrong, expanding quickly across the world with its famous golden arches seen in virtually every country and city in the world. However, it failed to spot quickly enough the growing worry amongst consumers about health and healthy eating - a quick look at their menu of burgers, chips and other fatty foods should have sounded the company alarm bells, but it took some bad publicity in the form of books, films and court cases to turn consumers against the offer and in 2002 the company made a quarterly loss for the first time in its history.

Not a company to give in easily, McDonald's has come back quickly and strongly with a new offer. The standard fare is still available, albeit with reduced fat, but the highlight is now firmly on its healthier options of salads and fruit and it has stopped its 'super sizing'. Incidentally McDonald's is now the world's largest purchaser of apples! With the change in strategy has come a return to profit with the company's US customers warming to its new menu and its European customers returning to its restaurants. You can't help feeling that if McDonald's had spotted the change in consumer behaviour soon enough and been brave enough to do something about it then the four or five years of turmoil would have been avoided.

Birds Eye

Birds Eye were somewhat quicker to act when they realised something was wrong, by keeping their finger firmly on the consumer's pulse they realised that attitudes and opinions were changing quickly and that their current range of frozen products would not match consumer requirements in the very near future. They took the brave decision to reformulate and re-brand all their products to fit in with consumers' healthier eating behaviour. Under the banner of 'We don't play with your food' the company invested time and money to ensure that their new range lived up to consumers' new expectations. The new Birds Eye promise was threefold - a guarantee not to add artificial colours, flavourings or preservatives; a commitment to the highest standards of food safety; and a pledge to use only sustainable practices. The results since the change speak for themselves.

Walkers

In our own industry, Walkers has introduced a very successful healthier snack aimed specifically at children. With bad press about crisps and snacks in general, and with many parents becoming increasingly aware about the health and nutrition of their children, Walkers were quick to react and launch 'Potato Heads'. These are a crinkly crisp containing no artificial flavours, colours or preservatives and cooked in high Oleic Sunflower Oil which has 70% less saturated fat. Altogether a healthier product but one that kids love and their parents worry less about.

How do we monitor and react to changing consumer behaviour at the BPC?

As I have already said, we take the task of monitoring consumer behaviour very seriously and we do this in a number of ways.

Monitoring sales data

We establish what is happening to sales by segment. What is up? What is down? What are the underlying causes of any movements?

Keeping abreast of the latest published consumer research

Including anything that might be related to, or have an impact on our market - fresh produce, shopping behaviour, store formats, convenience foods, eating out, etc.

Ad hoc research

This is used to fill in knowledge gaps where published research does not exist. For example we are currently running a research programme that monitors shoppers attitudes to potatoes and health three times a year, as this is clearly an important issue for us and has been for several years now. We are also currently examining the difference in attitudes to food and specifically carbohydrates and potatoes, between the younger and older age groups as we have begun to witness a polarisation of attitudes and behaviour that could begin to affect our market quite seriously.

PR activity

PR is used in both an offensive and defensive way - offensively to create and foster positive attitudes to potatoes and potato products and thus help shape demand, and defensively to correct and defend the image of potatoes against attack.

What do we do with the knowledge gained?

We start with a very basic product. It doesn't come much simpler than the potato. It has been a staple part of our diet in the UK for hundreds of years. But in the last thirty or so years things have started to change. Things that have and are affecting the market for potatoes, and these things are mainly to do with changing consumer attitudes.

From our own research, which is backed up by other organisations including Unilever, we have established the three most important factors that affect the food market in the UK. - health, convenience and indulgence. It is a combination of these factors that shape demand for each and every consumer and we believe will do for the foreseeable future.

Health has become a very important issue made even more important recently by government backed schemes to reduce obesity levels and bring peoples' diets back into balance. Also as science evolves, new problems and opportunities arise. A recent article in the Times highlights this. It deals with the new science of 'nutrigenomics', which aims to work out how each of us responds to different foods. By tailoring our diets to our genetic makeup, nutrigenomics will offer dietary advice specific to particular gene types. If it works then 'DNA or Gene diets' will become the holy grail of the diet world. It will make it possible for example, to create a diet that knows about and is sensitive to your family history of cardiovascular disease or bowel cancer.

The market for convenience products is also growing. With peoples' busier lifestyles and more women in the workforce, the demand for convenience products whether that is sliced and diced vegetables or a full, ready to eat meal is increasing quickly. So is the urge to indulge occasionally in more exotic foods, which is fuelled by the general increase in affluence.

We need to understand these shapers and monitor their changing effect on consumers if we want to be ahead of the game which is clearly, where we want to be.

By keeping our industry bang up to date with research and changing attitudes we will remain as much as we can be - masters of our own destiny. Moulding our offer to best suit the changing market place.

Activation examples

Health messages

We make sure the industry is at the forefront in terms of conveying positive health messages about potatoes because, as we all know, they are too often portrayed in a bad light. This work could take a number of formats - by being actively involved with the health debate at government level; by a positive PR campaign; or simply by persuading retailers to carry health messages on pack. We know from our own research that health messages are relatively short lived in consumer minds and therefore it is vital that we continually drip this information to keep it in mind.

Packaging and information

Research again tells us the importance to consumers of good, vibrant packaging with simple, prominent information on variety, health attributes, usage or provenance. We are currently looking at labelling and presentation within the whole category to see if we can make it better and easier for the consumer at the point of purchase.

Adding value to a commodity product

One of the ways we can promote change and help the industry develop its offer is to encourage suppliers to add value to their products.
As a commodity product, if all we have to compete on is price then we are on a downward spiral. What we have to do is add value. I want to give two quick examples of what I mean here.

Eggs

Firstly, take eggs, a commodity product if ever there was one. Looking for eggs in your local UK supermarket fifteen years ago, about the only choice you could get was small, medium or large. Today, however, a look around a supermarket will reveal that, before you start worrying about size you can choose your preferred breed of hen; your favourite farming method and even the hens diet.
Stonegate, a leading producer in the UK, sees the future in free range, barn, organic and speciality eggs which account for over a half of its £110m sales and substantially more of its profit. According to Stonegate, to persuade consumers to trade up at the egg fixture you can either focus on the end product - selling the sizzle not the steak - or you can focus on the hens and how well you look after them. More importantly, you can print photos onto packs, showing consumers exactly what they can do with an egg and some inspiration. 'Eggs are the ultimate in fast food', they say, 'but they are also a fresh, natural product - and a very cheap meal'.

Milk

Secondly take milk, another commodity product. Two added value products have come on to the market recently. Firstly a new range of milks that have been enriched with Omega 3, a fish oil derived substance which is known to tackle heart disease and ensure healthy nails, hair and skin. It is produced by cows which are given a special fish oil blend of their normal feed. It doesn't look or taste any different to normal milk, but will help consumers to get their recommended share of Omega 3 which research tells us they are not getting through the traditional route of eating oily fish.

The second added value milk product was launched a year or so ago and is now widely available in the UK, which claims to help people with sleep problems establish a good pattern of sleep within two weeks. "Night Time" milk looks and tastes like regular cow's milk but has higher levels of melatonin than other milks. Melatonin being a natural substance found in mammals, including humans, which controls our body clock and helps us to sleep. All milk contains some melatonin, but Night Time milk has more than most because it is collected from the morning milking of a small number of Somerset dairy herds specially selected to provide milk that is rich in the substance. The recommended retail price of Night Time milk and Omega 3 milk is more than double that of ordinary milk, with a much healthier margin.

Two brilliant ideas - simple, natural, a creative way of adding value through a functional element of the product. The same sort of thing is happening in bread as well, in fact a few months ago a London supermarket was selling a speciality loaf for £9 - if that is not adding value I don't know what is.

While these products in their own right may not sell in vast quantities, they do provide a halo effect, bringing fresh publicity as well as relevant, new products to commodity markets, all of which plays a role in persuading people that there is more to the product than they thought. Perhaps it is not as boring as they thought it was after all.
It is this type of innovation within these industries that we have perhaps been slow to embrace in potatoes.

Conclusion

From our knowledge of demand we know that consumers in general, find potatoes boring and are crying out for more interesting things to do with them. If we stick with just selling potatoes per se, then sales trends show we are destined for a slow lingering death. The challenge for the industry is to keep potatoes and potato products as relevant a choice as ever for the consumer, so that potatoes share of plate grows rather than slows. The challenge for us, the British Potato Council, is to help our levy payers to do this by disseminating our market and consumer knowledge to our growers and processors and for being the voice of the industry.

So finally, to answer the question posed in the title of this presentation, like it or not, consumer behaviour changes us. Yes, occasionally it does happen the other way, and a product will come along that actually shapes demand and changes consumer behaviour, but these are few and far between and are generally the result of a step change caused by invention. Generally though, particularly in the food business, consumer behaviour changes us, it is a continuous process and it is constantly re-shaping our industry. The better we monitor it, understand it and react to it, the more chance we have to shape our own destiny.

Curriculum vitae – David F. Walker

With a background in agriculture David Walker has extensive business experience in the agrochemical industry, being previously Commercial Manager of Shell Chemical UK's Agriculture Division and most recently Managing Director of Cyanamid Agriculture Limited in the UK and Ireland.

A former Chairman of the British Agrochemical Association, David has wide experience of the challenges involved in trade associations and collaborations within market sectors.
Having been Chairman of the British Potato Council since July 1997 he now feels privileged to have been introduced to many sides of the potato industry - enough at least to know why growing potatoes is an enduring challenge and that the marketing of them induces great passion!

Company profile – British Potato Council

The British Potato Council (BPC) was created in 1997 by a Parliamentary Order to work with all sectors of the potato industry to improve the competitiveness of its products to consumers and customers. Specific functions include Research and Development, Market Information/Statistics and Marketing. Within the organisation's marketing activities there is an energetic seed and ware export activity.

Great Britain has a world-class potato industry exporting seed and table potatoes, new varieties, science, machinery, consultancy plus pre and post harvest technologies.

Breeding and seed production

Genomic resources in potato and possibilities for exploitation

E. Ritter[1], F. Lucca[1], I. Sánchez[1], J.I. Ruiz de Galarreta[1], A. Aragonés[1], S. Castañón[1], G. Bryan[2], R. Waugh[2], V. Lefebvre[3], F. Rousselle-Bourgoise[3], C. Gebhardt[4], H. van Eck[5], H. van Os[5], J. Taco[6], E. Bakker[5] and J. Bakker[5]
[1]NEIKER, Apartado 46, E-01080 Vitoria, España; eritter@neiker.net
[2]Scottish Crop Research Institute, Invergowrie, Dundee DD2 5DA, United Kingdom
[3]Institut National de la Recherche Agronomique (INRA), 84143 Monfavet Cedex, France
[4]Max-Planck-Institut für Züchtungsforschung, Carl-von-Linné Weg 10, Cologne, Germany
[5]Wageningen University, P.O. Box 386, 6700 AJ Wageningen, The Netherlands
[6]Keygene N.V., P.O. Box 216, 6700 AE Wageningen, The Netherlands

Abstract

This paper reviews the evolution of genomic R&D in potato and related wild species. In the 1980s and 1990s different molecular marker types were developed and applied in potato. Several genetic linkage maps were constructed and numerous markers for qualitative and quantitative characters were integrated in these maps, including applications at the tetraploid level. Comparative genome analyses within and between species revealed resistance gene clusters at several regions of the potato genome and certain common structures between related species. QTLs were found to correspond to genes of associated metabolic pathways (candidate genes). Molecular markers closely linked to qualitative traits or QTLs and candidate genes have been used to develop applications for marker assisted selection and breeding. A high density reference linkage map of 10000 markers was established in 2002 and an anchored physical and functional map is under construction. Novel tools for genome analysis and gene discovery are applied in potato, such as EST, cDNA-AFLP, microchip analyses and physical mapping. Several extended genomic resources are available on the WEB and recent R&D initiatives including genome sequencing indicate that potato is becoming a model crop species.

Introduction

Molecular markers are an essential component of a genomic project and have provided a major contribution to the genetic knowledge of many cultivated plant species including potato. In addition to their basic importance for genetic and evolutionary studies including biodiversity analyses, molecular markers are useful to construct linkage maps and to localise monogenic and polygenic traits, which allows the efficient introgression and selection of individuals with specific characteristics. Such markers enable early selection tests already in breeding material. Moreover, molecular markers play also a crucial role in the isolation and cloning of plant genes by map-based cloning. Due to the economic importance and due to its physiological characteristics, potato represents a promising target species as model plant for genomic analyses.

In other, strategic plant species with small genome sizes such as *Arabidopsis* or rice genomic research is far ahead. However, several genomic projects have been initiated meanwhile also in potato including the establishment of a consortium to sequence the potato genome and potato is becoming a model species as a crop plant.

In this review paper we give numerous details on the evolution of genomic R&D in potato and related species and provide a large list of references. In this way the interested reader can find easily detailed information of each study.

Molecular markers and linkage maps

Different molecular marker types have been developed and applied in potato. Initially in the 1960s and 1970s these were proteins and isoenzymes. Already in that time molecular markers were not only used for cultivar identification (Zwartz, 1966), but also for the analyses of physiological changes (Stegemann *et al.*, 1973). The first DNA markers described in potato were RFLPs (Gebhardt *et al.*, 1989a). With the invention of the PCR technique other dominant and co-dominant marker types were developed such as RAPD, AFLP (van Eck *et al.*, 1995), SSR (Milbourne *et al.*, 1998), ISTR, ISSR, SCAR and CAP markers (Oberhagemann *et al.*, 1999). These molecular markers were used for identification and purity control of varieties (Görg *et al.*, 1992), for the analyses of biodiversity, for phylogenetic studies in the genus *Solanum* (Debener *et al.*, 1990) and for other applications.

The first genetic linkage maps using RFLP markers were published in 1988 by Bonierbale *et al.* for the interspecific cross between *S. phureja* x (*S. tuberosum* x *S. chacoense*) and by Gebhardt *et al.* (1989b) for a diploid *S. tuberosum* progeny. This latter map included also known RFLP probes such as *PAL, Rubisco,* CoA *ligase and glutamine synthetase.* Afterwards these maps were more saturated and it was possible to align potato and tomato maps using common probes descending from tomato (Tanksley *et al.*, 1992; Gebhardt *et al.*, 1992). These studies revealed also several inversions located at the distal ends of different chromosomes between the two species. Other reduced linkage maps were constructed in different genetic backgrounds and were aligned using co-dominant markers such as RFLPs and particularly SSRs from Milbourne *et al.* (1998).

Qualitative characters and QTL analyses

In these maps qualitative characters were integrated such as monogenic resistances to PVY (Ry_{sto}, Brigneti *et al.*, 1997; Ry_{adg}, Hämäläinen *et al.*, 1997), PVX (*Rx1, Rx2,* Ritter *et al.*, 1991; *Nb,* De Jong *et al.*, 1997; Nx_{phu}, Tommiska *et al.*, 1998), nematodes (*Gro1,* Barone *et al.*, 1990; *H1,* Gebhardt *et al.*, 1993; *Gpa1,* Kreike *et al.*, 1994; *Gpa2,* Rouppe van der Voort *et al.*, 1997) and *Phytophthora infestans* (*R1,* Leonards-Schippers *et al.*,1992; *R3,* El-Kharbotly *et al.*, 1994; *R2,* Li *et al.*, 1998; *R6* and *R7,* El-Kharbotly *et al.*, 1996). These studies showed that resistance genes were clustered and mainly located at distal regions of the chromosomes V, IX, XI and XII. Resistance genes to different pathogens were organised in clusters. On the other hand different analyses have been performed to map quantitative trait characters (QTL) considering for example polygenic resistances to *P. infestans* (Leonards-Schippers *et al.*, 1994) and *G. pallida* (Kreike *et al.*, 1993; Caromel *et al.*, 2003). Other QTL analyses consider yield components (Schäfer-Pregl *et al.*, 1998), tuberisation (Van den Berg *et al.*, 1996a), dormancy (Van den Berg *et al.*, 1996b), tuber form (Van Eck *et al.*, 1994), or starch and reducing sugar contents

(Menéndez *et al.*, 2002). These analyses revealed that many QTLs were located near known genes with specific biological properties. For example QTLs for *Phytophthora infestans* resistance were located near the mapped *R1* gene and near other *Prp* genes, which are involved in defence reactions (Leonards-Schippers *et al.*, 1994). QTLs for reducing sugar contents were associated with enzymes of the carbohydrate metabolic pathway such as AGPase and sucrose invertase (Menéndez *et al.*, 2002).

Since cultivated potatoes are generally tetraploid, also applications for this ploidy level have been developed. Meyer *et al.* (1998) developed a model for linkage mapping and Badshaw *et al.* (1998) detected QTLs for nematode resistance in tetraploid potato. Simko *et al.* (2004) applied LDA (linkage disequilibrium analysis) to identify QTLs for *Verticillium dahliae* resistance in tetraploids.

Candidate genes and comparative genome analyse

The detected relationships between QTLs and functional genes of associated metabolic pathways lead us to the concept of "candidate genes" influencing a specific trait. Such genes should be considered initially, when searching for genes influencing specific characteristics. The compilation and integration of all these results generates a "functional map" of the potato (Gebhardt *et al.*, 1999), which contains certain "strategic regions" where genes of interest for different characters are accumulated. Markers associated to QTLs or to qualitative genes as well as the candidate genes themselves allow the "marker assisted selection" (MAS) in breeding programmes (Gebhardt and Valkonen, 2001). Particularly the design of allele specific primers for these markers allows to distinguish for example between resistant and susceptible genotypes to a specific pathogen (Oberhagemann *et al.*, 1999, Kasai *et al.*, 2000).

Comparative genome analyses showed also that resistance genes from different solanaceous species and against different pathogens were located in homologous genomic regions (Grube *et al.*, 2000). This is the case for example for the nematode resistance gene *Gro1* from potato, the gene *pvr1* providing resistance against a Pepper mottle potyvirus and the fungal resistance gene *py-1* against *Pyrenochaeta lycopersici* in tomato. Based on conserved regions between resistance genes in tobacco and *Arabidopsis*, Leister *et al.* (1996) designed specific primers to amplify homologous resistance genes in potato. The amplification products could be mapped for example to the same locations as the genes *Gro1* and *R7*, confirming in this way their general applicability to detect homologs of resistance genes (RGL, RGH).

Gebhardt *et al.* (2003) performed comparative mapping and genomic analyses between potato and *Arabidopsis* using probes from both species. Certain genomic sequences and structures are conserved in numerous syntheny blocks between the two species, although these can have repeats and different genomic locations in each species. Therefore, the large amount of genomic resources available in the model species *Arabidopsis* can be exploited to a certain degree in order to advance in the discovering of the potato genome.

High density reference map of potato

In order to improve the integration of available information and to provide a highly saturated reference map of potato, the project: "Construction and Exploitation of an ultra-high density map of potato" (FAIR5-PL97-3565) was initiated in 1998. This ultra-high density map allows to obtain markers closely linked to any QTL, and to align other maps in potato based on SSR and RFLP reference markers and co-migrating AFLPs (van Eck *et al.*, 1995), SSR and RFLP. On the other hand such a map is a pre-requisite to obtain a physical map in potato. In 2002 a genetic linkage map was obtained which contained nearly 10000 markers based on the analysis of around 400 AFLP primer combinations in a population of 130 progeny genotypes (Isidore *et al.*, 2004). Considering a genome length of 850 MB in potato, this means an average marker density of one marker for less than 100 kb. Details of map and markers can be seen at: http://www.dpw.wageningen-ur.nl/UHD/. In order to visualize the large amount of markers and to improve the quality of data scoring, the novel "bin" concept based on recombination patterns was developed, which considers not only the absolute numbers but also the particular genotypes in which recombination events occur. The UHD map is composed of 1115 bins and on average 8.6 markers belong to each bin, which are also indicated in the mentioned web page. In the frame of this project also different satellite maps were constructed involving different genetic backgrounds: *S. spegazzini*, *S gourlayi* and *S. tarijense* and QTLs for nematode resistance were determined and compared in these maps. Moreover, markers were developed for different QTL locations and compared between species. An initial BAC library was established and smaller contigs were built in specific regions of chromosomes IV, V and XII.

Novel tools for genome analysis and gene discovery

Traditionally, markers used for linkage mapping and QTL analyses were neutral markers and identify in general genomic DNA. They can be more or less linked to a QTL and their allelic configuration depends on the particular genetic background. Therefore, it would be desirable to detect directly the genes which influence a trait of interest and analyse and compare the effects of their different alleles. These types of markers could be directly applied in marker-assisted selection, independent of the genetic background and are useful to establish functional maps. Actually, several molecular techniques are available for this purpose. These techniques include cDNA-AFLP, microarray analyses, EST and candidate gene mapping and the construction of physical maps.

Several specific EST libraries are available in potato derived from specific tissues, developmental stages or after infections with pathogens and numerous clones have been sequenced and the putative function of the corresponding genes was identified (Ronning *et al.*, 2003).
The use of microchips allows to analyse the spatio-temporal expression of many genes at the same time by hybridising specific, labelled mRNA populations with cDNAs which are immobilised on the chip surface. The TIGR institute (http://www.tigr.org) offers commercially potato microarrays with over 10,000 cDNAs.

The cDNA-AFLP technique represents a robust method to visualize differential gene expression (Bachem *et al.*, 2000; Dellagi *et al.*, 2000; Durrant *et al.*, 2000), a tool for the isolations of genes (Bachem *et al.*, 2000; Bachem *et al.*, 2001; van der Biezen *et al.*, 2000; Durrant *et al.*,

2000) and can be used also for quantitative transcript profiling (Breyne *et al.*, 2003). The technique targets coding regions of the genome and does not require prior sequence information. Transcriptional modulation of gene activity in specific tissues, developmental stages or stress situations can be monitored genome-wide by comparing gene expression patterns of appropriate plant material samples. In addition cDNA-AFLP can also be applied to obtain a complete genetic linkage map (Brugmans *et al.*, 2002). The allelic variants of constitutively expressed genes segregate in the progeny and can therefore be used like other markers for linkage mapping. However, these markers have generally a concrete biological meaning. If they are associated to a particular known QTL, then they may represent a potential candidate gene for this trait.

The availability of a physical map reduces the efforts to clone any gene of interest and allows an efficient mapping of ESTs and cDNAs, independent of segregating polymorphisms. On the other hand it was shown that numerous resistance genes are located in delimited genomic locations (Bakker *et al.*, 2000). Contigs of such regions reduce significantly sequencing efforts. This aspect is particularly important considering the relative large genome size of potato compared to *Arabidopsis* or rice. Song *et al.* (2000) constructed a BACs library in *Solanum bulbocastanum* and Chen *et al.* (2004) obtained two libraries from *S. pinnatisectum*.
Several resistance genes have been isolated meanwhile in potato using BAC libraries or applying other techniques. Bendahmane *et al.* (1997) obtained the virus resistance gene *Rx1* from tetraploid potatoes and van der Vossen *et al.* (2000) isolated the nematode resistance gene *Gpa2* in a gene cluster of the same genomic region. Although these genes confer resistance to different pathogens, a high degree of sequence homology could be observed. Bendahmane *et al.* cloned also *Rx2* in 2000 and in 2002 Ballvora *et al.* isolated *R1*, the first resistance gene against *P. infestans*. They observed that this gene has as other resistance genes in plants typical characteristic elements such as a leucine zipper motive, an NBS and leucine rich repeats (LRR).The isolation of other resistance genes is on the pipe (Bakker *et al.*, 2003, Bradeen *et al.*, 2003, Tian *et al.*, 2003).

Physical and functional maps of potato

In 2003 the R&D Project APOPHYS: Development of a physical and functional map of potato (QLK5-CT-2002-01849) was initialised with the aim to generate novel sources of molecular markers for obtaining new cultivars with multiple resistances and quality traits. Within this project a physical map is constructed based on a library of 70000 BAC clones. On the other hand also a complete transcriptome map is under construction using the cDNA-AFLP technique in the UHD map progeny, with the aim to associate specific cDNAs to published QTLs. Markers are generated for resistance QTLs against biotic stresses (*Phytophthora infestans*, *G. pallida*, *G. rostochiensis*, *M. hapla*, RGL), water stress, maturity, dormancy, tuberisation and starch and sugar contents. Candidate genes are identified for these characters and their allelic variants are analysed in *Solanum* germplasm and commercial cultivars in order to detect associations between particular alleles and specific gene expression. All information with respect to markers, QTLs, Candidate genes and BAC clones are unified in a single functional consensus map of potato.

Genomic resources on the WEB

In order to maximize the diffusion of results and the exploitation of the generated resources, it is necessary to establish user-friendly databases on the Internet which allow all kinds of complex queries. An example of such a database is the USDA-ARS SolGenes database developed at Cornell University (http://ukcrop.net/cgi-bin/WebAce/webace?db=SolGenes). The SolGenes database considers potato, tomato, pepper and other wild relatives. The database includes applications which allow to display chromosomal regions and their corresponding markers and QTLs. These are referenced by literature, probe catalogues, and restriction enzyme loci. Also studies on cultivar identification, QTLs, field trials and results of marker assisted selection are available and the database has links to important germplasm collections.

Another important public source for genomic resources represents the project "Potato Functional Genomics", financed by NSF (National Science Foundation, USA) in which several universities and the mentioned TIGR institute collaborate. The TIGR institute has generated over 60000 ESTs and maintains an expression database of genes involved in biotic and abiotic stress response of *Solanaceae*. All public potato ESTs and GenBank accessions have been used to generate the *Solanum tuberosum* Gene Index (StGI). The aim of this database is to provide a collection of non redundant genes and to provide data on expression patterns, cellular functions and evolutionary relationships. StGI applications allow complex queries based on codes, nucleotide or amino acid sequences, or starting from gene products, metabolic pathways or cDNA libraries.

Recent genomic R&D initiatives in potato

In 2003 the "Centre of Biosystems Genomics" (Wageningen, The Netherlands) initiated the "Potato Genome Project" with the collaboration of universities, R&D institutes and companies (http://www/biosystemsgenomics.nl/research02.html). Different subprojects are related to environmental quality, consumer and processing quality and to the exploitation of the genetic variability in this species. One aim considers the understanding, the control and the utilization of resistance genes which are available in germplasm collections and another approach considers the identification and use of quality genes related to growth, development and processing. In the frame of this project specific microarrays are developed to analyse the transcriptome of the mentioned characters and to screen the corresponding proteins and metabolites as well as for analysing the genetic variability in *Solanum* germplasm. A similar project submitted within the 6th Framework Programme of the EU "BIOEXPLOIT" and involving over 40 industrial and research partners is presently under contract negotiation.

All these examples show that the potato crop is advancing rapidly towards a model species for genomic research and that the international scientific community is realising large efforts to advance in the knowledge and exploitation of the potato genome.

Recently also a Consortium from different countries (PGSC) has been established with the aim to sequence the whole potato genome. The work has started and details can be found at: http://www.genomics.nl/homepage/research/special_activities/pgsc.

References

Bachem, C., B. Horvath, L. Trindade, M. Claassens, E. Davelaar, W. Jordi and R. Visser (2001). A potato tuber-expressed mRNA with homology to steroid dehydrogenases gibberallin levels and plant development. Plant Journal, 25, 595-604.

Bachem, C., R. Oomen, S. Kuyt, B. Horvath, M. Claassens, D. Vreugdenhil and R. Visser (2000). Antisense suppression of a potato alpha-SNAP homologue leads alterations in cellular development and assimilate distribution. Plant Molecular Biology 443, 473-482.

Bachem, C.W.B., R.S. van der Hoeven, S.M. de Bruijn, D. Vreugdenhil, M. Zabeau and R.G.F. Visser (1998). Visualization of differential gene expression using a novel method of RNA fingerprinting based on AFLP: Analysis of gene expression during potato tuber development. Plant Journal, 9, 745-753.

Bakker, E., P. Butterbach, J. Rouppe van der Voort, E. van der Vossen, J. van Vliet, J. Bakker and A. Goverse (2003). Genetic and physical mapping of homologues of the virus resistance gene Rx1 and the cyst nematode resistance gene Gpa2 in potato. Theoretical and Applied Genetics, 106, 1524-1531.

Bakker, J., W. Stiekema and R. Klein-Lankhorst (2000). Homologues of a single resistance gene cluster in potato confer resistance to distinct pathogens: a virus and a nematode. Plant Journal, 23, 1-11.

Ballvora, A., M.R. Ercolano, J. Weiss, K. Meksem, C.A. Bormann, P. Oberhagemann, F. Salamini and C. Gebhardt (2002). The R1 gene for potato resistance to late blight (Phytophthora infestans) belongs to the leucine zipper/NBS/LRR class of plant resistance genes. Plant Journal, 30, 361-371.

Bendahmane, A., M. Querci, K. Kanyuka and D.C. Baulcombe (2000). Agrobacterium transient expression system as a tool for the isolation of disease resistance genes: application to the Rx2 locus in potato. Plant Journal, 21, 73-81.

Bendahmane, A., K.V. Kanyuka and D.C. Baulcombe (1997). High resolution and physical mapping of the Rx gene for extreme resistance to potato virus X in tetraploid potato. Theoretical and Applied Genetics, 95, 153-162.

Barone, A., E. Ritter, U. Schachtschabel, T. Debener, F. Salamini and C. Gebhardt (1990). Localization by restriction fragment length polymorphism mapping in potato of a major dominant gene conferring resistance to the potato cyst nematodo Globodera rostochiensis. Molecular and General Genetics, 224, 177-182.

Bonierbale, M., R.L. Plaisted and S.D. Tanksley (1988). RFLP maps based on a common set of clones reveal modes of chromosomal evolution in potato and tomato. Genetics, 120, 1095-1103.

Bradeen, J.M., S.K. Naess, J. Song, G.T. Haberlach, S.M. Wielgus, C.R. Buell, J. Jiang and J.P. Helgeson (2003). Concomitant reiterative BAC walking and fine genetic mapping enable physical map development for the broad-spectrum late blight resistance region, RB. Molecular and General Genomics, 269, 603-611.

Bradshaw, J.E., C.A. Hackett, R.C. Meyer, D. Milbourne, J.W. McNicol, M.S. Phillips and R. Waugh (1998). Identification of AFLP and SSR markers associated with quantitative resistance to *Globodera pallida* (Stone) in tetraploid potato (*Solanum tuberosum* ssp. *tuberosum*) with a view to marker-assisted selection. Theoretical and Applied Genetics, 97, 201-210.

Breyne, P., R. Dreesen, B. Cannoot, D. Rombaut, K. Vandepoele, S. Rombauts, R. Vandeerhaeghen, D. Inze and M. Zabeau (2003). Quantitative cDNA-AFLP analysis for genome-wide expression studies. Molecular and General Genomics, 269, 173-179.

Brigneti, G., J. García-Mas and D.C. Baulcombe (1997). Molecular mapping of the potato virus Y resistance locus Ry_{sto} in potato. Theoretical and Applied Genetics, 94, 198-203.

Brugmans, B., A. Fernandez del Carmen, C.W. Bachem, H. van Os, H.J. van Eck and R.G. Visser (2002). A novel method for the construction of genome wide transcriptome maps. Plant Journal, 31, 211-222.

Caromel, B., D. Mugniery, V. Lefebvre, S. Andrzejewski, D. Ellisseche, M.C. Kerlan, P. Rousselle and F. Rousselle-Bourgeois (2003). Mapping QTLs for resistance against Globodera pallida (Stone) Pa2/3 in a diploid potato progeny originating from Solanum spegazzinii. Theoretical and Applied Genetics, 106, 1517-1523.

Chen, Q., S. Sun, Q. Ye, S. McCuine, E. Huff and H.B. Zhang (2004). Construction of two BAC libraries from the wild Mexican diploid potato, Solanum pinnatisectum, and the identification of clones near the late blight and Colorado potato beetle resistance loci. Theoretical and Applied Genetics, 108, 1002-1009.

Debener, T., F. Salamini and C. Gebhardt (1990). Phylogeny of wild and cultivated *Solanum* species based on nuclear restriction fragment length polymorphisms (RFLPs). Theoretical and Applied Genetics, 79, 360-368.

De Jong, W., A. Forsyth, D. Leister, C. Gebhardt and D.C. Baulcombe (1997). A potato hypersensitive resistance gene against potato virus X maps to a resistance gene cluster on chromosome 5. Theoretical and Applied Genetics, 95, 246-252.

Dellagi, A., P. Birch, J. Heilbronn, G. Lyon and I. Toth (2000). cDNA-AFLP analysis of differential gene expression in the prokariotic plant pathogen *Erwinia carotovora*. Microbiology Reading, 146, 165-171.

Durrant, W., O. Rowland, P. Piedras, K. Hammond-Kosack and J. Jones (2000). cDNA-AFLP reveals a striking overlap in race specific resistance and wound response gene expression profiles. Plant Cell, 12, 963-977.

El-Kharbotly, A., C. Palomino-Sánchez, F. Salamini, E. Jacobsen and C. Gebhardt (1996). *R6* and *R7* alleles of potato conferring race-specific resistance to *Phytophthora infestans* (Mont.) de Bary identified genetic loci clustering with the *R3* locus on chromosome XI. Theoretical and Applied Genetics, 92, 880-884.

El-Kharbotly, A., C. Leonards-Schippers, D.J. Huigen, E. Jacobsen, A. Pereira, W.J. Stiekema, F. Salamini and C. Gebhardt (1994). Segregation analysis and RFLP mapping of the R1 and R3 alleles conferring race-specific resistance to Phytophthora infestans in progeny of dihaploid potato parents. Molecular and General Genetics, 242, 749-754.

Gebhardt, C., B. Walkemeier, H. Henselewski, A. Barakat, M. Delseny and K. Stuber (2003). Comparative mapping between potato (Solanum tuberosum) and Arabidopsis thaliana reveals structurally conserved domains and ancient duplications in the potato genome. Plant Journal, 34, 529-541.

Gebhardt, C. and J.P. Valkonen (2001). Organization of genes controlling disease resistance in the potato genome. Annual Review of Phytopathology, 39, 79-102.

Gebhardt, C., R. Schäefer-Pregl, P. Oberhagemann, X. Chen, C. Chalot-Balandras, E. Ritter, L. Concilio, E. Bonnel, J. Hesselbach and F. Salamini (1999). Function Maps of potato. EBPN, Proceedings of Phytosphere, 99.

Gebhardt, C., D. Mugniery, E. Ritter, F. Salamini and E. Bonnel (1993). Identification of RFLP markers closely linked to the H1 gene conferring resistance to Globodera rostochiensis in potato. Theoretical and Applied Genetics, 85, 541-544.

Gebhardt, C., E. Ritter, A. Barone, T. Debener, B. Walkemeier, U. Schachtschabel, H. Kaufmann, R.D. Thompson, M.W. Bonierbale, M.W. Ganal, S.D. Tanksley and F. Salamini (1992). RFLP maps of potato and their alignment with the homeologous tomato genome. Theoretical and Applied Genetics, 83, 49-57.

Gebhardt, C., C. Blomendahl, U. Schachtschabel, T. Debener, F. Salamini and E. Ritter (1989a). Identification of 2n breeding lines and 4n varieties of potato (Solanum tuberosum, ssp. tuberosum) with RFLP-fingerprints. Theoretical and Applied Genetics, 78, 16-22.

Gebhardt, C., E. Ritter, T. Debener, U. Schachtschabel, B. Walkemeier, H. Uhrig and F. Salamini (1989b). RFLP analysis and linkage mapping in Solanum tuberosum. Theoretical and Applied Genetics, 78, 65-75.

Görg, R., U. Schachtschabel, E. Ritter, F. Salamini and C. Gebhardt (1992). Discrimination among 136 Tetraploid potato varieties by fingerprints using highly polymorphic DNA markers. Crop Science, 32, 815-819.

Grube, R.C., E.R. Radwanski and M. Jahn (2000). Comparative Genetics of Disease Resistance Within the Solanaceae. Genetics, 155, 873-887.

Hämäläinen, J.H., K.N. Watanabe, J.P.T. Valkonen, A. Arihara, R.L. Plaisted, E. Pehu, L. Miller and S.A. Slack (1997). Mapping and marker-assisted selection for a gene for extreme resistance to potato virus Y. Theoretical and Applied Genetics, 94, 192-197.

Isidore, E., H. van Os, S. Andrzejewski, J. Bakker, I. Barrena, G. J. Bryan, B. Caromel, H. van Eck, B. Ghareeb, W. de Jong, P. van Koert, V. Lefebvre, D. Milbourne, E. Ritter, J. R. van der Voort, F. Rousselle-Bourgeois, J. van Vliet and R. Waugh (2004). Toward a Marker-Dense Meiotic Map of the Potato Genome: Lessons From Linkage Group I. Genetics, 165, 2107 - 2116.

Kasai, K., Y. Morikawa, V.A. Sorri, J.P. Valkonen, C. Gebhardt and K.N. Watanabe (2000). Development of SCAR markers to the PVY resistance gene Ry_{adg} based on a common feature of plant disease resistance genes. Genome, 43, 1-8.

Kreike, C.M., J.R.A. de Koning, J.H. Vinke, J.W. van Ooijen, C. Gebhardt and W.J. Stiekema (1993). Mapping of loci involved in quantitatively inherited resistance to the potato cyst nematode Globodera rostochiensis pathotype Ro1. Theoretical and Applied Genetics, 87, 464-470.

Kreike, C.M., J.R.A. de Koning, J.H. Vinke, J.W. van Oojen and W.J. Stiekema (1994). Quantitatively inherited resistance to *Globodera pallida* is dominated by one major locus in *Solanum spegazzinii*. Theoretical and Applied Genetics, 88, 764-769.

Leister, D., A. Ballvora, F. Salamini and C. Gebhardt (1996). A PCR-based approach for isolating pathogen resistance genes from potato with potential for wide application in plants. Nature Genetics, 14, 421-429.

Leonards-Schippers, C., W. Gieffers, F. Salamini and C. Gebhardt (1992). The R1 gene conferring race-specific resistance to Phytophthora infestans in potato is located on potato chromosome V. Molecular and General Genetics, 233, 278-283.

Leonards-Schippers, C., W. Gieffers, R. Schäfer-Pregl, E. Ritter, S.J. Knapp, F. Salamini and C. Gebhardt (1994). Quantitative resistance to Phytophthora infestans in potato: a case study for QTL mapping in an allogamous plant species. Genetics 137: 67-77.

Li, X., H.J. van Eck, J.N.A.M. Rouppe van der Voort, D-J. Huigen, P. Stam and E. Jacobsen (1998). Autotetraploids and genetic mapping using common AFLP markers: the *R2* allele conferring resistance to *Phytophthora infestans* mapped on potato chromosome 4. Theoretical and Applied Genetics, 96, 1121-1128.

Menendez, C.M., E. Ritter, R. Schäfer-Pregl, B. Walkemeier, A. Kalde, F. Salamini and C. Gebhardt (2002). Cold Sweetening in Diploid Potato: Mapping Quantitative Trait Loci and Candidate Genes. Genetics 162: 1423-1434.

Meyer, R.C., D. Milbourne, C.A. Hackett, J.E. Bradshaw, J.W. McNichol and R. Waugh (1998). Linkage analysis in tetraploid potato and association of markers with quantitative resistance to late blight (Phytophthora infestans). Molecular and General Genetics, 259, 150-160.

Milbourne, D., R.C. Meyer, A.J. Collins, L.D. Ramsay, C. Gebhardt and R. Waugh (1998). Isolation, characterisation and mapping of simple sequence repeat loci in potato. Molecular and General Genetics, 259, 233-245.

Oberhagemann, P., C. Chalot-Balandras, E. Bonnel, R. Schäfer-Pregl, D. Wegener, C. Palomino, F. Salamini and C. Gebhardt (1999). A genetic analysis of quantitative resistance to late blight in potato: Towards marker assisted selection. Molecular Breeding, 5, 399-415.

Ritter, E.., T. Debener, A. Barone, F. Salamini and C. Gebhardt (1991). RFLP mapping on potato chromosomes of two genes controlling extreme resistance to potato virus X (PVX). Molecular and General Genetics, 227, 81-85.

Ronning, C.M., S.S. Stegalkina, R.A. Ascenzi, O. Bougri, A.L. Hart, T.R. Utterbach, S.E. Vanaken, *et al.* (2003). Comparative analyses of potato expressed sequence tag libraries. Plant Physiology, 131, 419-429.

Rouppe van der Voort, J., P. Wolters, R. Folkertsma, R. Hutten, P. van Zandvoort, H. Vinke, K. Kanyuka, A. Bendahmane, E. Jacobsen, R. Janssen and J. Bakker (1997). Mapping of the cyst nematode resistance locus *Gpa2* using a strategy based on comigrating AFLP markers. Theoretical and Applied Genetics, 95, 874-880.

Schäfer-Pregl, R, E. Ritter, L. Concilio, J. Hesselbach, L. Lovatti, B. Walkemeier, H. Thelen, F. Salamini and C. Gebhardt (1998). Analysis of quantitative trait locis (QTLs) and quantitative trait alleles (QTAs) for tuber yield and starch content. Theoretical and Applied Genetics, 97, 834-846.

Simko, I., S. Costanzo, K.G. Haynes, B.J. Christ and R.W. Jones (2004). Linkage disequilibrium mapping of a Verticillium dahliae resistance quantitative trait locus in tetraploid potato (Solanum tuberosum) through a candidate gene approach. Theoretical and Applied Genetics, 108, 217-224.

Song, J., F. Dong and J. Jiang (2000). Construction of a bacterial artificial chromosome (BAC) library for potato molecular cytogenetic research. Genome, 43, 199-204.

Stegemann, H., H. Francksen and V. Macko (1973). Potato proteins: genetic and physiological changes, evluated by one- and two-dimensional PAA-gel-techniques. Zeitschrift für Naturforschung, 28, 722-732.

Tanksley, S.D., M.W. Ganal, J.P. Prince, M.C. de Vicente, M.W. Bonierbale, P. Broun, T.M. Fulton, J.J. Giovannoni, S. Grandillo, G.B. Martin, R. Messeguer, J.C. Miller, L. Miller, A.H. Paterson *et al.* (1992). High density molecular linkage maps of the tomato and potato genomes. Genetics 132: 1141-1160.

Tian, Z.D., J. Liu and C.H. Xie (2003). Isolation of resistance related-genes to Phytophthora infestans with suppression subtractive hybridization in the R-gene-free potato. Yi Chuan Xue Bao, 30, 597-605.

Tommiska, T.J., J.H. Hämäläinen, K.N. Watanabe and J.P.T. Valkonen (1998). Mapping of the gene Nxphu that controls hypersensitive resistance to potato virus X in Solanum phureja IvP35. Theoretical and Applied Genetics, 96, 840-843.

Van den Berg, J.H., E.E. Ewing, R.L. Plaisted, S. McMurry and M.W. Bonierbale (1996a). QTL analysis of potato tuberization. Theoretical and Applied Genetics, 93, 307-316.

Van den Berg, J.H., E.E. Ewing, R.L. Plaisted, S. McMurry and M.W. Bonierbale (1996b). QTL analysis of potato tuber dormancy. Theoretical and Applied Genetics, 93, 317-324.

Van der Biezen, E.A., H. Juwana, J.E. Parker and J.D.G. Jones (2000). cDNA-AFLP display for the isolation of *Pernospora parasitica* genes *expressed* during infection in *Arabidopsis thaliana*. Molecular Plant-Microbe Interactions, 13, 895-898.

Van der Vossen, E.A., J.N. van der Voort, K. Kanyuka, A. Bendahmane, H. Sandbrink, D.C. Baulcombe, J. Bakker, W.J. Stiekema and R.M. Klein-Lankhorst (2000). Homologues of a single resistance-gene cluster in potato confer resistance to distinct pathogens: a virus and a nematode. Plant Journal, 23, 567-576.

Van Eck, H.J., J. Rouppe van der Voort, J. Draaistra, P. van Zandvoort, E. van Enckevort, B. Segers, J. Peleman, E. Jacobsen, J. Helder and J Bakker (1995). The inheritance and chromosomal location of AFLP markers in a non-inbred potato offspring. Molecular Breeding, 1, 397-410.

Van Eck, H.J., J.M.E. Jacobs, P. Stam, J. Ton, W.J. Stiekema and E. Jacobsen (1994). Multiple alleles for tuber shape in diploid potato detected by qualitative and quantitative genetic analysis using RFLPs. Genetics, 137, 303-309.

Zwartz, J.A (1966). Potato varieties and their protein electropherogram characteristics. European Potato Journal, 9, 111-128.

Curriculum vitae – Enrique Ritter

Enrique Ritter studied Agriculture at the University of Bonn (Germany), from which he received his degree in 1978. In 1979 he joined the Institute for Agricultural Botany at the University of Bonn, from which, in 1984, he received a PhD degree in Agricultural Sciences. In 1986 he joined the Max-Planck Institute for Plant Breeding in Cologne as a visiting scientist in the Department of Francesco Salamini, developing models and programmes for linkage mapping and QTL analysis, particularly useful for allogamous species. In 1991 he went to Spain to NEIKER - The Basque Institute for Research and Development in Agriculture - to become Senior Researcher and Head of the Department for Plant Production and Protection. Since 2003 he is head of the Department of Biotechnoloy at NEIKER. He has over 15 years experience particularly in potato breeding, in molecular biology and genetics and large experience in the participation and in managing of R&D&T projects. He is working in different fields of theoretical and practical genetics and molecular biology. Dr. Ritter was Chairman of the Virology Section of the European Association for Potato Research (EAPR) from 1993 to 1999, President of the EAPR (2002-2005) and currently serves as Past President of the association.

Company profile – NEIKER

NEIKER - The Basque Institute for Agricultural Research and Development - is a public, non-profit company owned by the Basque Government and controlled by the Department of Agriculture and Fisheries. The Institute has two centres. One is located in Arkaute (Araba) and the other in Derio (Bizkaia). The research institute performs basic and applied research in agriculture (plants and animals). Between 110 and 130 scientists, technical assistants and other staff members, including undergraduate and graduate students are active in four technical departments and administration. These departments include: Agrosystems and Animal Production, Biotechnology, Plant Production and Protection and Animal Health. NEIKER perfoms R&D&T projects in the field of agricultural, livestock, forestry and environmental sciences. The institute provides also local and regional Governments, Management Centres, Professional Associations, farmers, Agro-food Companies and other public and private entities with consulting services. One of the main research areas of the Department of Biotechnology is potato, since this crop plays an important role in the local area for seed potato production. NEIKER is one of the only two national institutions, which have a potato-breeding programme, and several new potato varieties have been released in the last years. The breeding programme combines breeding at tetraploid and diploid level covering - besides classical breeding through crossings - all novel technological aspects such as protoplast fusion, genetic transformation and DNA marker technology. A broad germplasm collection including *Solanum* wild species is available for this purpose. Screening for resistance to important potato pathogens are performed routinely within the breeding programme. Molecular techniques are well established at the institute and are continuously updated and optimized for potato as well as for other crop species.

Breeding for quality improvement: market fitness and nutritional quality

T.R. Tarn
Agriculture and Agri-Food Canada, Potato Research Centre, P.O. Box 20280, Fredericton, NB, Canada

Abstract

Many utilization-related traits of the potato have been improved over the past 40 years. As the world's fourth most important food crop the potato is an important dietary source of starch and of ascorbic acid, yet very little attention has been given to improving these or other traits of potential nutritional significance. Following reference to the consensus composition of the potato, variation and potential for modifying and improving the nutritional value of starch, protein, ascorbic acid, anthocyanins and carotenoids is discussed. With good potential for improvement and with its widespread acceptability, the potato can make a greater contribution to nutrition in the future.

Keywords: starch, protein, anthocyanins, carotenoids, antioxidants

Introduction

In the last 40 years progress in potato breeding in North America has included improved appearance, dry matter, processing colour and grade (Douches *et al.*, 1996, Love *et al.*, 1998). Reviewing 40 years of potatoes research Struik *et al.* (1997) describe progress in understanding texture, enzymatic and non-enzymatic discolouration and effects of deep fat-frying. They also note the potato is an interesting crop for functional and novel foods. However, any progress on nutritional aspects of potato did not attract the attention of any of these reviewers.

Nearly a decade after these reviews, potato consumption is falling in developed countries because of dietary concerns about the poor nutritional quality of fast foods and the increasing frequency of obesity. This situation is resulting in a fresh look at the quality characteristics of the potato. It also has many breeders looking at nutritional quality for the first time; considering the potato is the fourth most important crop in the world, this attention to quality is long overdue. The intent of this presentation is to review some of what has been achieved and to consider what may be possible in the foreseeable future.

The baseline is the consensus document on the nutritional and other compositional characteristics of the potato prepared by the OECD (2002) and reproduced in Table 1. This paper reviews research on three of these components, starch, protein and ascorbic acid, and two other components, anthocyanins and carotenoids, not shown in the table. Like ascorbic acid, anthocyanins and carotenoids are antioxidants and the potential to increase their levels can add to the nutritional quality of potato.

Table 1. Key nutrients of potato (fresh weight basis) (OECD, 2002).

Component	Unit	Mean	Ranges
Dry matter	%	23.7	13.1 - 36.8
Starch	%	17.5	8.0 - 29.4
Protein	%	2.0	0.69 - 4.63
Fat	%	0.12	0.02 - 0.2
Dietary fibre	%	1.7	1 - 2
Crude fibre	%	0.71	0.17 - 3.48
Minerals (crude ash)	%	1.1	0.44 - 1.87
Sugars	%	0.5	0.05 - 8.0
Ascorbic acid + dehydroascorbic acid	mg/kg	100 - 250	10 - 540

Sources: Lisinska and Leszczynski, 1989, and Woolfe, 1987.

Starch

Starch is the main component of potato accounting for 75 to 80% on a dry weight basis or about 17.5% on a fresh weight basis. In terms of the human diet, this translates into as much as 15 to 20% of total starch intake (see Soh and Brand-Miller, 1999). This has significance because of two characteristics, glycaemic index and resistant starch, yet has received very minimal attention from a potato quality perspective.

The glycaemic index (GI) is an index for the ranking of foods according to the blood glucose response following their consumption relative to a standard such as white bread. There is increasing evidence that low GI foods are desirable for the management of diabetes and hyperlipidaemia, as well as the prevention of diabetes and cardiovascular disease (see Leeman *et al.* 2004). Potatoes have a variable but generally high GI (Soh and Brand-Miller, 1999, Leeman *et al.*, 2005). Soh and Brand-Miller (1999) examined three cultivars and several methods of preparation. They found no differences among the cultivars and a significant difference only between canned new potatoes and the highest GI cultivar. They also found average tuber size to be correlated with GI. On this basis they suggested that the lower GI of new potatoes may be attributed to differences in starch structure and that it may be possible to produce cultivars with modified starch structure and lower GI. However, using an in vitro starch hydrolysis procedure and looking at four early cultivars and small and large tubers of six maincrop cultivars Leeman *et al.* (2005) found no support for this. On the other hand they did find significant differences between two of the early cultivars and in the interaction between two cultivars and short term storage.

Starch has a digestible component and an indigestible component. The latter, resistant starch, is important to the microflora of the colon and there is some evidence that resistant starch from potato may be of special interest in relation to health of the colon (see Leeman *et al.*, 2005). Raw potato is high in resistant starch because the starch is encapsulated in granules. This changes when potato is cooked and the starch is gelatinized, and can be modified by further treatments (see Garcia-Alonso and Goñi, 2000). Leeman *et al.* (2005) found significant differences in amounts of resistant starch in boiled potatoes followed by different time-temperature cycles and concluded that potato products may be important contributors to resistant starch intake in an average western diet.

Protein

Potatoes are able to produce more protein per unit area that any other major crop with the exception of soybean (Dale and Mackay, 1994). Nutritionally the protein is of high nutritional value (OECD, 2002). The major proteins are albumin, globulin, prolamine and glutenin, with a further fraction containing glycoproteins, metaloprotein and phosphoproteins. Compared with cereals, potato protein contains a higher proportion of the essential amino acid lysine and a lower concentration of the sulphur containing amino acids methionine and cysteine. There has been an interest in potato protein content since the 1970s (see Dale and Mackay, 1994) and increasing the levels of potato tuber protein has been justified on two needs. First, there is a general lack of protein in diets in many developing countries and where potato is an important constituent of the diet increased potato protein levels can be important. Second, where the potato is an industrial starch source, protein is a valuable by-product for feed supplements.

In the last ten years there have been several approaches to increasing the content of specific amino acids. The methionine synthetic pathway has attracted the interest of a number of researchers and at least a couple have obtained increases in levels of methionine. Protoplasts of cv. Russet Burbank selected on a medium containing the amino acid analogue ethionine resulted in a methionine increased of 2.66 times in tubers from mature plants (Langille *et al.*, 1998). Plants of cv. Désirée expressing a threonine synthase antisense construct produced tubers with a methionine increase of up to 30-fold (Zeh *et al.*, 2001). However, the line showing the strongest inhibition showed severe growth retardation and tuber yield reduction though lines with low to medium inhibition showed only marginal changes of phenotype.

A 35-45 percent increase in total protein and an increase in most essential amino acids was achieved through the introduction of the amaranth seed albumin gene (AmA1) from *Amaranthus hypochondriacus* into a late blight resistant diploid potato (Chakraborty *et al.*, 2000a and b). The AmA1 protein is non-allergenic, rich in all essential amino acids and in composition corresponds well with WHO standards for optimum human nutrition. In lines with higher levels of gene expression tuber number was increased over 2-fold and yield up to 3.5-fold. Protein content ranged from 14.57 to 16.58 mg/g tuber compared to 11.15 for the original diploid clone.

Ascorbic acid

Ascorbic acid from potatoes may contribute up to 40% of the daily recommended intake of humans (OECD, 2002) and is frequently identified as a major nutritional asset. Levels are higher in the central pith region of the tuber than in the outer cortex (Munshi and Mondy, 1989).

There have been several surveys of cultivars and germplasm to explore the potential to breed for higher ascorbic acid levels. Sinden *et al.* (1978) surveyed 98 clones of Tuberosum germplasm and found significant differences among them. These authors also examined ten tuber-bearing

Solanum species and found ascorbic acid levels top range from 53 mg/kg FW in *S. bulbocastanum* to 254 mg/kg in *S. stoloniferum*. Davies *et al.* (2002), working with diploid and tetraploid families, derived from the diploid *S. phureja* and haploid and tetraploid cultivars, found mean ascorbic acid levels to range between 110 and 128 mg/kg FW for diploid families and between 110 and 153 mg/kg FW for tetraploid families. Dale *et al.* (2003) evaluated 27 near-commercial breeding lines from three European breeding programmes and six processing cultivars over three locations and found the combined ascorbic acid and dihydroascorbic acid range to be between 177 and 363 mg/kg FW. Love *et al.* (2004) surveyed 75 clones from 12 North American breeding programmes and found ascorbic acid to range from 115 to 298 mg/kg. Mean values were similar for clones from 11 of the breeding programmes while the twelfth programme was represented only by two high content clones.

Ascorbic acid levels are well known to drop during storage. However, Dale *et al.* (2003) and Love *et al.* (2004) found that during storage not all genotypes lose ascorbic acid at the same rate. Davies *et al.* (2002) obtained a similar result when they found a diploid Phureja-haploid Tuberosum hybrid that retained a twofold higher content of ascorbic acid following storage.

Heritability estimates vary from 0.45 for narrow sense heritability (Sinden *et al.*, 1978) to 0.71 for broad sense heritability (Love *et al.*, 2004), while the interaction between genotype and environment is low (Dale *et al.*, 2003, Love *et al.*, 2004). Taken together, this information supports breeding and selecting for increased levels of ascorbic acid in potato. Dale *et al.* (2003) refine this further by suggesting that since most potatoes are consumed following storage the objective should be to increase content post-storage.

Anthocyanins

Anthocyanins are the most important group of water soluble pigments in plants. In potatoes these flavonoids are responsible for the red, blue and purple colours of tuber skin and flesh. These pigments have value as natural food colourants (Rodriguez-Saona *et al.*, 1998) and antioxidants and in the latter role have the potential to add to the nutritional value of the potato. Red colouration is caused by the presence of acylated glycosides of pelargonidin and purple by the presence of acylated glycosides of petunidin and peonin (Brown *et al.*, 2003). In the Andes of South America pigmented flesh is common in germplasm of the primitive cultivated diploid and tetraploid groups (Brown, 2005). However, pigmented flesh has been rare in the modern cultivated potato though this may be changing with the interest in the antioxidant value of the pigments.

Thirty three red-fleshed breeding lines and cultivars were studied by Rogrigues-Saona *et al.* (1998) who found monomeric anthocyanin content to range from 2 to 40 mg/100g FW. Lewis *et al.* (1998a) found anthocyanin content to range from 1 to 11 mg/100g FW in red-fleshed clones and from 0.3 to 23 mg/100g FW in purple-fleshed clones depending on the clone and season. White-fleshed cultivars with pink skin had lesser amounts while anthocyanins were absent in white-fleshed cultivars with white skin. Anthocyanins were also absent in the flesh of seven wild *Solanum* species and in one clone of the primitive cultivated species *S. stenotomum* studied (Lewis *et al.*, 1998b). Evaluating anthocyanins in breeding populations selected for coloured skin and flesh, Brown *et al.* (2003) reported total anthocyanins to range from 5.5 to 35.0 mg/100g FW. In a further study of similar material Brown *et al.* (2005) reported that yellow-fleshed clones had negligible amounts of anthocyanins while clones with red and yellow flesh combined had 9.5 to 17.9 mg/100g FW of total anthocyanins. In comparison they found purple-fleshed clones to have 17 to 20.1 mg/100g FW and red-fleshed clones to have 19.8 to 37.8 mg/100g FW.

The inheritance of tuber flesh colour is summarized by Brown (2005). A single gene, *D*, controls the synthesis of red pigment, and a single gene, *P*, controls blue pigment. Gene *I* epistatically determines the presence and absence of pigment even in the presence of the other pigment genes. However, the completeness of anthocyanin distribution may be under complex genetic control. Based on segregation for anthocyanin content in their breeding populations Brown *et al.* (2003) consider there is a possibility of enhancing anthocyanin levels in potato. Anthocyanin content decreases during the growth and maturation of tubers, while levels increase with greater day-length and cooler temperatures (Reyes *et al.*, 2004). Since anthocyanins are water soluble they leach out of tuber flesh during boiling but are retained during baking and microwaving.

Carotenoids

Yellow-fleshed potatoes, popular in many parts of the world, obtain their colour from the xanthophyll carotenoids. Less well known is deep yellow or orange flesh that results from the presence of zeaxanthin. Potato carotenoids are attracting attention because high serum levels of lutein are correlated with a reduced risk of age-related macular degeneration in humans.

During development carotenoid levels are reported to decrease as tuber size and dry matter increase (Haynes *et al.*, 1994, Morris *et al.*, 2004), though this was not the case for the intense yellow-fleshed cultivar Inca Dawn (Morris *et al.*, 2004). However, in storage total carotenoid levels are reported to be stable (Morris *et al.*, 2004) though there is a shift in carotenoid components.

Several authors have reported on carotenoid content. An early and detailed report of 13 German cultivars, with flesh colours ranging from white to intense yellow, identified violaxanthin, lutein and lutein-5,6-epoxide as the main xanthophylls (Iwanzik *et al.*, 1983). In this study total carotenoids varied from 27 to 73 μg/100g FW in the white-fleshed cultivars to 171 to 342 μg/100g FW in the intense yellow-fleshed cultivars. Similar results were obtained by Lu *et al.* (2001) from a diploid population derived from *S. phureja* and *S. stenotomum*, though their yellow-fleshed clones contained up to 1,435 μg/100g FW of total carotenoids, and by Nesterenko and Sink (2003) from a group of 15 selections from the Michigan State University breeding programme. The latter researchers included one orange-fleshed selection that contained zeaxanthin and 878 μg/100g FW total carotenoids. Brown *et al.* (2005) found carotenoid levels of 509 to 795 μg/100g FW in breeding lines with dark yellow flesh, and 109 to 327 μg/100g FW in breeding lines combining red and yellow flesh. Working with a diploid population derived from *S. phureja* and *S. stenotomum*, Brown *et al.* (1993) focused on orange coloured flesh. They found the orange colour to be associated with large amounts of zeaxanthin, and total carotenoid levels to range between 1390 and 2,175 μg/100g FW, the highest levels reported for potato and about one-sixth the total carotenoid content of carrot.

Yellow flesh is controlled by a single dominant gene, *Y*, with yellow flesh dominant over white, *y*. Brown *et al.* (1993) found orange flesh to be controlled by an allele, *Or*, at the same locus. This locus has been mapped to chromosome 3 (Bonierbale *et al.*, 1988). The genotype × environment interaction, though significant (Haynes *et al.*, 1996), is too small to be of practical importance in selecting for this trait. Potato has also been genetically transformed to increase zeaxanthin content. Using both sense and antisense constructs, Romer *et al.* (2002) were able to inhibit zeaxanthin conversion to violaxanthin in the cultivars Baltica and Freya. Both approaches resulted in higher levels of zeaxanthin, up to 134-fold increase, and most of the transformants also showed increases in total carotenoids, up to 5.7-fold. Their highest recorded total carotenoid level was 60.8 µg/g DW, equivalent to approximately 1,400 µg/100g FW. In a different modification of the carotenoid biosynthetic pathway Ducreux *et al.* (2005) introduced the gene encoding phytoene synthase into two yellow-fleshed cultivars cv. Désirée (*S. tuberosum*) and cv. Mayan Gold (*S. phureja*). This resulted in carotenoid levels in Désirée increasing from 5.6 µg/g DW to 35 µg/g DW with β-carotene increasing from negligible amounts in the controls to 11 µg/g DW. In cv. Mayan Gold levels increased from 20 µg/g DW to 78 µg/g DW (equivalent to approximately 1,800 µg/100g FW) in the most affected transgenic line.

Discussion

The potato is the world's fourth most important crop. Depending on the country, potato can contribute up to 20% of the total dietary starch intake and up to 40% of the ascorbic acid intake. In the diet of the Maori of New Zealand the purple-fleshed cultivar Urenika is an important functional food because of its anthocyanin content (Cambie and Ferguson, 2003). The importance of the crop and its often high rates of consumption mean that changes in tuber nutritional quality can have a significant dietary impact.

Starch intake can have important implications for glycaemic index and colon health. The high glycaemic index (GI) of potato is well known and from a dietary perspective that is an undesirable characteristic. Little information is available about the relationship between the characteristics of potato starch and GI, and about the variation in starch characteristics in potato germplasm so that it is difficult to forecast the changes that may be possible. Research in this area will provide an improved understanding of these relationships and indicate what the potential is for reducing the GI of potato or producing products with lower GI, and of increasing the proportion of resistant starch.

The ability to improve potato protein content has already been demonstrated. Where dietary protein intake is low and potatoes can be produced, the potato offers a valuable option to improve nutrition.

Antioxidants are recognized as important nutrients because of their ability to scavenge free oxygen radicals in the human body. Diets rich in antioxidants are associated with a lower incidence of atherosclerotic heart disease, some cancers, age related macular degeneration and severity of cataracts (see Brown, 2005). The dietary importance of ascorbic acid in the potato is already established (OECD, 2002). In addition, anthocyanin and carotenoid levels come close to or overlap those of other foods known for the value of their antioxidants (Tables 2 and 3).

Table 2. Anthocyanin content of potato and selected foods.

Food	Anthocyanin content Mg/100g FW	Reference
Potato, red flesh	2 - 40	Rodrigues-Saona *et al.*, 1998
Potato, red flesh	1 - 11	Lewis *et al.*, 1998a
Potato, red flesh	19.8 - 37.8	Brown *et al.*, 2005
Potato, red and yellow flesh	9.5 - 17.9	Brown *et al.*, 2005
Potato, purple flesh	0.3 - 23.2	Lewis *et al.*, 1998a
Potato, red and purple flesh	5.5 - 35.0	Brown *et al.*, 2003
Potato, purple flesh	17.0 - 20.1	Brown *et al.*, 2005
Cabbage, red	25	Clifford, 2000
Blueberry	82 - 420	Clifford, 2000
Grape, red	30 - 750	Clifford, 2000
Strawberry	15 - 35	Clifford, 2000

Table 3. Lutein + zeaxanthin and total carotenoid content of potato and selected vegetables.

Food	Lutein + zeaxanthin µg/100g FW	Total carotenoids µg/100g FW	Reference
Cv. Russet Burbank, white	Not available	58	Brown *et al.*, 2005
Cv. Superior, white	20	64	Lu *et al.*, 2001
Cv. Yukon Gold, yellow	28	111	Lu *et al.*, 2001
Potato, white	15.7 - 40.2	27.4 - 73.5	Iwanzik *et al.*, 1983
Potato, white	19.8 - 51.1 [1]	37.6 - 86.9 [1]	Nesterenko and Sink, 2003
Potato, white	Not available	40 - 101	Brown *et al.*, 2005
Potato, yellow	Not available	101 - 250	Brown *et al.*, 2005
Potato, yellow	34.8 - 81.6	107.3 - 260.3	Nesterenko and Sink, 2003
Potato, yellow	62 - 575	283 - 1435	Lu *et al.*, 2001
Potato, intense yellow	37.6 - 88.0	171.2 - 342.7	Iwanzik *et al.*, 1983
Potato, dark yellow	Na	509 - 795	Brown *et al.*, 2005
Potato, orange	559.1	878.2	Nesterenko and Sink, 2003
Beans, snap, green, raw	640	1085	Holden *et al.*, 1999
Broccoli, raw	2445	3225	Holden *et al.*, 1999
Brussels sprouts, raw	1590	2421	Holden *et al.*, 1999
Cabbage, raw	310	375	Holden *et al.*, 1999
Carrots, raw	0	13458	Holden *et al.*, 1999
Cassava, raw	0	8	Holden *et al.*, 1999
Squash, winter, acorn, raw	38	258	Holden *et al.*, 1999
Sweet potato, raw	0	9180	Holden *et al.*, 1999

[1]One outlier has values of 119.0 and 265.0, respectively

Research to-date provides good evidence for the potential to increase these levels further. If such highly coloured yellow, red and purple-fleshed cultivars gain more than what, in most countries, is at best a small niche market, then the dietary impact of such nutritionally enhanced potatoes can be significant.

Antioxidant values may be compared using oxygen radical absorbance capacity (ORAC) values. Brown (2005) has shown that ORAC values of light red-fleshed and purple-fleshed potatoes exceed those of white potatoes by 183% and 186% respectively, and that dark red potatoes exceed white potatoes by 330%. These values place highly pigmented potatoes in the same range of ORAC values as common vegetables such as broccoli, red bell pepper and spinach known as antioxidant sources.

The achievement of these potential improvements will require new collaborations between food scientists and researchers knowledgeable in potato genetics and genomics. Even moderate success can have a significant dietary and health impact and provide a strong boost for an already important world crop.

References

Bonierbale, M.W., R.L. Plaisted and S.D. Tanksley (1988). RFLP maps based on a common set of clones reveal modes of chromosomal evolution in potato and tomato. Genetics, 120, 1095-1103.

Brown, C.R. (2005). Antioxidants in potato. American Journal of Potato Research, 82, 163-172.

Brown, C.R., C.G. Edwards, C.-P. Yang and B.B. Dean (1993). Orange flesh trait in potato: Inheritance and carotenoid content. Journal of the American Society of Horticultural Science, 118, 145-150.

Brown, C.R., R. Wrolstad, R. Durst, C.-P. Yang and B. Clevidence (2003). Breeding studies in potatoes containing high concentrations of anthocyanins, American Journal of Potato Research, 80, 241-250.

Brown, C.R., D. Culley, C.-P. Yang, R. Durst and R. Wolstad (2005). Variation of anthocyanin and carotenoid contents and associated antioxidant values in potato breeding lines. Journal of the American Society of Horticultural Science, 130, 174-180.

Cambie, R.C. and L.R. Ferguson (2003). Potential functional foods in the traditional Maori diet. Mutation Research. Fundamental and Molecular mechanisms of Mutagenesis, 523/524, 109-117.

Chakraborty, S., N. Chakraborty and A. Datta (2000a). Nutritional quality improvement of transgenic potato by expressing a seed albumin protein from *Amaranthus hypochondriacus*. Pages 236-241, In: S.M.P. Khurana, G.S. Shekhawat, B.P. Singh and S.K. Pandey, eds. Potato, Global Research and development, Vol.1. Proceedings on the Global Conference of the Potato. New Delhi, India, December 1999.

Chakraborty, S., N. Chakraborty and A. Datta (2000b). Increased nutritive value of transgenic potato by expressing a nonallergenic seed albumin gene from *Amaranthus hypochondriacus*. Proceedings of the National Academy of Sciences, 97, 3724-3729.

Clifford, M.N. (2000). Anthocyanins - nature, occurrence and dietary burden. Journal of the Science of Food and Agriculture, 80, 1063-1072.

Dale, M.F.B., D.W. Griffiths and D.T. Todd (2003). Effects of genotype, environment, and postharvest storage on the total ascorbate content of potato (*Solanum tuberosum*) tubers. Journal of Agricultural and Food Chemistry, 51, 244-248.

Dale, M.F.B. and G.R. Mackay (1994). Inheritance of table and processing quality. In Potato Genetics, J.E. Bradshaw and G.R. Mackay (eds.). Wallingford, Oxon, UK. CAB International, pp. 285-315.

Davies, C.S., M.J. Ottman and S.J. Peloquin (2002). Can germplasm resources be used to increase the ascorbic acid content of stored potatoes? American Journal of Potato Research, 79, 295-299.

Douches, D.S., D. Maas, K. Jastrzebski and R.W. Chase (1996). Assessment of potato breeding progress in the USA over the last century. Crop Science, 36, 1544-1552.

Ducreux, L.J.M., W.L. Morris, P.E. Hedley, T. Shepherd, H.V. Favies, S. Millam and M.A. Taylor (2005). Metabolic engineering of high carotenoid potato tubers containing enhanced levels of β-carotene and lutein. Journal of Experimental Botany, 56, 81-89.

García-Alonso, A. and I. Goñi (2000). Effect of processing on potato starch: in vitro availability and glycaemic index. Starch/Stärke, 52, 81-84.

Haynes, K.G., W.E. Potts, J.L. Chittams and D.L. Fleck (1994). Determining yellow-flesh intensity in potatoes. Journal of the American Society of Horticultural Science, 119, 1057-1059.

Haynes, K.G., J.B. Sieczka, M.R. Henninger and D.L. Fleck (1996). Clone x environment interactions for yellow-flesh intensity in tetraploid potatoes. Journal of the American Society of Horticultural Science, 121, 175-177.

Holden, J.M., A.L. Eldridge, G.R. Beecher, I.M. Buzzard, S. Bhagwat, C.S. Davis, L.W. Douglass, S. Gebhardt, D. Haytowitz and S. Schakel (1999). Carotenoid content of U.S. foods: an update of the database. Journal of Food Composition and Analysis, 12, 169-196.

Iwanzik, W., M. Tevini, R. Stute and R. Hilbert (1983). Carotinoidgehalt und - zusammensetzung verschiedener deutscher Kartoffelsorten und deren Bedeutung für die Fleischfarbe der Knolle. Potato Research, 26, 149-162.

Langille, A.R., Y. Lan and D.L. Gustine (1998). Seeking improved nutritional properties for the potato: ethionine-resistant protoclones. American Journal of Potato Research, 75, 201-205.

Leeman, A.M., L.M. Bårström and I.M.E. Björck (2005). In vitro availability of starch in heat-treated potatoes as related to genotype, weight and storage time. Journal of the Science of Food and Agriculture, 85, 751-756.

Lewis, C.E., J.R.L. Walker, J.E. Lancaster and K.H. Sutton (1998a). Determination of anthocyanins, flavonoids and phenolic acids in potatoes. I: Coloured cultivars of *Solanum tuberosum* L. Journal of the Science of Food and Agriculture, 77, 45-57.

Lewis, C.E., J.R.L. Walker, J.E. Lancaster and K.H. Sutton (1998b). Determination of anthocyanins, flavonoids and phenolic acids in potatoes. II: Wild, tuberous *Solanum* species. Journal of the Science of Food and Agriculture, 77, 45-57.

Lisinska, G. and W. Leszczynski (1989). Potato Science and Technology. Elsevier Applied Science. London.

Love, S.L., J.J. Pavek, A. Thompson-Johns and W. Bohl (1998). Breeding progress for potato chip quality in North American cultivars. American Journal of Potato Research, 75, 27-36.

Love, S.L., T. Salaiz, A.R. Mosley and R.E. Thornton (2004). Stability of expression and concentration of ascorbic acid in North American potato germplasm. HortScience, 39, 156-160.

Lu, W., K. Haynes, E. Wiley and B. Clevidence (2001). Carotenoid content and color in diploid potatoes. Journal of the American Society of Horticultural Science, 126, 722-726.

Morris, W.L., L. Ducreux, D.W. Griffiths, D. Stewart, H.V. Davies and M.A. Taylor (2004). Carotenogenesis during tuber development and storage in potato. Journal of Experimental Botany, 55, 975-982.

Munschi, C.B. and N.I. Mondy (1989). Ascorbic acid and protein content of potatoes in relation to tuber anatomy. Journal of Food Science, 54, 220-221.

Nesterenko, S. and K.C. Sink (2003). Carotenoid profiles of potato breeding lines and selected cultivars. HortScience, 38, 1173-1177.

Organization for Economic Co-Operation and Development (OECD) (2002). Series on the Safety of Novel Foods and Feeds, No. 4. Consensus Document on Compositional Considerations for New Varieties of Potatoes: Key Food and Feed Nutrients, Anti-nutrients and Toxicants.

Reyes, L.F., J.C. Miller, Jr. and L. Cisneros-Zevallos (2004). Environmental conditions influence the content and yield of anthocyanins and total phenolics in purple- and red-flesh potatoes during tuber development. American Journal of Potato Research, 81, 187-193.

Rodrigues-Saona, L.E., M.M. Giusti and R.E. Wrolstad (1998). Anthocyanin pigment composition of red-fleshed potatoes. Journal of Food Science, 63, 458-465.

Römer, S., J. Lübeck, F. Kauder, S. Steiger, C. Adomat and G. Sandmann (2002). Genetic engineering of a zeaxanthin-rich potato by antisense inactivation and co-suppression of carotenoid epoxidation. Metabolic Engineering, 4, 263-272.

Sinden, S.L., R.E. Webb and L.L. Sandford (1978). Genetic potential for increasing ascorbic acid content in potatoes. American Potato Journal, 55, 394-395.

Soh, N.L. and J. Brand-Miller (1999). The glycaemic index of potatoes: the effect of variety, cooking method and maturity. European Journal of Clinical Nutrition, 53, 249-254.

Struik, P.C., M.F. Askew, A. Sonnino, D.K.L. MacKerron, U. Bång, E. Ritter, O.J.H. Statham, M.A. Kirkman and V. Umaerus (1997). Forty years of potato research: highlights, achievements and prospects. Potato Research, 40, 5-18.

Woolfe, J.A. (1987). The potato in the human diet. Cambridge Press, Cambridge, UK.

Zeh, M., A.P. Casazza, O. Kreft, U. Roessner, K. Bieberich, L Willmitzer, R. Hoefgen and H. Hesse (2001). Antisense inhibition of threonine synthase leads to high methionine content in transgenic potato plants. Plant Physiology, 127, 792-802.

Breeding and diagnostic developments for better storage of potatoes to meet future industry needs

N.W. Kerby[1], M.F.B. Dale[2], A.K. Lees[2], M.A. Taylor[2] and J.E. Bradshaw[2]
[1]Mylnefield Research Services Ltd, Invergowrie, Dundee DD2 5DA, Scotland, UK
[2]Scottish Crop Research Institute, Invergowrie, Dundee DD2 5DA, Scotland, UK

Abstract

This paper describes recent research at the Scottish Crop Research Institute (SCRI) to: 1) combine the development of specific and quantitative diagnostic techniques with epidemiological studies to inform and facilitate decision making for the control of potato storage diseases; 2) studies investigating the control of tuber dormancy and 3) breeding strategies to reduce low temperature sweetening and the potential formation of acrylamide in potatoes.

Keywords: diagnostics, disease, dormancy, breeding, low temperature sweetening

Introduction

Efficient and effective storage is a priority for the fresh, processing and seed potato industries. Both the quality and quantity of marketable tubers need to be maintained throughout the storage periods to maximise profits and reduce wastage. Stored tubers are living organisms, which produce heat through respiration and lose moisture through evaporation causing wastage. Potatoes may be stored for periods of up to 10 months. Tuber quality is defined relative to the intended end-use of the raw product. Potatoes intended for processing into French fries and crisps (chips) must meet specific quality standards in terms of low reducing sugar concentrations, minimum dry matter levels and finished product colour. Final standards may change geographically depending on national or regional consumer preference. Potatoes destined for the fresh market or seed potatoes do not have the same quality standards. Storage should control sprouting and limit the development of diseases which are important in stored seed tubers to ensure that any contaminated seed do not act as a source of inoculum for disease transmission and soil infestation in the following season.

Crop losses in potato stores can be considerable and diseases of concern in storage include silver scurf (*Helminthosporium solani*), Fusarium dry rot, late blight (*Phytophthora infestans*), pink rot (*Phytophthora erythroseptica*), Phythium leak (various *Phythium* spp.) and soft rot (*Erwinia carotovora*). Bacterial soft rot is potentially the most damaging of these diseases. Potato storage diseases are often difficult to control for two reasons: a) there are very few post-harvest chemicals registered for use on potatoes and b) the storage environment and conditions can favour the spread of certain diseases. The control of temperature, humidity and airflow will help in managing many of these diseases in storage and good store management is used in combination with biocides and fungicides. Valid and robust disease diagnostics are required to facilitate better disease control and store management strategies. The development of superior cultivars with resistance to storage diseases would also enhance profitability and reduce wastage with a reduced reliance on chemical control.

The manipulation of the potato tuber life cycle, in order to improve the timing of tuberisation, tuber size distribution and dormancy characteristics is a major economic target (reviewed in Fernie and Willmitzer, 2001). As it is often necessary to store potato tubers for periods beyond that of natural dormancy (generally 1-15 weeks), sprouting is controlled commercially either by storage at low temperatures or by the use of chemical sprout suppressants such as chloropropham (CIPC) or a combination of both approaches. The EU has proposed a maximum residue level of 10 parts per million, but many potato buyers and product manufacturers are calling for lower or even zero residues. There is a trade off between low temperature storage, reducing sugar accumulation, sprouting and the quality of the end-product. With lower storage temperatures sprouting is delayed, but reducing sugars, mainly glucose and fructose, accumulate more rapidly. This paper reviews some of the relevant research at SCRI, which is sponsored by our private sector partners, the British Potato Council and UK Government departments in the important areas of storage, storage diseases and breeding to improve the quality of stored tubers.

Using diagnostics to inform decision making for the control of *Fusarium* dry rot during storage

The potato industry requires high quality disease free seed, but few scientifically valid testing procedures for determining seed health have been available to date, with most disease assessments based on visual examinations. Diagnostic tests for a wide range of potato pests and diseases, including the blackleg pathogen *Erwinia carotovora* ssp. *atroseptica* (Toth *et al.*, 2003), *Helminthosporium solani* (= Silver scurf) (Cullen *et al.*, 2001) and *Fusarium* dry rot causing species (Cullen *et al.*, 2005), have been developed at SCRI in recent years. SCRI research will improve decision making for the management of potato diseases using predictive diagnostics, the deployment of relevant and robust sampling techniques and knowledge of the epidemiology of pathogens of potato to establish risk assessment criteria in collaboration with industry.

Diagnostic assays have been developed and used to study the epidemiology and control of dry rot, an economically important storage disease of potato tubers that can be caused by several *Fusarium* spp. All commercially grown potato cultivars are considered susceptible to the disease and contaminated seed or rotting tubers are significant sources of inoculum for the transmission of *Fusarium* spp., resulting in soil infestation. In order to validate control and management strategies and ensure disease-free stocks, the potato industry requires robust and reliable diagnostics to provide a rapid means of identifying and quantifying pathogen inoculum levels in diseased stocks and in soil.

Specific and sensitive quantitative diagnostics, based on real-time (TaqMan) PCR and PCR ELISA, have been developed at SCRI to detect dry rot causing *Fusarium* spp. (*F. avenaceum, F. coeruleum, F. culmorum,* and *F. sulphureum*). Experiments examined the effect of *Fusarium* species (listed above) and inoculum concentration on dry-rot symptom development over time. The objective of the SCRI research, some of which is reported here, is to develop and use diagnostic assays to investigate the extent of dry rot inoculum in potato stocks, the factors affecting disease development during storage and the relationship between inoculum load and disease risk.

Methodology

Tubers of three potato cultivars (Morene, Hermes and Spunta) with differing resistance to dry rot were collected from commercial seed stores as first generation stocks and again in the following year as second generation stocks. Tuber stocks were monitored by randomly sampling 10% of the total number of one tonne boxes per stock, according to a statistical design devised by Biomathematics & Statistics Scotland. DNA was extracted from peel collected from twelve tubers, randomly sampled from all areas within each box and tested for dry-rot causing *Fusarium* spp. All methods are fully described by Cullen *et al.* (2005).

Results

The results show that significantly higher disease scores were observed on tubers inoculated with *F. sulphureum* and *F. coeruleum* than with *F. avenaceum* and *F. culmorum*. Tubers inoculated with *F. sulphureum* had more evident disease symptoms at the lowest inoculum concentration (10^4 spores/ml) after 8 and 12 weeks storage than those inoculated with any other species, whereas *F. coeruleum* produced the highest disease scores at the highest inoculum (10^6 spores/ml) concentration. The threshold inoculum levels for significant disease development were set at 10^4 spores/ml for *F. sulphureum* and 10^5 spores/ml for *F. coeruleum*. The relationship between inoculum concentration and disease score for *F. sulphureum* is given in Figure 1. Overall, there was no significant difference in disease severity recorded between the cultivars Spunta and Morene.

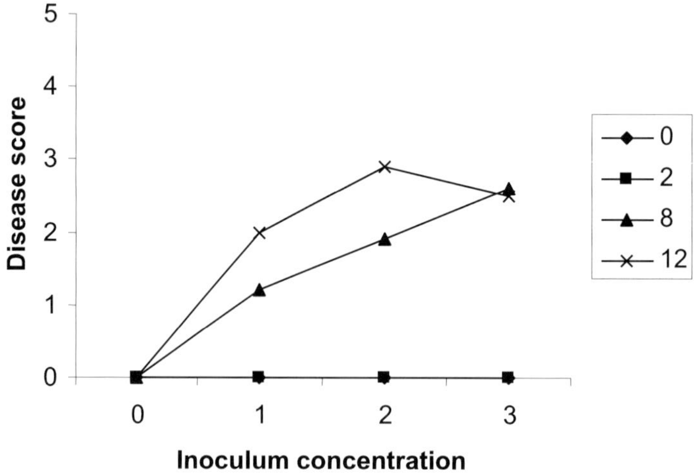

Figure 1. Relationship between disease score (0-5 scale, 0 = no symptoms, 5 = severe symptoms) on cultivar Spunta measured after 0, 2, 8 and 12 weeks storage and initial Fusarium sulphureum inoculum concentration (0 = 0 spores, 1 = 10^4, 2 = 10^5 and 3 = 10^6 spores/ml).

Real-time PCR testing of tuber samples revealed that all four potato seed stocks sampled from commercial stores were contaminated with *F. avenaceum, F. sulphureum, F. culmorum,* and/or *F. coeruleum* and there was a general trend towards increased *Fusarium* contamination in the second generation of seed sampled.

SCRI studies have generated information about the levels of *Fusarium* dry-rot causing species that exist on commercial seed stocks and the levels that are responsible for causing disease under different conditions. This data forms part of a much larger ongoing, industry led, study which will provide the information needed to interpret the results of diagnostic assays for risk assessment and disease control purposes for a wide range of fungal and bacterial diseases of potato.

Control of potato tuber dormancy

An understanding of the processes that lead to stop-start cycles in the growth of the potato tuber apical meristem is important to control tuberisation and dormancy. For example, although tuber apical buds exhibit the phenomenon of endodormancy (that is, meristem growth is repressed under apparently favourable conditions for growth), control of the length of the dormancy period is difficult. Premature dormancy break can lead to deterioration in quality during potato tuber storage (Wiltshire and Cobb, 1996) due both to disease-related and to physiological processes. Sprouting is accompanied by changes that are detrimental to processing, including increases in reducing sugar content, respiration, water loss and glycoalkaloid content (Burton, 1989). The length of the post-harvest dormancy period depends on both the genetic background of the cultivar and the prevailing environmental conditions during tuber development (Kotch *et al.*, 1992). As the potato industry diversifies its product range to satisfy consumer demand for salad potatoes or yellow-fleshed *Solanum phureja* cultivars, tuber dormancy has become even more of a problem. Some *S. phureja* cultivars have extremely low levels of tuber dormancy and thus cannot be supplied to the consumer all year. There is a growing awareness that CIPC treatment is expensive and leaves chemical residues in the food product. These economic and food safety concerns have stimulated interest in the mechanisms underpinning dormancy bud break.

As with many plant developmental processes, roles for the phytohormones in the control of dormancy have been investigated (reviewed in Claassens and Vreugdenhil, 2000; Fernie and Willmitzer, 2001; Suttle, 2004). Recent results have demonstrated that continuous ethylene treatment can be used as an effective sprout suppressant in commercial stores (Prange *et al.*, 2005), whereas short-term ethylene treatment can prematurely terminate tuber dormancy (Rylski *et al.*, 1974). Thus, although some effects of the phytohormones have been described, limited information exists on the mechanisms underpinning bud dormancy in potato and particularly on the genetic and molecular processes involved. In particular only limited forward and reverse genetic approaches have been applied to the investigation of the roles of the plant phytohormones, although such approaches are now technically feasible.

Several research avenues are currently being explored to enhance our understanding of the control of tuber dormancy. A multidisciplinary approach recently developed at SCRI led to the concept of the potato tuber life cycle being controlled by cycles of meristem activation and deactivation, mediated via symplastic association and disassociation of the tuber apical bud (Viola *et al.*, 2001). Thus on dormancy release, the apical bud regains symplastic connection with the tuber and growth resumes. Subsequent work examining dormancy release in buds of the mature tuber identified molecular markers in potato tuber buds that were induced or repressed specifically on release from endodormancy, in some cases prior to any visible sign of growth (Faivre-Rampant *et al.*, 2004 a,b).

Other studies have demonstrated that potato tuber sprouting can be controlled by manipulation of carbohydrate metabolism (Sonnewald, 2001). Tuber sprout growth is initially supported by energy captured from sucrose breakdown. As inorganic pyrophosphate is a necessary co-factor for sucrose breakdown, removal of pyrophosphate by expression of a bacterial pyrophosphatase in transgenic tubers increases sucrose content and prevents its use as an energy supplier. Consequently, sprout growth is significantly inhibited when sucrose is limited but accelerated when sucrose supply is increased. Transgenic approaches have also started to address the role of the plastid-derived isoprenoid hormones (particularly, cytokinins, gibberellins and abscisic acid) in the control of potato tuber dormancy. For example, over-expression of the gene encoding the first step of the isoprenoid biosynthetic pathway (catalysed by 1-deoxy-D-xylulose 5-phosphate synthase) leads to accelerated sprouting, probably due to an increase in the ratio of cytokinins to gibberellins (M.A. Taylor, unpublished). This provides an understanding of the steps of the pathway that should be targeted for molecular breeding approaches, such as allele mining of populations that exhibit variation in the trait of interest.

Acrylamide

In 2002 the Swedish National Food Administration announced that acrylamide could be found in foods containing starch that were cooked at high temperatures. Acrylamide is formed in the Maillard reaction as part of the cooking process and can never be completely eliminated. Acrylamides are a family of chemicals known to cause damage to the nervous system of humans at very high levels of exposure and are also potential human carcinogens, though there is no direct evidence that they cause cancer in humans. Acrylamide in foods is largely derived from heat-induced reactions between the R-amino group of the free amino acid asparagine and carbonyl group(s) of reducing sugars (fructose and glucose) at a rate strongly increasing with temperatures increasing from 120 to 170 °C. The levels in the food vary depending on the food type, cooking temperature and time. The major foods containing acrylamide have been found to be French fries and crisps, coffee, pastries, biscuits, breads, rolls and toasts.

Amrein *et al.* (2003) showed that the acrylamide levels are strongly correlated with the content of reducing sugars and asparagine. Since asparagine levels are high and exhibit limited variation, the far broader variation of the reducing sugars essentially determines the final levels of acrylamide formation. Reducing sugars are also important substances determining the quality of potato products like chips and French fries. The storage conditions are even more important than the cultivar. When stored below 8-10 °C, reducing sugar levels in the tubers increase. Biedermann *et al.* (2002a, b) found that for the cultivar Erntestolz, 15 days of storage at 4 °C increased the potential acrylamide levels by a factor of 28, i.e. the total sum of the reducing sugars found in the fresh weight increased from 80 to 2250 mg/kg (Biedermann *et al.*, 2002a, b).

The potato variety has a strong influence on the content of acrylamide in potato crisps and French fries. There is a good correlation between the content of acrylamide and glucose levels in potatoes, but no correlation with the content of asparagine is found. The strong relationship between reducing sugar levels and the formation of acrylamide in the variety Agria is demonstrated in Figure 2 below (Amrein *et al.*, 2003).

The amount of acrylamide depends on the nature of the processed product, e.g. coarse cut French fries have lower amounts of acrylamide than fine cut French fries.

Breeding efforts in many of the SCRI programmes to produce processing varieties for a number of markets described previously include a strong element of selection for storage at low temperatures (e.g. 4 °C) for prolonged periods without accumulating significant levels of reducing sugars. Given the significant relationship observed in Figure 2 it is evident that the reduction in reducing sugar levels in new varieties approaching the market may reduce potential acrylamide levels in processed products.

SCRI is also looking to candidate gene approaches to investigate the potential to suppress genes encoding enzymes involved in the biosynthesis of asparagines. A number of other approaches to decrease acrylamide levels in processed products available to processors include lower processing temperatures resulting in lower acrylamide formation, blanching to lower reducing sugar levels and the use of asparaginase to catalyse the hydrolysis of asparagines.

Breeding strategies for better storage

An increasing proportion of the European potato crop is being processed (25-30%) into value added products, while, in the UK, this figure has now increased to approximately 50% (British Potato Council, 2004). These products have stringent quality targets and require higher standards of certain attributes of the original raw product/germplasm entering the factories.

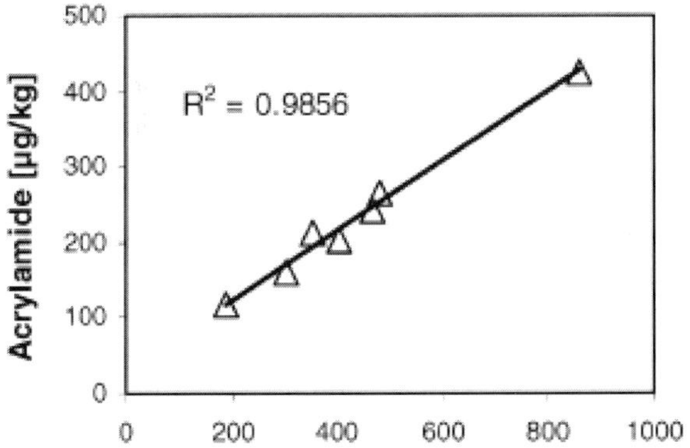

Figure 2. Concentrations of reducing sugars and potentials of acrylamide formation for seven tubers from the same lot of Agria potato. (Amrein et al.*, 2003).*

Potatoes for processing in Great Britain are harvested in autumn and stored throughout the winter and spring to maintain continuity of supply to the factories until fresh produce starts to become available again in summer. Tubers have to be stored between 2-4 °C to maintain them in good condition, to prevent sprouting and minimise disease problems. However, at these temperatures starch reserves are normally mobilised and there is a rapid accumulation of sugars, particularly of reducing sugars. When the tubers are fried a complex Maillard reaction occurs between the reducing sugars and some amino acids and results in a dark bitter-tasting product. Consequently, the manufacturers have to store tubers at relatively high temperatures (>6.5 °C), thus adding to their costs and exacerbating disease problems.

In 1982, SCRI research identified sources of low temperature, sugar stability (i.e. resistance to cold-induced sweetening) amongst its diverse genepool of the principal cultivated potato, *Solanum tuberosum*. Crosses between these clones and other 'normal' breeding clones and cultivars showed that this trait was heritable and culminated in the submission of two sugar stable clones to official trials as potential varieties. Cultivar Brodick was National Listed in 1990 and cultivar Eden in 1991, and the former enjoyed some commercial success. Subsequent research showed that progenies (families) with superior low temperature storage characteristics could be identified very early in a breeding programme. By determining the sugar stability (assessed as acceptable fry colour) of tuber progenies during storage at 4 °C after their first field grown year, it is possible to select those progenies which have the greatest likelihood of containing superior clones within two years of sowing true seedlings in the glasshouse. By concentrating selection for superior clones within these progenies, SCRI research demonstrated that potential cultivars for processing could be identified more efficiently than by selection of clones with some processing potential in a more general breeding programme. Industry funding allowed this new breeding procedure to be put into practice and resulted in five new cultivars being National Listed from 1999 to 2001 (Montrose, Harborough Harvest, Golden Millennium, Scarborough and Tay) (Mackay *et al.*,1997). These new cultivars all had Brodick as one parent and three had Eden as the other parent. Most recently, research has shown how the use of mid-parent values and progeny tests can increase the efficiency of potato breeding for combining processing quality with the disease and pest resistance now required for economically and environmentally sustainable potato production (Bradshaw *et al.*, 2003). The five new cultivars, and other germplasm, are therefore being used as parents in industry funded breeding programmes to achieve these new objectives.

More than one cycle of crossing and selection is required to combine many desirable traits into a new cultivar and this can only be done more quickly through new molecular breeding methods made possible by recent advances in genetics and genomics. A big impact on the rate of progress would be the identification of superior clones genotypically as seedlings in the glasshouse and the use of modern methods of rapid multiplication to progress them to commercialisation, whilst also using them as new parents in further hybridisations. This would require molecular-marker assisted selection, or preferably direct recognition of the desired allele at a genetic locus. Prospects for doing this for cold temperature sweetening have increased now that a potato molecular-function map for carbohydrate metabolism and transport has been published (Chen *et al.*, 2001; Menendez *et al.*, 2002; Dale and Bradshaw, 2003).

Given the timescale and complexity of conventional breeding, further improvement of successful cultivars using *Agrobacterium*-mediated transformation systems is an attractive proposition, despite consumer concerns in some countries about genetically modified potatoes. Transgenic approaches to cold temperature sweetening are certainly possible based on an understanding of primary carbohydrate metabolism, as shown by the following two examples. First, as stored tubers maintain the potential for starch synthesis, encouragement of this should limit the availability of hexose phosphates for sucrose biosynthesis, and hence reducing sugar formation. Thus, Stark *et al.* (1992) increased tuber starch content and lowered the levels of reducing sugars by expressing an *E. coli* glgC16 mutant gene which encodes for the enzyme ADPglucose pyrophosphorylase and increases the production of ADPglosose, which in turn becomes incorporated into the growing starch granule. Second, and perhaps the most obvious target, Greiner *et al.* (1999) were able to minimise the conversion of sucrose to glucose and fructose by expressing a putative vacuolar invertase inhibitor protein from tobacco, called Nt-inhh, in potato plants under the control of the CaMV35S promoter. However, for commercialisation, demonstration of the economic, environmental and health benefits of genetically modified potatoes will be crucial in convincing sceptical consumers of their value. More information on the control of sugar accumulation, starch biosynthesis and potato quality can be found in the review by Davies (1998).

Conclusions

To improve the storage of potatoes SCRI is: developing robust potato disease diagnostics to improve store management and enhance profitability; conducting research on tuber developmental processes that impact on storage; and developing superior cultivars that can be stored at low temperatures without the deleterious effects of accumulating reducing sugars.

Acknowledgements

This work was funded by: the British Potato Council, Department Environment, Food and Rural Affairs, Scottish Executive Environment and Rural Affairs Department, SAPPIO LINK.

References

Amrein, T. M., S. Bachmann, A. Noti, M. Biedermann, M. F. Barbosa, S. Biedermann-Brem, K. Grob, A. Keiser, P. Realini, F. Escher and R. Amadó (2003). Agricultural Food Chemistry, 51 (18), 5556 -5560.

Biedermann, M., S. Biedermann-Brem, A. Noti and K. Grob (2002a). Methods for determining the potential of acrylamide formation and its elimination in raw materials for food preparation, such as potatoes. Mitt. Lebensm. Hyg. 93, 653-667.

Biedermann, M., A. Noti, S. Biedermann-Brem, V. Mozzetti and K. Grob (2002b). Experiments on acrylamide formation and possibilities to decrease the potential of acrylamide formation in potatoes. Mitt. Lebensm. Hyg. 93, 668-687 (2002).

Bradshaw, J.E., M.F.D. Dale and G.R. Mackay (2003). Use of mid-parent values and progeny tests to increase the efficiency of potato breeding for combined processing quality and disease and pest resistance. Theoretical Applied Genetics, 107:36-42.

Burton, W.G., (1989). The Potato, 3rd Edition, Logman, London.

Chen, X., F. Salamini and C. Gebhardt, (2001). A potato molecular-function map for carbohydrate metabolism and transport. Theoretical Applied Genetetics, 102, 284-295.

Claassens, M.M.J. and D. Vreugdenhil, 2000. Is dormancy breaking of potato tubers the reverse of tuber initiation? Potato Research, 43: 347-369.

Cullen, D.W., A.K. Lees, I.K. Toth and J.M. Duncan (2001). Conventional PCR and real-time quantitative detection of *Helminthosporium solani* in soil and on potato tubers. European Journal of Plant Pathology, 107: 387-398.

Cullen, D.W., I.K. Toth, Y. Pitkin, N. Boonham, K. Walsh, I. Barker and A.K. Lees (2005). The use of quantitative molecular diagnostic assays to investigate Fusarium dry rot in potato stocks and soil. Phytopathology (In Press)

Dale, M.F.B. and J.E. Bradshaw, (2003). Progress in improving processing attributes in potato. Trends in Plant Science, 8: 310-312.

Davies, H.V., (1998). Prospects for manipulating carbohydrate metabolism in potato tuber. Aspects of Applied Biology, 52: 245-254.

Faivre-Rampant, O, L. Cardle, D. Marshall, R. Viola and M.A. Taylor (2004a). Changes in gene expression during meristem activation processes in Solanum tuberosum with a focus on the regulation of an auxin response factor gene. Journal of Experimental Botany, 55, 613-622.

Faivre-Rampant, O, G.J. Bryan, A.G. Roberts, D. Milbourne, R. Viola and M.A. Taylor (2004b). Regulated expression of a novel TCP domain transcription factor indicates an involvement in the control of meristem activation processes in *Solanum tuberosum*. Journal of Experimental Botany, 55, 951-953.

Fernie, A.R. and L. Wilmitzer, 2001. Molecular and biochemical triggers of potato tuber development. Plant Physiology, 127: 1459-1465.

Greiner, S., T. Rausch, U. Sonnewald and K. Herbbers (1999). Ectopic expression of a tobacco invertase inhibitor homolog prevents cold-induced sweetening of potato tubers. Nature Biotechnology, 17, 708-711.

Kotch, G.P., R. Ortiz and S.J. Peloquin, 1992. Genetic analysis by use of potato haploid populations. Genome, 35, 103-108.

Mackay, G.R., D. Todd, J.E. Bradshaw and M.F.B. Dale (1997). The targeted and accelerated breeding of potatoes. SCRI Annual Report 1996/97, pp. 40-45.

Menendez, C.M. E. Ritter, R. Schafer-Pregl, B. Walkemeier, A. Kalde, F. Salamini and C. Gebhardt, (2002). Cold sweetening in diploid potato: mapping quantitative trait loci and candidate genes. Genetics, 162, 1423-1434.

Prange, R.K., B.J. Daniels-Lake, J.C. Jeong and M. Binns (2005). Effects of ethylene and 1-methylcyclopropene on potato tuber sprout control and fry color. American Journal of Potato Research, 82, 123-128.

Rylski, I., L. Rappaport and H.K. Pratt (1974). Dual effects of ethylene on potato dormancy and sprout growth. Plant Physiology, 53, 658-662.

Sonnewald, U., (2001). Control of potato tuber sprouting. Trends in Plant Science, 6, 333-335.

Sowokinos, J.R. (2001). Biochemical and molecular control of cold-induced sweetening in potatoes. American Journal of Potato Research, 78, 221-236.

Stark, D.M., K.P. Timmerman, G.F. Barry, J. Preiss, G.M. Kishore (1992). Regulation of the amount of starch in plant tissues by ADP glucose pyrophosphorlase. Science, 258, 287-292.

Suttle, J.C., (2004). Physiological regulation of potato tuber dormancy. American Journal of Potato Research, 81, 253-262.

Toth, I.K., L. Sullivan, J.L. Brierley, A.O. Avrova, L.J. Hyman, M. Holeva, L. Broadfoot, M.C.M. Pérombelon and J. McNicol (2003). Relationship between potato seed tuber contamination by Erwinia carotovora ssp. atroseptica, blackleg disease development and progeny tuber contamination. Plant Pathology, 52, 119-126.

Viola, R., A.G. Roberts, S. Haupt, S. Gazzani, R.D. Hancock, N.Marmiroli, G.C. Machray and K.J. Oparka, 2001. Tuberization in potato involves a switch from apoplastic to symplastic phloem unloading. Plant Cell, 13,385-398.

Wiltshire, J.J.J. and A.H. Cobb, 1996. A review of the physiology of potato tuber dormancy. Annals of Applied Biology, 129: 553-569.

Seed potato systems in Latin America

Marcelo Huarte
National Potato Program Coordinator, Balcarce Experiment Station, INTA, Ruta Nac 226 km 73,5, (7620) Balcarce, Argentina. huarte@balcarce.inta.gov.ar

Abstract

Seed potato production systems are characterized by combining more than one seed production techniques and/or distribution and commercialization procedures. In Latin America there are six main seed production systems: the Traditional seed system. Systems based on imports and mass multiplication, the Clonal multiplication system, the Rapid multiplication system, the In vitro, minituber and microtuber production system and some New systems like SAH (Autotrophic Hydroponic System). A summary of their characteristics is given. Most countries show several systems at the same time, differences among growers (size and technology) and varieties (andigena or tuberosum) are crucial for the implementation of each system. Future trends in the seed market are discussed.

Keywords: seed systems, clonal multiplication, in vitro multiplication, minitubers, Autotrophic Hydroponic System

Introduction

In order to understand potato seed production in Latin America a characterization of two distinctive potato systems is needed: the Andean and the non Andean ecosystems. The former one is characterized by small growers, many of them under subsistence production with *Solanum tuberosum* ssp. *andigena* varieties in fragile agricultural systems. This system is typical of Colombia, Ecuador, Peru and Bolivia. The non Andean potato growers are large scale market oriented, cropping in flatlands mostly and using high technology and *S. tuberosum* ssp. *tuberosum* varieties. This system is typical of Argentina, Brazil, Chile, Mexico, Uruguay and Venezuela. Seed potato production systems are to some extent related to these differences. Although globalization and larger involvement of multinational processing companies are influencing the region, a greater government participation is observed in the Andean region, whilst greater involvement of private firms is seen in the non Andean region in the whole chain of seed production.

Seed potato systems

Several seed production systems coexist in the Latin American region and even within a given country. The Andean countries have a larger proportion of informal and traditional systems, whereas the non Andean countries have more formal and commercial structures than the former. In all countries formal certification or officially controlled seeds coexist with varying proportions of informal seed business. At present, six distinct systems can be identified in the region:

1. Traditional.
2. Imports and mass multiplication.

3. Clonal multiplication.
4. Rapid multiplication.
5. In vitro, minituber and microtuber production.
6. New systems (SAH).

The **traditional system** dates from the pre-Hispanic colonization and consists in supplying seed from the highlands where low aphid and other pests' pressure is characteristic. Also there is some evidence that taking back unhealthy potatoes from the lowlands to the highlands a "cure" is operated by the competition of the healthy plants on the unhealthy ones that ends in almost no yield from the latter. This logic has been known by the Andean growers for centuries and has been passed by oral tradition from one generation to the other (Crissman and Uquillas, 1990; Thiele, 1997). This system has maintained some of the old varieties but many have been lost. Some improvements have occurred:

- the introduction of IPM practices for the control of Andean potato weevil and tuber moth;
- the visual selection of tubers;
- the application of certification systems, although in some cases there has been reported little advantage compared to the informal system.

The traditional system has the lowest technology input compared to the others and there is no basic material produced. Its adaptation to the Andean region is based on tradition and local beliefs of the ancestral culture of potato cropping. In Colombia there has been a negligible advantage in yield for some varieties that were produced under modern seed systems compared to seed of the same varieties produced under the traditional system in the high altitudes. Therefore this system is still a valid economic alternative.

The **imports and mass multiplication system** has been in place since the early part of the 20th century in countries like Argentina, Chile and Brazil where seed supply from Northern Hemisphere countries was possible for large commercial growers. These growers have little knowledge of seed production, they normally choose adequate fields and seasons for multiplying the imported seed and they apply heavy inputs on the crop (insecticides, fungicides and fertilizers). The crops are mostly handled by mechanized systems for land preparation, planting, spraying, hilling, irrigation and harvest. The growers renew the seed of susceptible varieties almost on a yearly basis. The healthy seed arrives most of the time dormant in order to be planted in the maincrop areas at high latitudes; it is generally planted in double cropping areas with more virus pressure, prior to planting in the maincrop areas. Sometimes, imported seed is planted in seed areas and contaminates local healthy stocks because most imported seed stocks have not ELISA testing or any other post-control to guarantee its health status. The growers depend on foreign imports but that dependency is closely related to the exchange rates, and generally the costs of imported seed are high. Therefore the system has proven to be of little sustainability in countries with local capacity for seed production. The commercial struggle between exporting and importing countries is far from being over: the former claim competitive advantages and importing countries look forward not to be obliged to import large quantities of seed every year. Offer and demand in a changing exchange rate environment play an important role for this system to be established. This system is certainly important in the introduction of new varieties (Huarte *et al.*, 1984; Huarte and Inchausti, 1994). The technology level for this system is almost nil in terms of seed production, although the technology applied to the crop is very high, comparable to that applied in Northern Hemisphere countries. The system costs depend on the price of the imported seed and on the exchange rate. This system has been well adapted to countries in which tuberosum varieties are grown and therefore allow imports from Northern Hemisphere seed exporters with such varieties. Most of the growers that imported seed changed to the *in vitro*/minituber or the SAH systems when they were forced to stop importing because of devaluation of the local currency. Table 1 gives some insight on the potato production and exchange in Latin America and selected countries. Although no discrimination between seed or ware potatoes is given, Colombia, Argentina and Chile are net exporters of seed and ware potatoes; conversely, Brazil, Mexico and Cuba are net potato importers. The overall import value reveals that still extraregional imports are very important.

The **clonal multiplication** has been in use in The Netherlands and Canada since the early fifties. Argentina and Chile has also used this system since long time ago. The system has been improved by the use of ELISA tests for indexing tuber units or plant units. Also mechanization, where applicable, has been used to improve the planting efficiency of the clonal units. Restricted seed areas in which only basic seed can be planted have contributed to extend the number of clonal generations with good health status that can be grown starting from prebasic seed stocks. Recently, CIP has proposed a Speed Seed system, similar to that applied in Argentina in the 1980s, for use among small growers. A high number of tubers are indexed for virus testing with ELISA and only healthy tubers are planted. The system can be used by small growers with government aid for training and antisera supply. Clonal multiplication has proven to carry a high proportion of fungal and bacterial diseases in soil-back-to-soil produced tubers. Present technology level is intermediate and the system has not been widely distributed. A cost benefit analysis for this system has favoured more modern alternatives as seen below. Costs are intermediate and are highly influenced by the amount of ELISA testing, the degree of mechanization for harvest and planting and the number of clonal generations needed to reach the final production volume.

Table 1. Potato production, area and yield, exports and imports in Latin America and selected countries.

Region/country	1995-1997			2003		Rank vs 20 other food crops
	Production (1000 t)	Area (1000 ha)	Yield (t/ha)	Exports (1000 US$)	Imports (1000 US$)	
Latin America (n=24)	**14,486**	**1,080**	**13**	**9531**	**74700**	**7**
Colombia	2,770	170	16	4815	0	2
Brazil	2,701	182	15	37	1716	7
Peru	2,355	240	10	0	0	1
Argentina	2,155	98	22	829	58	4
Mexico	1,278	63	20	31	11445	6
Chile	1,001	66	15	328	770	3
Bolivia	734	131	6	0	48	2
Ecuador	473	66	7	13	15	5
Cuba	337	14	24	77	6623	1

Sources: CIP and FAO.

Rapid multiplication from healthy mother plants or tubers, generally in greenhouses, is a valid alternative to speed up propagation of small scale seed growers. Aged healthy tubers, normally remnant from greenhouse production of minitubers or from mother plants, can be used to produce a great amount of sprouts, which in turn can be placed in pots to root and grow in a clean environment, and can then be transplanted to a greenhouse. It is evident that this system can be combined with both the clonal and the *in vitro* systems, and has proven to be very low cost. In Argentina, this system has been transferred to farmer's schools and to small cooperatives of subsistence farmers. Technology level can be considered as intermediate, similar to the clonal multiplication.

The ***in vitro*-greenhouse minituber system** is the most widespread system in Latin America. Both private enterprises (generally large sacale) and state owned facilities apply this method and large amounts of minitubers are produced in greenhouses (about 5 million in total). The *in vitro* facilities and inputs are somewhat expensive and the rate of multiplication for plantlets (1 plantlet is propagated 5 to 8 times in a month) renders a much higher number of propagules than the previously described methods. The aseptical conditions in the lab are a bonus, although some labs have gone into bacterial or mite contamination. Some seed growers have encountered slow growth of *in vitro* plantlets transplanted to the greenhouse together with some proportion of losses due to transplant stress (Rigato *et al.*, 2000b). The available microtubers produced *in vitro* have proven to be quite uneven in sprouting, therefore rendering quite uneven crop stand in the greenhouse. The technology level and inputs required can be considered as high, compared to the other methods, and therefore is the highest system cost wise. These reasons have moved many of the growers to change from this system to the SAH system.

The **Hydroponic Autotrophic System (SAH),** developed in Argentina to enhance the plantlet production for greenhouse or field transplants, is a new methodology that is being adopted by many companies and official government units. SAH profits from the *in vitro* aseptic conditions but doesn't have its drawbacks in growth rate and contamination or transplant stress as no sugar is used in the growth media. SAH therefore makes use of the autotrophic capacity of the plant to grow in a friendlier environment where gas exchange is more adequate and similar to the natural conditions (Rigato *et al.*, 2000a). The rate of plantlet multiplication has proven to be about ten times faster than *in vitro* and the costs of production are about three times less than *in vitro*. The system has been transferred to 8 private labs in Argentina (80% of the country's labs), one company in Mexico and 5 official labs in Bolivia, Chile, Colombia, Ecuador, and Venezuela. So far, and after five years of intensive utilization of this system, no drawbacks have appeared. The technology level required is intermediate.

A summary of the importance of each system in the Andean and Non Andean regions, a global assessment of their cost and the technology level required is presented in Table 2. In all systems, costs and importance are estimated in a global way due to the lack of reliable statistical sources.

Table 2. Cost, technology level and importance in Andean and Non Andean regions of six potato seed production systems.

Seed Production Systems	Cost	Technology level	Importance in Agro-ecological regions	
			Andean	Non Andean
Traditional	Low	Low	Very important	Almost not existent
Imports and mass multiplication	High	Low	Almost not existent	Still important but declining
Clonal multiplication	Intermediate	Intermediate	Not important	Not important
Rapid multiplication	Low	Intermediate	Some importance	Some importance
In vitro, minituber and microtuber production	High	High	Important and growing	Very important
SAH	Low	Intermediate	In official labs	Growing among private companies (Argentina & Mexico)

General discussion

Adequate seed areas have been identified in most countries of Latin America. These areas may have some degree of restriction with respect to the category of seed that can be multiplied inside them in order to guarantee a healthier production. A generalized safer and cleaner production, with increases in monitoring and testing is foreseen. Some countries, like Argentina, rely on heavy post-control testing together with the conventional certification procedures. Combinations among seed systems are also an alternative to decrease costs of production and to render more efficient systems adapted to each particular situation; minituber production together with rapid sprout multiplication is an excellent alternative when small quantities are required; minitubers smaller than 10 g remultiplied in greenhouses is a common technique too; clonal indexing together with in vitro and minituber production is a combination used in a few labs in order to guarantee variety identity. Certification has proven to be an adequate means to achieve healthy seed production in most countries, especially in the non-Andean ones, although there is still a high proportion of informal seed used. Informal seed frequently passes through the same control procedures as the certified seed (100% of informal seed has post-control testing in Argentina); therefore its quality is also high and thus it is difficult to eradicate.

Healthy seed is present in all countries of Latin America. Chile and Argentina are the main seed exporters and Brazil, Uruguay and Venezuela are the main seed importers. Colombia has an important role in exporting Andean potatoes but also imports for processing. Canada, The Netherlands, France and Germany are the main sources of foreign seed and varieties. A great challenge for local growers and official institutions is to decrease the larger proportion of extra-regional seed trade as compared to the intra-regional trade (see Table 1). A larger trade among in-region countries is foreseen but also partnership with Northern Hemisphere companies: the two-Hemisphere business can supply the adequate seed with the best physiological age to extra-regional seed importer countries. A society between a seed company in Europe or North America and a seed company in Argentina or Chile can supply seed almost all the year round;

considering the harvest and storage periods in both regions, that seed will have the best physiological age to be planted in many other countries that presently receive dormant seed only from one of the Hemispheres.

Increase in consumption could determine a better adjustment to the slow but steady increase in yields and therefore seed prices could maintain an adequate level as well as ware potatoes. Seed price levels could slide down due to global decrease of processing industry paid prices for the raw product. Concentration processes in the market economy are being witnessed; fewer and larger growers remain in this highly competitive market. Conversely, a high variability in varieties and an atomized production scope is observed in the subsistence economy, where small farmers struggle for subsistance.

The overall scope reveals a trend towards self seed supply in Latin America, given the seed production infrastructure existent in most countries. Countries like Paraguay, the Guyanas, and some Caribbean and Central American countries, will still rely heavily on importation.

Acknowledgements

Special thanks to the organizers of Potato 2005 for inviting the author to the event.

References

Crissman, C.C. and J. E. Uquillas, 1989. Seed Potato Systems in Ecuador: A case study. Lima, International Potato Center. 70 pp.

Huarte, M.A., A.O. Mendiburu and O.A Garay,. 1984. Factores que afectan la difusión de cultivares de papa argentinos (Factors that affect the Argentine cultivar adoption). In: XII Reunión Anual de ALAP. Paipa, Colombia, May 1984. pp. 161-170.

Huarte, M. and M. Inchausti, 1994. La producción de papa en la República Argentina y su relación con el MERCOSUR (Potato production in Argentina and its relationship to MERCOSUR). In: III Simposio de Integración Frutihortícola del Cono Sur. Montevideo, Uruguay, June 22-25, 1994.

Rigato, S., A. Gonzalez y M. Huarte, 2000a. Producción de plántulas de papa a partir de técnicas combinadas de micropropagación e hidroponía para la obtención de semilla prebásica (Potato plantlet production by combined micropropagation and hydroponic techniques for the obtainment of prebasic seed). In: XIX Congreso de la Asociación Latinoamericana de la Papa, February 28th-March 3rd, 2000, La Habana, Cuba. Proceedings, p. 155.

Rigato, S., A. Gonzalez y M. Huarte, 2000b. Determinación de la tasa de multiplicación en minitubérculos de plántulas de la variedad Frital INTA originadas por micropropagación y por multiplicación de brotes de tubérculos cultivados en un sistema hidropónico. XIX Congreso de la Asociación Latinoamericana de la Papa, February 28th-March 3rd, 2000, La Habana, Cuba. Proceedings, p. 156.

Thiele, G., 1997. Sistemas informales de Semilla de Papa en los Andes: por qué son importantes y qué hacer con ellos? Lima, Internacional Potato Center, Social Sciences Department. Work Document # 1997-1, 56 pp.

Curriculum vitae – Marcelo Huarte

Marcelo Huarte studied Agronomy at Mar del Plata National University, from which he received the Ingeniero Agrónomo degree in 1975. That same year he joined the Research Staff of the Agronomy Department of the Balcarce Agricultural Experiment Station of the National Institute for Agricultural Technology (INTA) to work on potato breeding under Dr Americo Mendiburu's direction. In 1979, he engaged in graduate work at Cornell University, USA, with Dr. Robert Plaisted as advisor and he received a PhD degree in 1983 with a thesis on potato breeding. When he returned to the Balcarce Station he took responsibility in conducting the potato breeding programme from which he released over ten varieties with national and international distribution and permanent scientific publication. In 1991 he became the National Potato Subprogram Coordinator, his present position, advising research and development projects in different regions of the country. He has participated in teaching at the National University of Mar del Plata lecturing on Quantitative Genetics, Breeding and Potato Production. He has been project evaluator for UN, Swiss Cooperation Agency and IBD in several countries in Latin America. He is current member of the Steering or Advisory Committees of GILB, PREDUZA, ALAP and WPC. He has been elected President of ALAP in 2004 until 2006.

Company profile – The National Institute for Agricultural Technology (INTA)

The National Institute for Agricultural Technology (INTA) from Argentina has 42 experiment stations, 240 extension and technology transfer units, and 12 research institutes, that enables it to contribute substantially to production and managerial change processes that the Argentine farming sector needs in order to compete in the world's new production and marketing scenario. General research and extension strategies are oriented towards generating, adjusting and transferring technology for the development of production systems of all regions of Argentina. INTA´s main outputs are agricultural information and know how, technical services training and education, for many different social and productive sectors of Argentina´s life. Postgraduate training projects in agricultural sciences have been performed by INTA since 1963, with the cooperation of several universities. The training courses are directed to professionals of Argentine and other countries. In this way INTA has contributed to the development of solid scientific knowledge, with emphasis on domestic problems and circumstances. Degrees are granted by several universities. INTA has a great deal of cooperation agreements with countries and international agencies such as the Interamerican Institute for Agricultural Cooperation (**IICA**), **FAO**, the Interamerican Development Bank (IDB), the World Bank, the **CGIAR** centres, the German Cooperation Agency, the Institut National de la Recherche Agronomique (**INRA**), among others. Also, INTA has gained considerable experience with regional research and development institutions, as a partner in PROCISUR, closely connected now with the goals of MERCOSUR. Through joint venture agreements with private companies innovations generated by INTA are developed by specialized firms, so that they can reach the market. Risks and benefits are shared, combining INTA's scientific expertise with industrial management and marketing expertise. Main joint venture projects are done with seed, agricultural machinery and veterinary products industries. Visit our web page at www.inta.gov.ar or contact us at propapabalc@balcarce.inta.gov.ar.

Technology driving change in the seed potato industry

David McDonald
Managing Director Technico Pty. Limited, 226 Argyle Street, Moss Vale NSW 2577, Australia

Introduction

The objective of this paper is to provide a brief outline on the changing dynamics occurring in the global seed potato sector and its effects on future potato markets.

In the last decade we have witnessed the emergence of a number of factors which are impacting on the world potato markets and the seed export trade to these markets.

Some of these factors include:
- The emergence of both India and China as high growth potato consumption markets.
- The booming potato processing market in developing countries.
- Changing consumer habits being partly driven by:
  - An aging population, more diet conscious in developed countries.
  - A high proportion of under 25 years population in developing countries driving a greater consumer trend toward processed potato products.
- Increasing potato disease challenge throughout the world.
- High oil prices driving greater transport costs.

Technology impacts and opportunities

Know-how is increasingly playing a bigger role in the markets and as the traditional potato producing markets in developed countries have recorded declining production in recent years, there has been greater demand experienced in the emerging markets of Asia, Latin America, Africa and Middle Eastern countries. This is putting greater pressure on the establishment of reliable seed potato programmes of the right varieties in the emerging markets.

For decades many of the developing countries have turned to developed countries for supply of their seed potato products given the lack of know-how and technology to implement a quality seed programme in their own country. Climatic conditions prevailing in many countries would make a multiple generation (4 or 5 year seed programme) impossible and since conventional seed programmes have traditionally used multiple year programmes this has resulted in most seed needing to be imported.

The emergence of rapid multiplication seed technologies is making in-roads to this long seed supply chain and today large volumes of economically priced seed potato product can be produced in as few as two field generations.

Technology is currently being commercialised that can efficiently produce low cost minituber seed products in protected environments at a fraction of the cost that minitubers can be produced in any of the developed countries.

Markets such as China and India, who are embarking on structural change and improvement to their traditional seed production programmes are already embracing rapid seed potato multiplication techniques. As with most technology changes, both China and India are moving to the most modern technology available and as such are now developing seed programmes that can and are producing affordable seed potato products in as few as two field generations. This is reducing time to market, agricultural risk and improving quality by reducing the number of field multiplications in open field conditions. It is also allowing new processing industries to emerge with a speedy access to processing varieties to meet their production requirements.

Seed potato trade is also undergoing some changes with traditional export markets being challenged by high oil prices, fluctuating exchange rates and higher disease quarantine requirements within these export markets. Importing markets are looking for feasible alternatives and technology changes are providing an opportunity for affordable, limited generation seed schemes to be introduced as a viable alternative.

The emergence of low cost rapid multiplication technology is providing the following benefits to the industry:

- Enabling the rapid multiplication and introduction of new varieties:
  - benefiting variety developers - speed to the market
  - for processing companies in need of faster implementation of their seed and commercial programmes.
- Lower transport costs:
  - benefiting importing countries by being able to access high quality seed (produced in protected environments) which need only 100 kg planting material per hectare compared to much higher volumes (3-4 metric tonnes per hectare) using traditional imported seed products, thus creating substantial savings in transport and storage costs.
- Ability to implement/Quality improvement:
  - Countries that have not previously been able to consider a domestic seed programme because of the need for multiple field generations are now better positioned to be able to do this with cost effective and quality seed being produced in just two field generations.

Conclusion

The combination of new seed production technologies, low production cost environments, protected production facilities and soil-less production methods is changing global industry dynamics by slashing years off traditional seed potato production practices.

Managing Intellectual Property portfolios in potato

R. Korenstra
European Trademark, Design and Plant Variety Rights Attorney, Algemeen Octrooi- en Merkenbureau, P.O. Box 645, 5600 AP Eindhoven, the Netherlands

Abstract

Intellectual Property (I.P.) protection is effected by a system of *sui generis* legislations. The protection of achievements, like plant breeding, can be distinguished from the protection of identifying signs. The first system, for plant breeding called Plant Breeders' Rights, protects new varieties, the second system is focused on Trademarks and Designs and protects the goodwill of the companies.
This paper provides a brief description of these systems, including the conditions for registering new varieties, trademarks and designs. Last, but not least, the exploitation of the rights will be reviewed, including infringement and valuation.

Keywords: Intellectual Property, trademarks, designs, plant variety rights, royalties, license-agreements

Introduction

In agriculture Intellectual Property Rights are the most neglected rights both in production and in marketing activities. Companies are investing significant amounts in their production, harvesting and storage facilities as well in processing techniques. Breeding companies are investing many years of efforts in obtaining new varieties which benefits their customers. However, a well considered investment in an Intellectual Property portfolio will not only satisfy the profits of a company in the longer term, but also strengthen the competitiveness in the world market.

Plant variety rights

Introduction

New varieties of potatoes which produce improved yields, higher quality or provide better resistance to plant pests and diseases are a key element and a most cost-effective factor in increasing productivity and product quality in agriculture, whilst minimizing the pressure on the natural environment. Many other modern technologies of plant production need to be combined with high-performing varieties in order to deploy their full potential. The tremendous progress in agricultural potato productivity in various parts of the world is largely based on improved varieties.

Potato breeding has wider economic and environmental benefits than just increasing food production, including for developing countries. The development of new improved varieties with, for example, higher quality increases the value and marketability of crops in the global market of the 21st century. In addition, breeding programmes for potato plants can be of substantial economic importance for an exporting country. The breeding and exploitation of new varieties is a decisive factor in improving rural income and overall economic development.

The process of potato breeding is long and expensive; however, it can be very quick and easy to reproduce a variety. Clearly, few breeders would spend many years of their life, making substantial economic investment, in developing a new variety if there was no means of being compensated for this commitment. Hence, sustained breeding efforts are only possible if there is a chance to reward investment. It is, therefore, important to rely on an effective system of plant variety protection, with the aim of encouraging the development of new varieties of plants, for the benefit of society.

The system of plant variety protection was established in 1961 when there was recognition of the intellectual property rights of plant breeders in their varieties on an international basis. It is a form of intellectual property protection which has been specifically adapted for the process of plant breeding and has been developed with the aim of encouraging breeders to develop new varieties of plants.

In Plant Variety Rights legislation a breeder is defined as: the person who bred, or discovered and developed a variety, or the person who is the employer of the aforementioned person, or the successor in title.

Conditions
A variety shall be granted protection if it is:
- new;
- distinct;
- uniform;
- stable.

The grant of protection shall not be subject to any further conditions, provided the variety is designated by an acceptable denomination and the applicant complies with all the formalities and pays the required fees.

Novelty
To be eligible for protection, a variety must not have been sold, or otherwise disposed of, in the territory of the member of the Union concerned for more than one year prior to the application for a breeder's right, or more than 4 years in another territory in which the application has been filed.

Distinctness
Criterion: A variety is deemed to be distinct if it is clearly distinguishable from any other variety whose existence is a matter of common knowledge at the time of filing of the application. A variety whose existence is a matter of common knowledge (a "variety of common knowledge") must fall within the definition of a variety, but this does not necessarily require fulfillment of the Distinctness, Uniformity and Stability (DUS) criteria required for grant of a breeder's right.

Uniformity
Criterion: A variety is deemed to be uniform if, subject to the variation that may be expected from the particular features of its propagation, it is sufficiently uniform in its relevant characteristics. The uniformity requirement has been established to ensure that the variety can be defined as far as is necessary for the purpose of protection. Thus, the criterion for uniformity does not seek absolute uniformity and takes into account the nature of the variety itself. Furthermore, it relates only to the relevant characteristics for the protection of the variety.

Stability
Criterion: A variety is deemed to be stable if its relevant characteristics remain unchanged after repeated propagation or, in the case of a particular cycle of propagation, at the end of each such cycle. As with the uniformity requirement, the criterion for stability has been established to ensure that the identity of the variety, as the subject matter of protection, is kept throughout the period of protection. Thus, the criterion for stability relates only to the relevant characteristics of a variety.

The distinctness, uniformity and stability (DUS) criteria are often grouped together and referred to as the "technical criteria." They are most easily understood by considering the criteria together with the way in which they are examined.

Denomination
Anyone who offers material of the protected variety for sale or markets propagating material of the variety is obliged to use a denomination, even after the expiration of the breeder's right of that variety. The denomination is chosen by the breeder of the new variety but it must conform with some criteria. In summary:
- it must be different from all other denominations used for the same, or a closely related, species;
- it must not be liable to mislead or cause confusion concerning the nature of the variety or identity of the breeder;
- it must enable the variety to be identified;
- no rights in the denomination shall hamper its free use as the variety denomination (even after expiry of the breeder's right);
- prior rights of third persons must not be affected and such rights can require a change of the variety denomination;
- it may not consist solely of numbers, unless this is an established practice.

The breeder must submit the same denomination to all other countries and, unless this is considered to be unsuitable within a particular territory, this same denomination will be registered. A trademark, trade name or other similar indication may be associated with the denomination for the purposes of marketing or selling, but the denomination must be easily recognizable.

Scope of the rights
The following acts in respect of the propagating material of the protected variety require the authorization of the breeder:

a) production or reproduction (multiplication);
b) conditioning for the purpose of propagation;
c) offering for sale;
d) selling or other marketing;
e) exporting;
f) importing;
g) stocking for any of the purposes mentioned in (a) to (f), above.

The scope of the breeder's right with respect to the propagating material is extended to harvested material, where this has been obtained through the unauthorized use of propagating material of the protected variety, unless the breeder has had reasonable opportunity to exercise his right in relation to the propagating material.

In addition to the protected variety itself, the scope of the breeder's right also covers:
a) varieties which are essentially derived[1] from the protected variety, where the protected variety is not itself an essentially derived variety;
b) varieties which are not clearly distinguishable from the protected variety; and
c) varieties whose production requires the repeated use of the protected variety.
[1]The purpose of the provision on essentially derived varieties is to ensure that sustainable plant breeding development is encouraged by providing effective protection for the classical plant breeder and by encouraging cooperation between classical breeders and breeders employing techniques, such as genetic engineering.

Exceptions
The breeder's right does not extend to:
- acts done privately and for non-commercial purposes. This exception means that, for example, subsistence farming is excluded from the scope of the breeder's right;
- acts done for experimental purposes; and
- acts done for the purpose of breeding other varieties and, for the purpose of exploiting these new varieties provided the new variety is not a variety essentially derived from another protected variety (the initial variety).

This exception, for the purpose of breeding other varieties, is a fundamental aspect of the system of plant variety protection and is known as the "breeder's exemption". It recognizes that real progress in breeding - which, for the benefit of society, must be the goal of intellectual property rights in this field - depends on access to the latest improvements and new variation. Access is needed to all breeding materials in the form of modern varieties, as well as landraces and wild species, to achieve the greatest progress and is only possible if protected varieties are available for breeding.

The breeder's exemption optimizes variety improvement by ensuring that germplasm sources remain accessible to all the community of breeders. However, it also helps to ensure that the genetic basis for plant improvement is broadened and is actively conserved, thereby ensuring an overall approach to plant breeding which is sustainable and productive in the long term. In short, it is an essential aspect of an effective system of plant variety protection system which has the aim of encouraging the development of new varieties of plants, for the benefit of society.

In addition the breeder's right can be restricted in relation to any variety in order to permit farmers to use for propagating purposes, on their own holdings, the product of the harvest which they have obtained by planting, on their own holdings, the protected variety or other variety covered by the protection. This optional provision is known as the "farmers' privilege." This provision recognizes that there has been a common practice of farmers saving their own seed, i.e. seed is produced on a farm for the purpose of re-sowing on the same farm and not for the purpose of selling the seed.

Trademarks

Introduction

A trademark is any word, name, symbol or device, slogan, package design or combination of these that serves to identify and distinguishes a specific product from others in the market place or in trade. Even a sound colour combination, smell or hologram can be a trademark under some circumstances.

A trademark provides protection to the owner of the mark by ensuring the exclusive right to use it to identify goods like potato varieties, or to authorize another to use it in return for payment. The period of protection varies, but a trademark can be renewed indefinitely beyond the time limit on payment of additional fees. Trademark protection is enforced by the courts, which in most systems have the authority to block trademark infringement.
In a larger sense, trademarks promote initiative and enterprise worldwide by rewarding the owners of trademarks with recognition and financial profit. Trademark protection also hinders the efforts of unfair competitors, such as counterfeiters, to use similar distinctive signs to market inferior or different products or services. The system enables people with skill and enterprise to produce and market goods in the fairest possible conditions, thereby facilitating international trade.

Registration

An application for registration of a trademark must be filed with the appropriate national trademark office. The application must contain a clear reproduction of the sign filed for registration, including any colours, forms, or three-dimensional features. The application must also contain a list of goods or services to which the sign would apply. The sign must fulfill certain conditions in order to be protected as a trademark or other type of mark. It must be distinctive, so that consumers can distinguish it as identifying a particular product, as well as from other trademarks identifying other products. It must neither mislead nor deceive customers or violate public order or morality. The rights applied for cannot be the same as, or similar to, rights already granted to another trademark owner. This may be determined through search and examination by the national trademark offices, or by the opposition of third parties who claim similar or identical rights.

Designs

Introduction

An industrial design is the ornamental or aesthetic aspect of an article. The design may consist of three-dimensional features, such as the shape or surface of an article, or of two-dimensional features, such as patterns, lines or colour. Industrial designs are what make an article attractive and appealing; hence, they add to the commercial value of a product and increase its marketability. In the potato business especially (new) packaging can be protected by industrial designs.

When an industrial design is protected, the owner - the person or entity that has registered the design - is assured an exclusive right against unauthorized copying or imitation of the design by third parties. This helps to ensure a fair return on investment. An effective system of protection also benefits consumers and the public at large, by promoting fair competition and honest trade practices, encouraging creativity, and promoting more aesthetically attractive products. Protecting industrial designs helps economic development, by encouraging creativity in the industrial and manufacturing sectors, as well as in traditional arts and crafts. They contribute to the expansion of commercial activities and the export of national products. Industrial designs can be relatively simple and inexpensive to develop and protect. They are reasonably accessible to small and medium-sized enterprises as well as to individual artists and craftsmen, in both industrialized and developing countries.

Registration

In most countries, an industrial design must be registered in order to be protected under industrial design law. As a general rule, to be registrable, the design must be *"new"* or *"original"*. Different countries have varying definitions of such terms, as well as variations in the registration process itself. Generally, "new" means that no identical or very similar design is known to have existed before. Once a design is registered, a registration certificate is issued. Following that, the term of protection is generally 5 years, with the possibility of further periods of renewal up to, in most cases, 20 years.

Depending on the particular national law and the kind of design, an industrial design may also be protected as a work of art under copyright law. In some countries, industrial design and copyright protection can exist concurrently. In other countries, they are mutually exclusive: once the owner chooses one kind of protection, he can no longer invoke the other. Under certain circumstances an industrial design may also be protectable under unfair competition law, although the conditions of protection and the rights and remedies ensured can be significantly different.

I.P. licensing

Introduction

A license is an agreement between an I.P. owner (the "licensor") and another person (the "licensee") in which the licensor permits the licensee to use its I.P.-rights in commerce. Usually, a license is expressed in a written contract specifying the scope of the license and will identify:

- the definition of the I.P. -rights;

- the licensor and the licensee;
- the right(s) to be licensed (including the territory in which the rights are being licensed).

Control is needed because a license represents the I.P. owner's reputation of a certain level of quality, and consumers tend to rely on this reputation in making purchasing decisions. If a licensor does not exercise sufficient control over the quality offered by the licensee, the rights may, in some countries become vulnerable to attack by the licensee or a third party.

Conditions

A license will provide for quality control by the licensor by including provisions such as:

- Usage - The licensor may specify the manner in which the rights will be used and on advertising and promotional materials;
- Quality Control Monitoring - A licensor may require access to a licensee's facilities, seed production, finished products, personnel and records to monitor the licensee's adherence to the licensor's quality standards.

Other significant issues are:

- Royalty - When a licensor grants a license in return for royalty payments from its licensee, a royalty amount is usually set forth explicitly in the license.
- License Term - An I.P. license usually states a fixed term for the license and the conditions under which the license may be (a) renewed for an additional period of time or (b) terminated for breach of the license conditions.
- Exclusivity - I.P. rights may be licensed exclusively to a single licensee or licensed non-exclusively to more than one licensee. In a non-exclusive licensing arrangement, the licensor retains rights to use the I.P. -rights itself, or to license itself, or to itself, or to license it to others, or both.

In some countries, there is no legal requirement that trademark licenses be recorded with the national trademark office. Such recording will simply provide notice to the public of the existence of the license agreement. In other countries, however, a license must be recorded to be effective against third parties.

Infringements

Introduction

Infringement is the commercial use of the proprietary rights by another with respect to the potato varieties which is "likely to cause confusion" with respect to actual or potential customers. Many factors are considered in determining the existence of a "likelihood of confusion" including the similarity of the varieties and their commercial connotation, the similarity or relationship of the respective varieties, the commonality of trade channels, the sophistication of purchasers, the fame and strength of the trademarks related to the varieties, the number and nature of similar marks in use, and the existence of actual confusion. No one factor is necessarily controlling and in general infringement is evaluated on a case by case basis based upon the totality of circumstances.

The focus of infringement analysis is on the confusion of actual or potential customers. Accordingly, an infringer's product needs only be sufficiently related to the proprietary varieties so that it is likely that both are promoted to and/or used by common customers.

Generally, the existence of a likelihood of confusion is not required to establish either dilution or unfair competition which is not based on infringement. Dilution focuses primarily upon the strength of the I.P. rights and remedies are usually limited to injunctive relief. Although unfair competition may be based on infringement, unfair competition is often based upon false or misleading statements in marketing and advertising materials. However, in many cases of dilution and false advertising, infringement also exists. Remedies against infringements include injunctive relief to prohibit infringement, the impoundment and destruction of plant material and for instance packaging bearing infringing trademarks, an infringer's profits, the owner's actual damages and court costs. Attorney's fees may be awarded in exceptional cases.

I.P. valuation by "brands"

Introduction

A brand is a plant variety right, a trademark, a design (or a combination thereof) that, through promotion and use, has acquired significance in distinguishing the source or origin of the variety offered by others in the marketplace. Plant variety rights are representing exclusive potato varieties while a trademark is often thought of as a word, phrase or logo and designs as the appearance or shape of the varieties or their packaging. Brands are intellectual property and are part of the assets or "goodwill" of a company and may be bought or sold like any other asset or property.

Valuation

The value of an I.P. portfolio is often defined as the amount of money another party is prepared to pay for it. Sometimes this is easily ascertainable after a company purchases a portfolio and the associated goodwill without any other assets; however, in many situations, determining a value for a portfolio can be significantly more complicated. Another easily identified value is the difference between the amount paid to buy a company and the value of the fixed assets of that company. That difference represents the "goodwill" being purchased, and this goodwill is usually reflected in the I.P. portfolio of the company. The true value is ascertainable only when there is a willing purchaser and willing seller who reach agreement in the marketplace. A much more difficult question is how to value when there is no current offer to sell or purchase.

Companies need to value their I.P. portfolio's for many purposes. In the simplest of cases, the reasons to ascertain valuation include the following: (1) to find a buyer, (2) to value the assets in the balance sheet, (3) as security for a loan, or (4) for tax reasons.

Potential buyers also need to value the portfolio of another company before they make an offer to buy that company or its assets. Portfolio valuation is a useful tool used in merger and acquisition planning, joint venture negotiations, secured borrowing transactions, bankruptcy administration and proceedings, litigation support services, fair trading investigations, tax planning, marketing budgeting, new product and market development analysis, advertising agency evaluation, external investor relations and internal communications.

It is difficult to precisely define the "goodwill" of a business, and it is equally difficult to accurately assess the true value of any I.P. portfolio. Nevertheless, some economic and mathematical methods are commonly used for this purpose. Some valuation models examine the price tag for the sale of similar portfolios in the marketplace. Other models estimate the likely profits over a period of time (whether the profits be from direct sales, license fees or other revenue generated from the brand), and then use an applicable discount rate to calculate the net present value of the future profit stream. Portfolios are also generally valued higher if the owner has obtained appropriate legal (I.P.) protection, such as plant variety rights and trademark registrations. For companies that choose to value their I.P. in their balance sheet, it is important that a consistent methodology is followed so that they can be easily analyzed and compared with one another.

Different businesses are likely to have different expectations in terms of the longevity of the I.P. portfolio, competitive threats and other risks in the marketplace, opportunities for growth and sustainability of profits. These and other variables make it very difficult to pinpoint the exact amount of the value at any particular time; however, the valuations produced can be used as a useful guide as an approximation of the true value. I.P. rights, such as plant variety rights, trademarks and designs always are a substantial part of the balance sheet.

Curriculum vitae – Drs. Ronald Korenstra

Ronald Korenstra (MScBA) studied at the Faculty of Economics at the Rotterdam Erasmus University and Business Administration at the Delft Interfaculty of Business Management (IIB) from which he received his Masters' degree in 1980. After a short period being a staff member of an employers' confederation he became member of the management team of one of the leading floricultural companies in the Netherlands. In this position he was responsible for the research and development section specifically the management of the Plant Variety Rights and Trademarks port folio and the exploitation of multiple license agreements world-wide. In 1985 the export management and the management of the foreign subsidiaries was included in his responsibilities.

In the early ninety's Ronald Korenstra decided to focus completely on Intellectual Property matters and he started his own consulting company providing filing application services and licence agreement consultancy to a growing number of clients in the floriculture and horticulture sectors. This company merged in 1998 into ALGEMEEN OCTROOI- EN MERKENBUREAU and a department for Plant Variety Rights was established. By the years Ronald Korenstra extended his profession also to trademark, design and copyright law and some years later he was appointed as Certified Trademark and Design Attorney at the Office of Harmonization of the Internal Market, directed by the European Union, Alicante (Spain). Within ALGEMEEN OCTROOI- EN MERKENBUREAU Ronald Korenstra manages a practice for associates and clients in the agricultural, horticultural and floricultural business. He was deputy board member of CIOPORA International and member of an International Breeders' Organisation.

During his professional career he always like to exchange know-how in his profession and therefore he gives lectures at (foreign) business schools, the international agricultural centre of the University of Wageningen and several private institutions. Ronald Korenstra is a member of the Benelux Association of Trademark and Designs and President of AMPO, the European Association of Plant Variety Rights' representatives.

Company profile – Algemeen Octrooi- en Merkenbureau

Algemeen Octrooi- en Merkenbureau was established in 1967 through a merger of several patent attorney offices. It now has a staff of more than 70, and is one of the major Industrial Property law firms in the Netherlands. The full range of services related to the protection of industrial property rights is provided to associates and clients in accordance with the philosophy of the firm, which is based on professional skill, mutual confidence and friendly relationship.

The main activities of Algemeen Octrooi- en Merkenbureau comprise the filing and prosecution of national, European and international Patent applications, Benelux, European, national and international Trademark and Design applications, and national, European and world wide application for Plant Variety Rights. Additionally to these procedural activities, the attorneys of Algemeen Octrooi- en Merkenbureau are providing infringement opinions and advice concerning license agreements. Algemeen Octrooi- en Merkenbureau is unique in that it has a comprehensive department for the agricultural, horticultural and floricultural business. In these fields the company is a professionally skilled business partner with a long standing experience and it provides an extensive international network specialized in Intellectual Property matters.

Decision support systems

Present role and future potential of decision support systems in managing resources in potato production

A.J. Haverkort
Wageningen University and Research Centre, Plant Research International, P.O. Box 16, 6700 AA Wageningen, The Netherlands, anton.haverkort@wur.nl

Abstract

Sustainable potato production calls for profitable production systems that are healthy for man and environment. Resources are used less efficiently when amply available and when other resources than the targeted one are in short supply. This is demonstrated in the laws of von Liebig and Mitscherlich. Decision support systems (DSS) generate a recommendation based on sensing, environmental data and a mechanistic model. They are aimed at applying resources as timely as useful and at as low as possible rates. The paper shows a few examples of such DSS. They generate data by sensing and they use generic data such as weather and soil data and produce data in the decision rule they generate. If intelligently combined with what the grower records together with the feed back from the processing industry they yield valuable databases. These databases can be used with the aid of data mining techniques to further fine-tune existing DSS and help reduce the administrative workload of growers if stakeholders are granted access to the database.

Keywords: environment, nitrogen, water, nematodes, crop protection, administrative workload

Introduction

Consumers want food to be safe, of good quality and produced in a sustainable way: in terms of a license to produce from society. This means in terms of triple P: fair pay and working conditions for People, optimal use of resources of our Planet and Profit for stakeholders and shareholders. Different sciences occupy themselves with the different aspects of sustainability: social sciences with the people, economics with the profit and plant and environmental sciences with the planet. The most important resources in potato production are land and its resource use efficiency is expressed as yield in g per unit resource m^2 (so g m^{-2}), labour (g h^{-1}), water (g l^{-1}), nutrients g g^{-1}), energy (g J^{-1}) and crop protection agents (g g^{-1}). Usually the more of a resource is made available for growth, the less efficient its use, the first quantities being the most efficient. Mitscherlich (1909) described this for the resource nutrients (Figure 1). The more nitrogen given to plants the higher the yields but with each increased dose the corresponding yield increase was less: Law of the Diminishing Increments. Justus von Liebig as explained by e.g. Van der Ploeg *et al.* (1999) stated that yield is proportional to the amount of the most limiting nutrient (Figure 1). If the deficient one is supplied yields increase up to the point that some other nutrient becomes the most limiting: the Law of the Minimum (LM). Van der Berg (1998) and authors prior to him challenged Liebig's law in the Multiple Limitation Hypothesis (MLH) based upon cost-benefit analysis. If growth is limited by one resource the plant allocates more effort into acquiring the limiting resource and less to the other ones

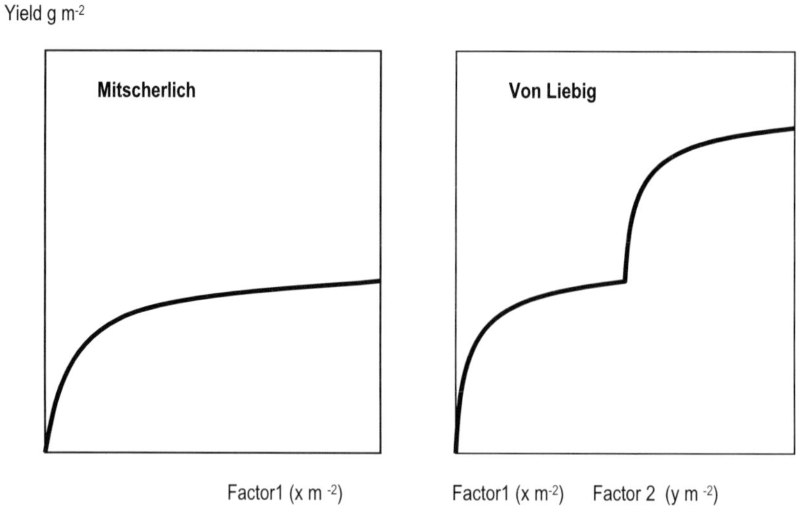

Figure 1. Schematic representation of Mitscherlich's Law of the Diminishing Increments and Von Liebig's Law of the Minimum.

creating a dynamic balance. Rubio *et al.* (2003) tested this hypothesis with the aquatic plant *Lemna minor* subjected to increasing amounts of N, P, K and Mg and found that neither LM nor MLH adequately predicted plant responses.

For modern agriculture the exact response of combinations of resources is less important than the realisation that the more of a resource a grower applies the less efficient its use and the more so if other than the applied resource is more limiting yield. Hence to optimally use resources they should be used as sparse as possible and in equilibrium with the other. This calls for instruments that allow growers to achieve an integration of managerial decisions.

The objective of this paper is to show the need for decision support systems to boost the optimal use of resources - to help growers to achieve two aims: increase profit and improve sustainability - and the perspectives of their future use. Therefore first the existing decision support systems (DSS) are highlighted. Examples are all taken from a recent book: Decision support systems in potato production: bringing models to practice (MacKerron and Haverkort, Eds., 2004). DSS can be grouped in those concerning integrated pest, crop and farm management. DSS on pest management are aimed at the application of biocides (fungicides, insecticides and herbicides) only when needed and at as low doses as needed. DSS on integrated crop management deal with the timing and dose of factors limiting growth such as minerals and water. Integrated management at the farm level is aimed at optimising soil health and fertility through crop rotation, organic matter amendments, erosion control and catch crops of minerals.

Decision support systems (DSS)

Most of the model based decision support systems deal with diseases and pests such as planners assisting in the control of late blight, nematodes, aphids, Colorado beetle and weed. They all measure something in the field, use a quantitative model and generate an advice.

Late blight caused by an oomycete *Phytophthora infestans* is the greatest threat of potato. Control of the disease amounts to about 20 percent of the production costs. If an outbreak takes place the foliage of that field has to be killed immediately. Several DSS exist and they are all based on the accurate estimation of the time that the leaves are wet. Spores of the oomycete need 8 hours in water to germinate and penetrate the leaf. The duration of wetness depends on the relative humidity of the air, the air temperature and wind speed. The period may be longer if it rains or when the crop is irrigated overhead, these two factors also enhance run off of the most recent fungicide application. Other factors that influence the risk of an outbreak is the time lap since the last fungicide application, the rate of (unprotected) leaf appearance, the proximity of the nearest infection site and the relative abundance of the canopy. All these factors are taken into consideration by most of the late blight control planners (Figure 2) such as PLANT-plus (Raatjes *et al.*, 2004). The outcome - advice - of the DSS may

```
--- Calculated infections -------------------------------------------------
      stage from              until              level
      D    12/06/00 12:00 - 19/06/00 18:00        33
      C    19/06/00 18:00 - 21/06/00 10:00         0
      B     21/06/00 10:00 - 22/06/00 20:00       10
      A     22/06/00 20:00 - 25/06/00 12:00      311

      advice threshold:  50 consideration 200 execution

======Advice================================  sub-model 3

      Application of Contact Fungicide necessary

---------Weather forecast ----------------------------------------------------
            Temperature    Rainfall  Radi-  RH   Wind
      Day  date   av.  max.  min. chance mm   ation  min. dir.  m/s
      fr 23/06/00  16.4  20.0  12.7   50   0.1   25.430   55  NW    4.9
      sa 24/06/00  16.2  19.8  11.5   60   0.6   27.220   50  W     3.8
      su 25/06/00  16.7  21.9   9.8   30   0.1   21.530   55  NW    3.6
      mo 26/06/00  16.0  21.8   9.0   20   0.0   24.850   45  WSW   4.2
      tu 27/06/00  15.0  17.3  12.4   40   3.6   16.615   60  W     5.2
-------- Spraying conditions ---------------------------------------------------
           04 05 06 07 08 09 10 11 12 13 14 15 16 17 18 19 20 21 22 23
      fr 23-06                      - - - - - - o - - o + +
      sa 24-06 + + + o ++ + o o - - - - - o - - - o + +
      su 25-06 ++ ++ ++ o + o o o - o - - - - - - - o + +
      mo 26-06 - -- -- + + o o - - - - - - - - - - o o o
      tu 27-06 -- o o o o o o o -- o - - - -- -- - - - -
----------------------------------------------------------------------------------
```

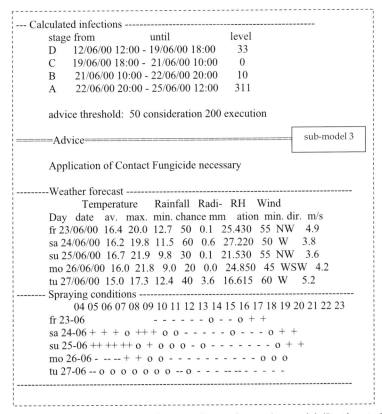

Figure 2. Example recommendation output of Dacom disease forecasting model (Raatjes et al., *2004).*

be: no treatment needed, consider a treatment or a treatment is necessary. Some DSS may suggest a contact or a systemic fungicide. Figure 2 shows an example of the output of a DSS.

For potato cyst nematodes (PCN) a comprehensive DSS is being developed in the Netherlands "Nemadecide" which is treated elsewhere in this volume. Like the Scottish approach (Elliott *et al.*, 2004, Figure 3) it projects PCN populations and yield decline at various management strategies. Populations will increase more rapidly when soils are lighter, rotation is shorter and the variety planted is more susceptible and less tolerant of PNC. Figure 3 exemplifies such figures with a starting point of 0.001 eggs per g soil. The DSS based on such projections recommend growers to change rotation patterns, suggests proper varieties, how and when to use nematicides and also when to sample the soil again for PCN.

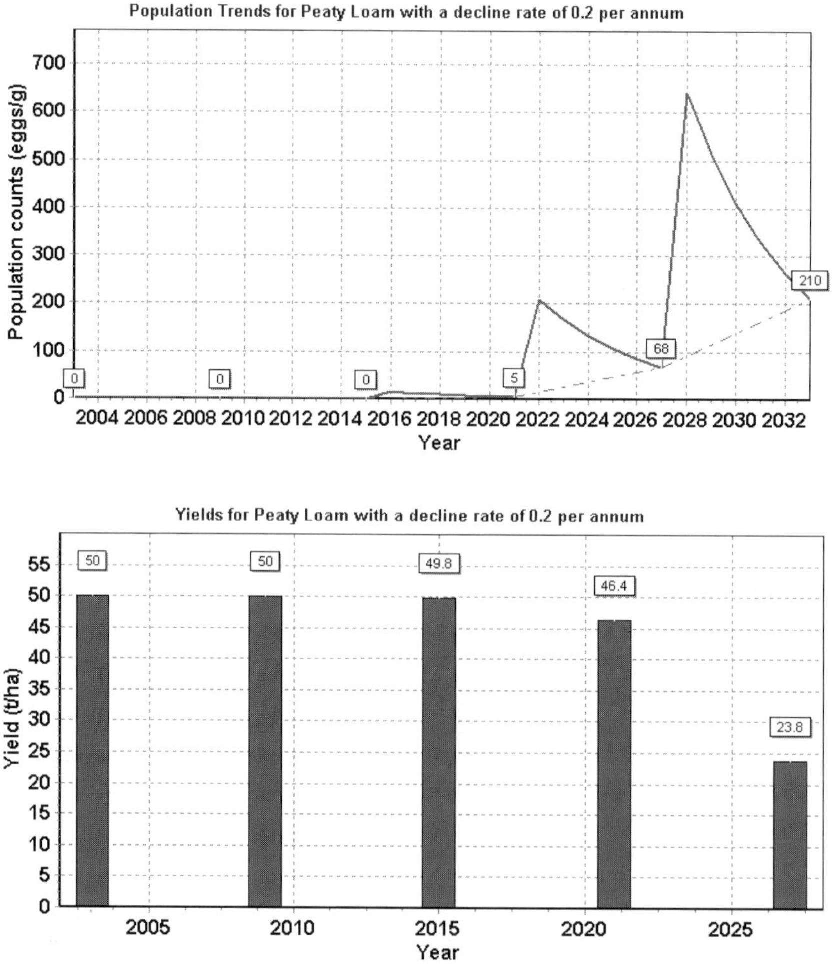

Figure 3. Very low population densities of PCN Globodera pallida (0.001 eggs/g soil) rise rapidly when exposed to susceptible potato varieties on commercial rotations of 6 years (Elliott et al., *2004).*

Kempenaar and van den Boogaard (2004) showed a concept of a DSS assisting growers in reducing the amount of herbicide without the risk of not killing the weeds. This concept MLHD for minimum lethal herbicide dose consists of farmers having to check which species of weeds (divided in 4 classes of sensitivity to herbicides) are present in the field and in what stages they are (cotyls, 1, 2, 4 6 or more leaves). The programme then recommends a certain dose of herbicide. After 2 days the grower checks if the photosynthesis apparatus still functions or not by sensing chlorophyll fluorescence - indicative of disrupted photosystem (PS) 2 - or hampering of PS1. This sensing is done with a small hand held device (Figure 5) that gives readings between 0 (optimal photosynthesis) and 100 (plants are dead). Figure 4 shows an example of the outcome of black nightshade treated with two doses of Sencor® and a control. A dose of 188 g of the herbicide effectively controlled the weed whereas a dose of 38 g would have given a recommendation to treat again.

The amount of nitrogen to be applied to a potato crop equals the amount of nitrogen absorbed by the crop and added to it the amount soil mineral N not available to the crop minus soil mineral N available at the start of the cycle minus the amount of nitrogen resulting from mineralization of soil organic matter. Pre-planting mineral N is often assessed. When during

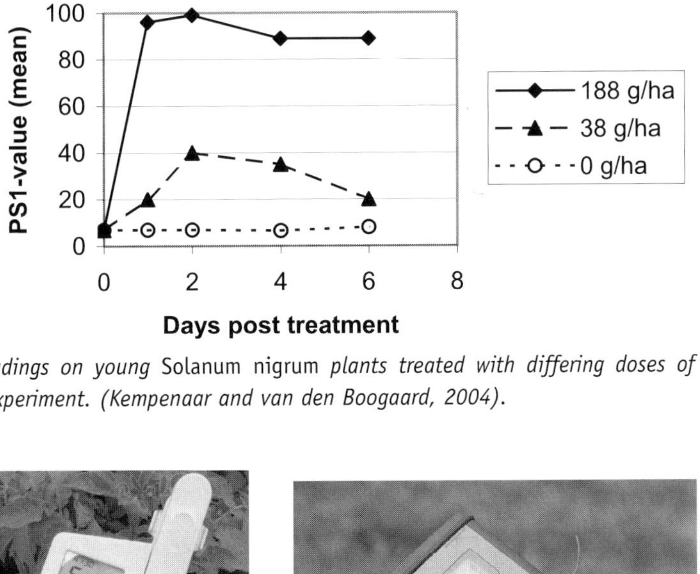

Figure 4. PS1-readings on young Solanum nigrum *plants treated with differing doses of Sencor in a greenhouse pot experiment. (Kempenaar and van den Boogaard, 2004).*

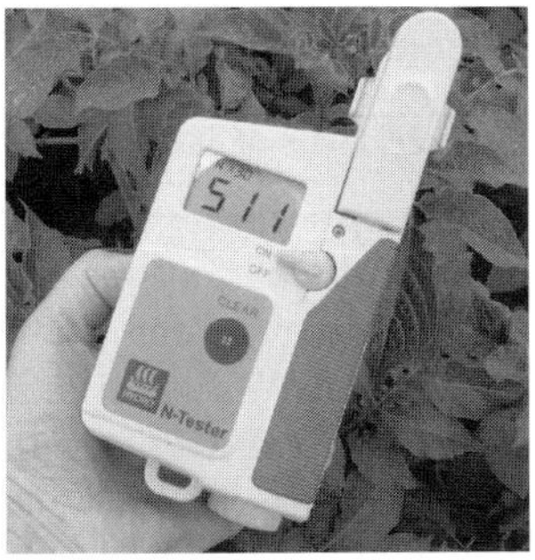

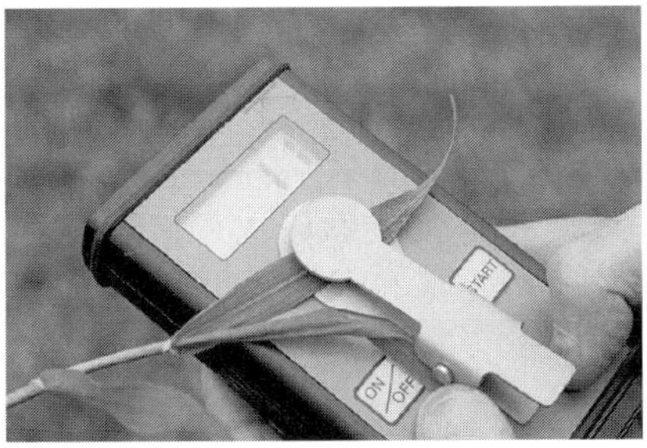

Figure 5. Examples of hand held sensors used in DSS. Left Chlorofyll sensor in nitrogen DSS and right a sensor used in a herbicide DSS.

crop growth the soil is sampled as well (to follow mineralization) it can be made obvious that the soil N value risks becoming too low for the crop's need. Then an advice may be given to apply nitrogen. This is a procedure followed by some DSS on nitrogen management. Recent developments assume that pre-planting residual N and mineralization during growth account for about half to two thirds of the crop needs. The remainder "third" - if at all - then needs to be applied depending on the crop's mid-season N-status. Figure 6 shows the procedure whereby a chlorophyll meter assists in assessing the crop's N-status.

Several DSS packages exist that combine a number of decision systems in one automated package (MacKerron *et al.*, 2004). An example is MAPP (Figure 7) the management advisory package for potatoes. This package includes a series of existing models such as the forecasting of potential and water constrained yields that are time-dependent, weather-driven, semi-mechanistic simulation models. The package further includes models of tuber multiplication and tuber size distribution and contains a DSS to assist in scheduling irrigation.

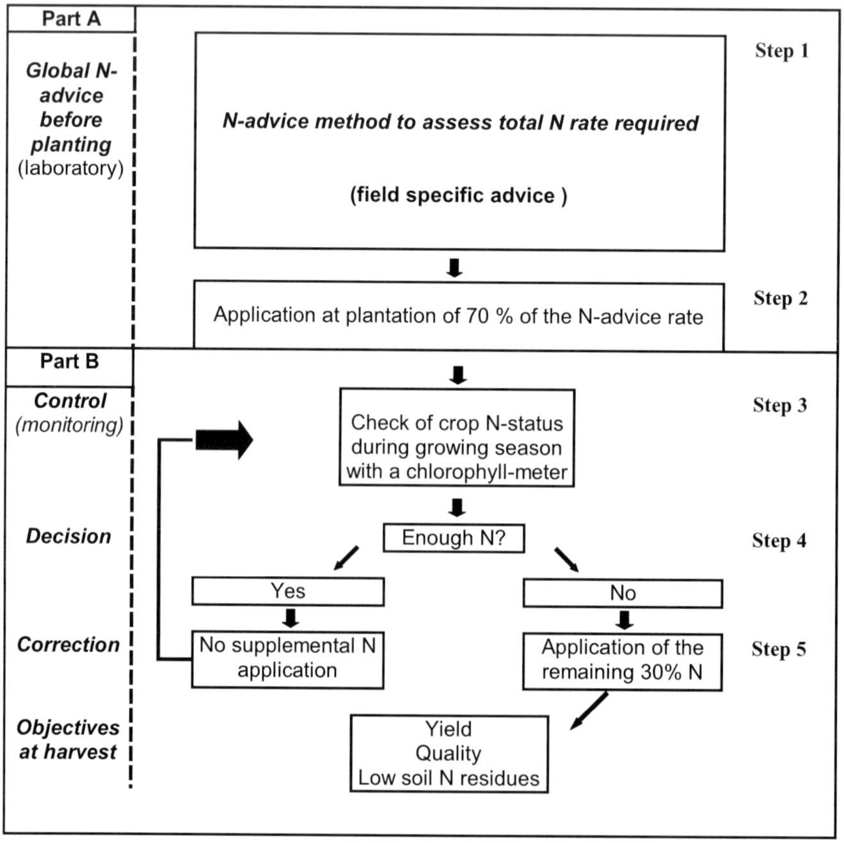

Figure 6. Steps in the strategy for management of nitrogen fertilization of the potato crop (Goffart and Olivier, 2004).

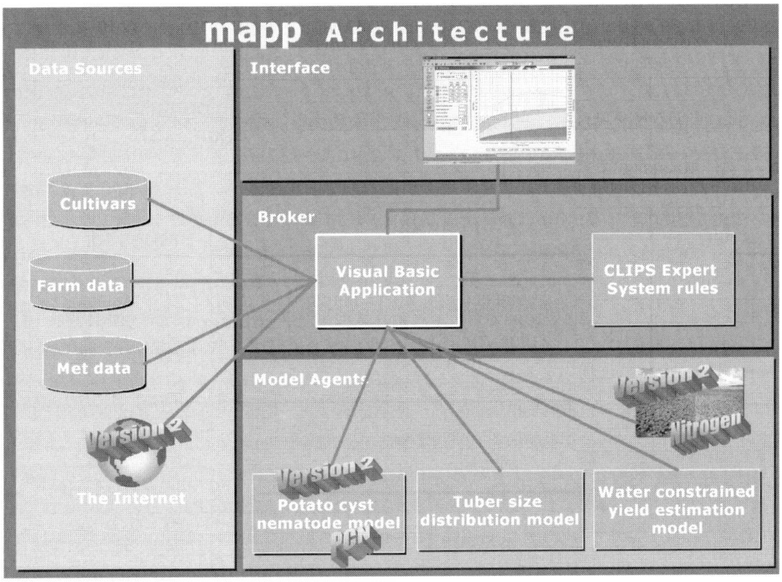

Figure 7. Diagrammatic structure of MAPP (MacKerron et al.*, 2004).*

Numerical challenges

Primary producers of agricultural produce generate increasing quantities of data in their field. Some data are generic such as meteorological and soil data. Other data the grower collects based on sensing and decision support systems shown above. Some data are generated by the grower such as seed rate, time of planting and harvesting. Finally feed back data from the processing industry (e.g. yield and quality) are available to the grower. Figure 8 schematically represents such flow of data between stakeholders.

Beside the grower other parties are interested in the database of the field: governments want to know if production complied with regulations, buyers want production data for tracing and tracking and extensionists and consultants want particular data to improve the quality of the advise. Growers themselves want the data to be able to compare with others (benchmarking) and other fields and years. Research and development are facing the challenge to identify the relevant databases, standardise data and to set up the base to allow partial access to them and to capitalise on data to develop techniques for data mining, benchmarking and self learning capacities (Figure 9).

Research is on-going into the type of data that exist such as mandatory data that growers have to make available to governmental bodies such as mineral balances, EU registration for subsidised crops and the presence of quarantine diseases. There are semi-obligatory data such as information in labels e.g. EUREPGAP and information that traders and processors want for reasons of tracing and tracking. Finally there are the voluntary data such as collected in decision support systems and the edaphysical ones. Collecting all relevant data, to standardise them, to develop an adequate system of storage and retrieval and to develop tools for

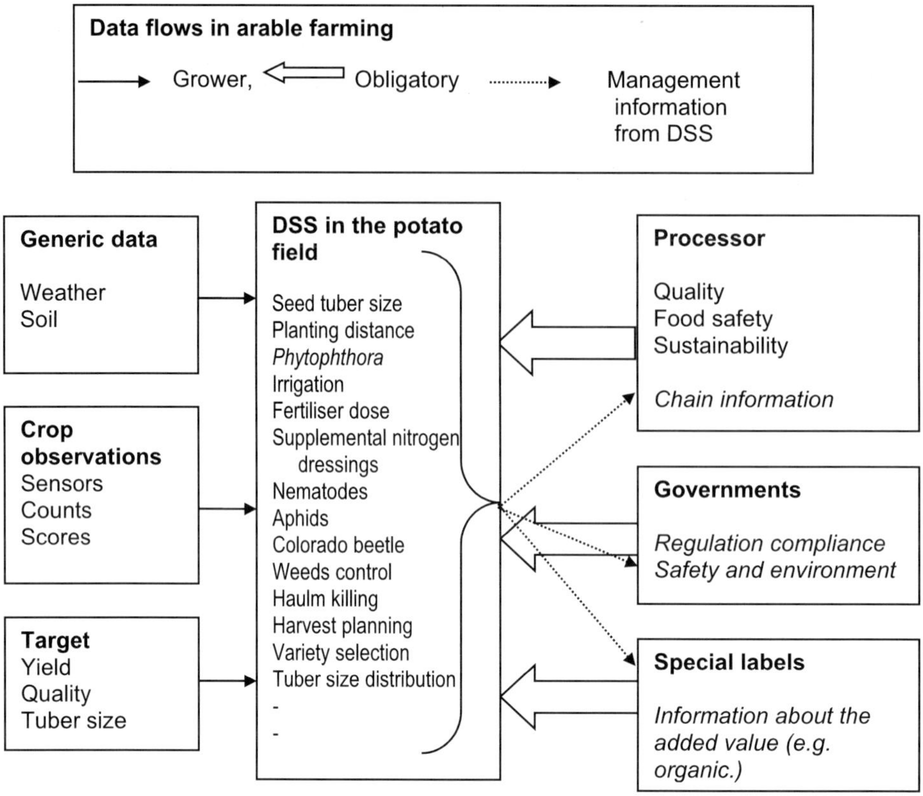

Figure 8. Schematic representation of data flows to, in and out a potato field, potentially all in a single database.

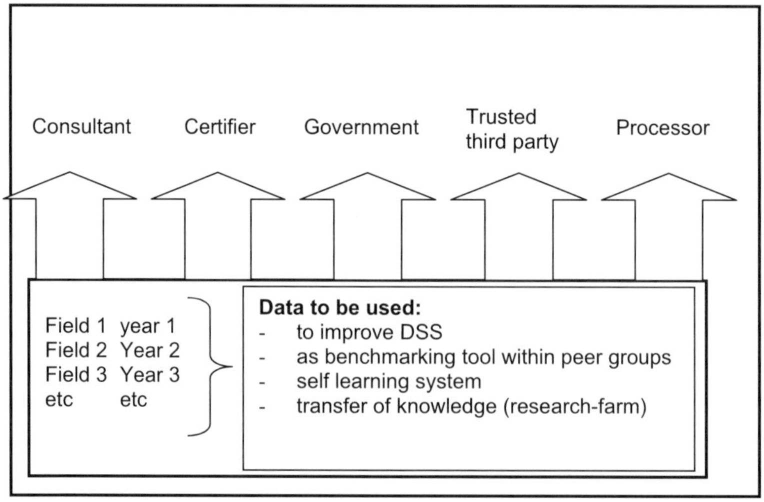

Figure 9. Schematic representation of access given to various stakeholders to the database shown in Figure 1.

benchmarking and self learning is subject of current research and development. Progress is being made in potato production in the Netherlands to gain insight in type of data and how to arrive at benefits for growers and other stakeholders in the potato industry and governments. One major benefit is the reduced administrative duties of growers towards governments and procurers. Giving automated partial access to interested parties considerably increases the efficiency of the production process.

With increased size of farms and reduced farmer's knowledge of individual fields the database acts as a transfer of knowledge. A coherent complete database of a particular field considerably adds to its value. The DSS in this way not only allow the grower to reduce his or her administrative workload and optimise the rate of return of the inputs but by furthering the laws of the minimum and of the diminishing increments also contribute to the enhancement of sustainability of potato production

References

Elliott, M.J., D.L. Trudgill, J.W. McNicol and M.S. Phillips (2004). Projecting PCN population changes and potato yields in invested soils. In: D.K.L. MacKerron and A.J. Haverkort (Eds.) Decision support systems in potato production: bringing models to practice. Wageningen Academic Publishers: pp. 142-152.

Goffart, J.P. and M. Olivier (2004). Management of N-fertilization of the potatop crop using total N-advice software and in-season chlorophyll-meter measurements. D.K.L. MacKerron and A.J. Haverkort (Eds.) Decision support systems in potato production: bringing models to practice. Wageningen Academic Publishers: pp. 68-83.

Kempenaar, C and R. van den Boogaard (2004). MLHD, a decision support system for rational use of herbicides: developments in potatoes. In: D.K.L. MacKerron and A.J. Haverkort (Eds.) Decision support systems in potato production: bringing models to practice. Wageningen Academic Publishers: pp. 186-196.

Mitscherlich, E.A. (1909). Das Gesetz des Minimums und das Gesetz des abnehmenden Bodenertrages. Landwirtschafliches Jahbuch der Schweiz, 38, 537-552.

Raatjes, P., J. Hadders, D. Martin and H. Hinds (2004). PLANT-Plus: turnkey solution for disease forecasting and irrigation management. In: D.K.L. MacKerron and A.J. Haverkort (Eds.) Decision support systems in potato production: bringing models to practice. Wageningen Academic Publishers: pp. 168-185.

Rubio, G., J. Zhu and J.P. Lynch (2003). A critical test of the two prevailing theories of plant response to nutrient availability. American Journal of Botany, 90, 143-152.

Van der Berg, H., (1998). Multiple nutrients limitation in unicellulars: reconstructing Liebig's Law. Mathematical Biosciences, 149, 1-22.

Van der Ploeg, R.R., W. Böhm and M.B. Kirkham (1999). On the origin of the theory of mineral nutrition of plants and the Law of the Minimum. Soil Science Society of America Journal, 63, 1055-1062.

Curriculum vitae – Anton J. Haverkort

Anton J. Haverkort - born on a basic seed potato farm - studied agronomy at Wageningen University from which he received a MSc in 1978. He then worked for ten years for the International Potato Centre in Peru, Turkey, Rwanda and Tunisia. He received a PhD degree from the University of Reading (UK) in 1985 based upon his work on management of resources in potato production in the tropical highlands of central Africa. The interaction of the influence of temperature and sunlight on growth and development was for the first time analyzed in field trials. It supplied the much needed parameter values in potato models. After he returned to the Netherlands Dr Haverkort joined Wageningen University and Research Centre as a senior researcher and research leader specializing in potato. He led a research programme on the interaction of pests and diseases and growth of potato - biotic stress - and the development of decision support systems based upon mechanistic models. He is the author of over 50 scientific papers and hundreds of columns en conference papers. He wrote and co-edited several books among which the "Decision Support Systems in Potato Production: bringing models to practice". He frequently travels all over the world as a consultant for governments and the agro-food industry. He is a member of the editorial board of the European Journal of Agronomy and is secretary general of the European Association of Potato Research.

Company profile – Plant Research International B.V.

Plants are the basis of life. They convert sunlight into an inexhaustible supply of food and renewable raw materials. In addition, plants play a stabilising role in agriculture and natural ecosystems. They also perform an essential landscape function and have enormous ornamental value. Plant Research International - within Wageningen University is the ideal choice for clients and partners that place a premium on top quality research. We combine leading-edge scientific research with innovation and an acute eye for business. In addition, we offer you next-generation facilities and technologies, plus a genuine global outlook. These attributes are further strengthened by our close ties to the market, which means the organisation is fully prepared today for the research questions of tomorrow. The research facilities at Plant Research International are among the most modern in the world. They range from laboratories with robotic genomics equipment to greenhouses and growing rooms that are customised to research requirements. We work among others on genomics, breeding, sustainable production systems, nutrient cycling and crop protection, all from the molecular to the systems level. Plant Research International has divided its activities and technologies into four dedicated business units: Bioscience, Biointeractions and plant health, Biodiversity and breeding, Biometry and Agrosystems research. It studies crop growth, develops models and automated decision support systems, designs production systems and analyzes their repercussions on the environment.

Thanks to the combination of know-how and experience across the entire spectrum of plant research, Plant Research International can offer unique new perspectives for governments and private companies, for agriculture and horticulture, and for rural and environmental development.

Plant Research International B.V. P.O. Box 16, 6700 AA Wageningen, The Netherlands, Telephone: +31 317 47 70 17, info.plant@wur.nl

Calibration of a crop growth simulation model to study irrigation scheduling effects on potato yield

R. Rocha-Rodríguez, J.A. Quijano-Carranza and J. Narro-Sánchez
Instituto Nacional de Investigaciones Forestales, Agrícolas y Pecuarias, Campo Experimental Bajío, 38010 Celaya, México

Abstract

Dynamic simulation models integrate the knowledge of different sciences and disciplines and permit to study the behaviour of systems as a function of time. Simulation models can be used to support decision making on very different kinds of systems. In agricultural research, system dynamics applications have been used to understand the course of plant growth and development process. Crop Growth Simulation Models solve the interactions between Soil-Plant and Weather simultaneously and represent an important tool to understand the changing nature of agricultural production. The Simulation Model of Potato Production, MSPEC-PAPA, was developed to calculate daily dry matter production of potato under both potential production, and water limited conditions. The model was calibrated to simulate the growth of the cultivar Alpha in a light clay soil in the location of San Francisco del Rincón, Guanajuato state, México. The comparison between farmer's irrigation scheduling (52,200 kg.ha^{-1}) and the simulated one (58,400 kg.ha^{-1}) showed an opportunity to increase at least by 10% the crop yield through increasing water use efficiency and optimizing the potato process of production.

Keywords: simulation, potential production, water balance, irrigation

Introduction

In México, potato is grown in 61,027 ha, with about 64% under irrigation conditions and the remaining 36% under rainfed conditions, although the uncertainty of water availability from rainfall is causing significative reductions in rainfed land cultivated with this crop. The main potato producer states are Sinaloa, Chihuahua, Sonora, Veracruz, Michoacán, México state, Guanajuato, and Puebla, (Rocha *et al.*, 2004). The weather conditions of the country permit that potato crop be cultivated throughout the year, and in areas near the sea level (Sonora and Sinaloa) until regions of high altitude (at least 3500 m in Puebla). In Guanajuato potato is cultivated in two seasons, one of them with planting dates from January 15 to February 28, and the other from August 1 to September 10. Although some studies have demonstrated that potential yield of potato in Guanajuato is about 80 ton.ha^{-1}, the regional mean yield under farmer conditions remains about 25 ton.ha^{-1}.

There are several causes for that gap, like seed health management, pests and weeds management, nutrition, water supply, and cultivars, but this study focuses just on one of them: the inflexibility of irrigation management, i.e., the use of a scheme of water supply to the crop on fixed intervals of time and with the same lamina, according, in a general way to the crop stage. It is obvious that a strategy like this would cause that water supply be under or over crop requirements from one season to another, because it doesn't consider the effect of the variability of weather conditions on soil moisture and plant needs. In the region of Guanajuato more than 95% of the farmers use pressurized irrigation systems. Economically speaking, irrigation management has not a significative effect on crop costs, because it represents about 3 to 5% of total crop costs. The only criterion to regulate the water amount to the crop is applied at the end of the growing season, when the producers reduce the water amount as a strategy to increase the tuber commercial quality. The reasons that most of the farmers have to use a fixed irrigation scheduling are mainly operative, and linked to the pressure of planting and managing great surfaces (50 ha or more).

In this study, a simulation approach was intended to generate information about the reductions on fresh tuber yield related to the lack of flexibility in the water management. This is an important feature to consider in evaluating the possibility of using a simulation based water management methodology, to increase potato crop productivity.

Materials and methods

The Simulation Model of Potato Potential Production, MSPEC-PAPA, was developed in the early 1990s. The MSPEC-PAPA, is a dynamic model that calculates daily dry matter production of potato under both potential production, and water limited conditions. This model was constructed to estimate potential production and support management decisions of potato production in the different environments of Guanajuato, México. The modelling approach is based on the principles and concepts established by Van Keulen and Wolf (1986), and applied for maize modeling by Quijano *et al* (1998).

The main inputs of MSPEC-PAPA are: Daily data of maximum and minimum temperature, rainfall, and solar radiation. The required soil data include clay, silt, and sand percentages, and soil depth. Crop parameters to characterize growth and development are: the required heat units to complete the vegetative phase and tuber formation and the pattern of dry matter distribution as a function of the stage of development. The MSPEC-PAPA simulates the development rate and tuber growth as a function of time, the rate of growth is determined by solar radiation and leaf area index and mainly limited by the soil water content and actual transpiration.

In the early 1990s a collaborative work with the local potato growers association was initiated to evaluate potential production in the region and the actual efficiency in the soil and weather resources management. Through this work several trials were established in farmer's plots to evaluate the effect of irrigation and nutrition management on the level of production, which showed that there is a significant effect of these factors on potato yield, although data were only collected regarding the growth of biomass and final yield. The results presented in this paper come from the first trial where the data also included the soil water content measuring. This experiment was conducted on a farmer plot in the location of San Francisco del Rincón, in the state of Guanajuato, México in 2001. The soil analysis indicated that a clay loam texture soil was present in the trial plot. Hydraulic properties were calculated based on a methodology proposed by van Keulen and Wolf (1986). The water content at field capacity and at wilting point was determined as 0.3742 $cm^3.cm^{-3}$ and 0.2281 $cm^3.cm^{-3}$, respectively.

The planting date was January 29, and plant density was 45,000 plants per ha for a seed size of 180 g and 55,000 plants per ha for the 90 g seed size. Climatic data were obtained from an automatic weather station located at a distance of 1 km of the trial location. The treatments consisted of two seed sizes, 90 g and 180 g evaluated under the farmer irrigation scheduling. Dry matter per organ and leaf area index were measured in four repetitions every week from the crop establishment until the end of tuber growth.

Soil water sensors (tensiometer) were installed at a depth of 30 cm to measure water content on different places of the plot. Daily climatic data were used to feed the MSPEC-PAPA model to simulate both the farmer and the plant needs based irrigation scheduling, focusing on the water balance, dry matter accumulation and tuber yield.

Results

Based on plant growth demand, 540 mm distributed over 12 dates were calculated with the model (Model scheduling). The farmer scheduling consisted of the application of 480 mm distributed over 11 dates (farmer scheduling). Table 1 shows the time of application and lamina used by these different distributions.

A comparison of the observed (measured through sensors) and simulated soil water percentage during the growing season at the 30 cm depth is presented in Figure 1, observed data represent the average of four sensors, one located in each repetition. In general the MSPEC-PAPA has a good estimation of the variation patterns of soil water content. Nevertheless it seems to overestimate the consumption rates of water by the crop, there is a good agreement in the times when the water percentage rises after the irrigation. A simple linear correlation coefficient of 0.84 was obtained for the relationship between these series.

A comparison between the predicted yield of potato, using MSPEC-PAPA for both farmer and model schedulings and observed data for tuber fresh weight at the two seed sizes is given by Figure 2.

The estimated yield was 58, 400 kg.ha^{-1} and 52, 200 kg.h^{-1} for the Model irrigation scheduling and the farmer's one, respectively. The comparison between simulated and observed yield data showed a determination coefficient (R^2) of 0.94. The analysis of variance for potato yield showed that there weren't significative differences between the seed sizes evaluated.

Table 1. Time intervals and laminas of irrigation used by the farmer and estimated by the MSPEC-PAPA for a potato crop, cv. Alpha in San Francisco del Rincón, México. 2001.

Days after planting	0	28	37	46	54	63	69	76	83	90	97		Total lamina mm	Number of irrigations
Farmer schedule mm	130	50	46	38.1	41.9	32	32	30	30	30	25		485	11
Days after planting	0	10	19	32	40	47	51	60	74	80	87	93		
Model schedule mm	130	56.3	50.4	39.5	29.9	21.3	24.8	51.8	38.4	30.5	39.4	28	540.14	12

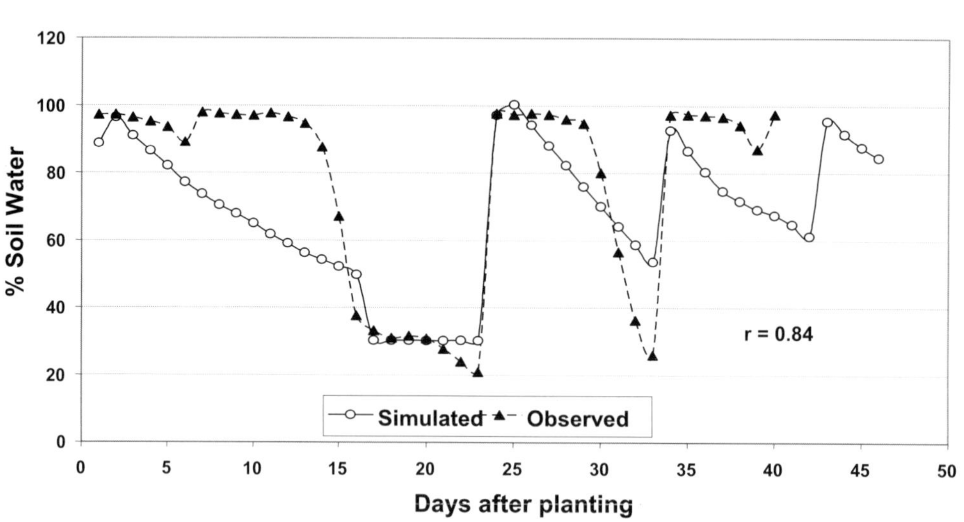

Figure 1. Simulated and observed soil water percentage during the growing season of a potato crop cv. Alpha in San Franciso del Rincón, Mexico. INIFAP. 2001.

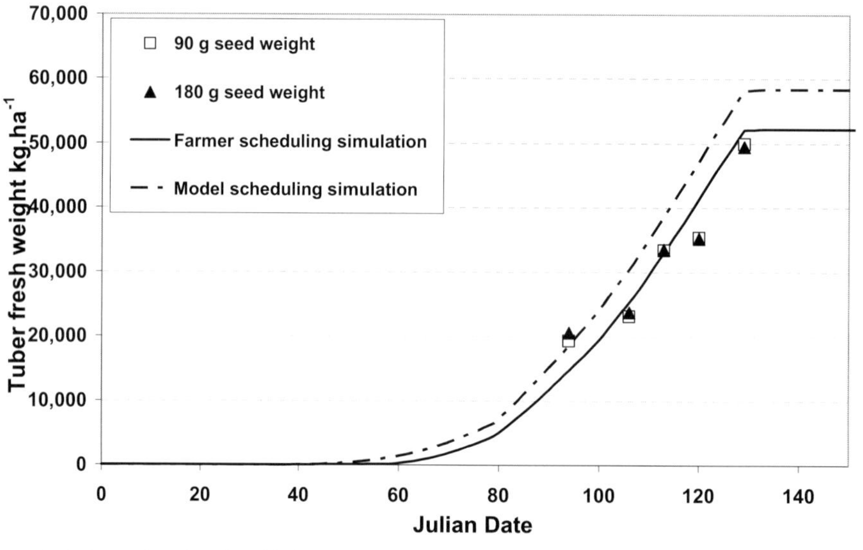

Figure 2. Estimated and real yield of potato cv. Alpha. San Francisco del Rincón, Guanajuato, Mexico. INIFAP. 2001.

Discussion

In general, the gap between the yield obtained under the model irrigation scheduling and the one currently realized by the farmer, indicates that there is an opportunity for an improvement of about 6,200 kg.ha^{-1}. This tuber yield differential can be explained by the periods of water stress that affected crop growth approximately between 15 and 20 days after planting and the 32 and 35 days after planting. In these periods, the water content dropped below the level of 60% which is considered as the lower limit to avoid plant water stress in potato (Rocha, *et al.* 2004). Simulation was carried out considering an optimal nutrient supply, which represents a common condition of production in the zone. The non-difference found between seed sizes (90 and 180 grams per seed), was probably due to a compensatory effect of the crop that produced a very similar performance in the number of stems per surface unit and the Leaf Area Index.

To this point, the main aim has been to organize a decision support system, integrated by a well-tested crop growth model, and data bases on soil, climate and crop management. The actual work consists in promoting the utilization of this system to analyze the possibilities of growing potato in different regions based on the potential production level, and the demand for supplementary applications of water and nutrients accordingly with the soil and climatic conditions. For both, the fresh market oriented farmers and the potato industry, the MSPEC-PAPA can be a powerful tool to evaluate the economic and environmental efficiency in terms of the millimetres of water necessary to be applied to maintain a economically profitable yield of potato in the different regions.

Conclusions

The results indicate that the MSPEC-PAPA can be confidently used as a tool to support the optimization of potato crop production.

Optimization of the water supply estimated with the use of the model, permits to obtain an increase of more than 10% in the yield of the crop.

A strict procedure to calibrate the model regarding to the components: cultivar, soil type, and the production conditions is required to gain precision in model estimates.

Acknowledgements

The authors want to acknowledge the Fundación Guanajuato Produce A.C. for partly financing this study, and the participant farmers for the facilities granted, especially to the "Granja Victoria" ranch.

References

Keulen, H. van and J. Wolf Eds. (1986). Modeling of agricultural production: weather soils and crops. Simulation Monographs, Pudoc Wageningen, The Netherlands, 479 pp.

Quijano, C. J. A. (1998). MSPEC.IM, Modelo de simulación del potencial ecológico de los cultivos. En Memoria del XVII Congreso nacional de Fitogenética. Acapulco, Guerrero. 195-249.

Rocha, R. R., J. A. Quijano C. and J. Narro S. (2004). El Cultivo de papa en Guanajuato. folleto para productores no. 1. INIFAP, Campo experimental Bajío, Celaya, Guanajuato, México, 47 pp.

Setting out the parameters of IRRINOV®, a method for irrigation scheduling

J.M. Deumier[1], F.X. Broutin[2] and D. Gaucher[3]
[1]ARVALIS - Institut du végétal, 6 Chemin de la Côte Vieille, 31450 Baziège, France, jm.deumier@arvalisinstitutduvegetal.fr
[2]ARVALIS - Institut du végétal, 2 Chaussée Brunehaut, Estrées-Mons, BP 70156, 80203 Peronne Cedex, France, fx.broutin@arvalisinstitutduvegetal.fr
[3]ARVALIS - Institut du végétal, Station Expérimentale, 91720 Boigneville, France, d.gaucher@arvalisinstitutduvegetal.fr

Abstract

The results of two irrigation trials carried out in 2002 and 2004 on the Charlotte cultivar in silt soil in Santerre are presented. These trials compare the effects of several irrigation patterns based mainly on the measurement of soil water potential. These results demonstrate the advantage of maintaining low soil water potential values from the emergence stage.

In 2002 this resulted in a significant increase in the number of tubers per plant (16.3 tubers per plant compared with 13.5 to 14 tubers per plant when the soil water potential values were higher), and in the yield in 35-55 mm marketable size (49.29 t/ha versus 44.20 t/ha for the other water regimes). In 2004, however, during a relative drought, particularly in the first part of the potato growing cycle, the number of tubers remained very low, around 12 tubers per plant. The poor quality of the seed tubers may be the cause of this. On the other hand, maintaining low soil water potential in the soil helped to limit the attack of common scab.

These trials enabled us to develop a method of potato irrigation - IRRINOV® - for use by farmers. This method, based on the measurement of soil water potential, was circulated in 2005.

Keywords: irrigation scheduling, firm-fleshed potato, soil water potential

Introduction

Since 2000, an irrigation scheduling method for farmers has been developed for wheat, barley, pea then for corn, and it is now available for potato. It is based on four principles:

- the determination in the fields of landmark stages to demarcate the period of irrigation and apply the irrigation pattern rules: emergence, tuber initiation, closure of the rows and beginning of plant senescence;
- the proposal of a "frequency-quantity" basic irrigation rhythm for each environment that covers the requirements for at least 8 years in 10;
- suggested tensiometric thresholds to decide when to trigger the first irrigation, when to alter the proposed rhythm during the growing season and when to stop the irrigation;
- a series of precise rules for the use of tensiometry.

Irrigation trials have been carried out since 2002 on the firm-fleshed cultivar Charlotte, with the aim of setting parameters for the method for this cultivar in the silt soil of Picardy and Pas-de-Calais. The results of the 2002 and 2004 trials are presented here and the method was circulated to farmers in 2005.

Materials and methods

The trials were carried out at the experimental station of Villers-Saint-Christophe (Aisne) on silt soil (15% clay, 80% silt, 5% sand and 1.5% organic matter). The cultivar grown was Charlotte: 38,300 plants/ha in 2002 and 43,300 plants/ha in 2004, with the row distance 0.75 m. Four water regimes were compared in a 4 block operation: 1 non-irrigated regime (T0) and 3 irrigated regimes (T1, T2, T3), where the irrigation took into account the soil water potential values.

The soil water potential was measured on 1 block by 6 Watermark® probes: 3 placed 30 cm below the top of the ridge and 3 placed 60 cm below. The threshold tension values were considered to be reached when the threshold tension figures on the following table were measured on 2 probes out of 3 at 30 cm depth and on 2 probes out of 3 at 60 cm depth.

Threshold tension values in kPa to implement the irrigation in 2002.

Stages		Emergence	Tuber initiation	Full Canopy	Start of senescence	Haulm killing
Number of indicative days after emergence		0	14	28	55	67
T1	Tension 30 cm	20	20	30	30	30
	Tension 60 cm	5	5	10	10	10
T2	Tension 30 cm	25	25	40	45	45
	Tension 60 cm	5	5	30	30	30
T3	Tension 30 cm	40	40	60	60	60
	Tension 60 cm	10	10	40	50	50

Threshold tension values in kPa to implement the irrigation in 2004.

Stages		Emergence	Tuber initiation	Full canopy	Start of senescence	Haulm killing
Number of indicative days after emergence		0	14	28	55	67
T1	Tension 30 cm	20	20	40	50	60
	Tension 60 cm	20	20	20	50	60
T2	Tension 30 cm	30	30	60	60	70
	Tension 60 cm	30	30	40	60	70
T3	Tension 30 cm	50	50	70	70	80
	Tension 60 cm	30	30	50	70	80

The irrigation depth and minimal frequency were respectively:

- 14 mm x 4 days from the "emergence" stage to the "full canopy" stage;

- 17 mm x 5 days after the "full canopy" stage;

i.e. a rhythm of around 3.5 mm/day.

We have established that this rhythm is necessary and sufficient to achieve a good yield for Bintje (Deumier and Vallade, 2002) 8 years out of 10, and we presume that the same is true for Charlotte.

The irrigation was carried out when the minimal periodicity since the date of the previous irrigation had been met and when the threshold tension values at 30 and 60 cm depth had been reached.

The growing stages are established for each treatment:
- emergence: 50% of plants with at least 3 leaves; measured on 2 rows of 9 m per block;
- tuber initiation: four plants of eight have at least half their early tubers, with a diameter double that of the stolon bearing them; measured on 8 plants in a block;
- full canopy and start of senescence are established by the ground coverage of the plants with a 90 cm x 75 cm frame cut out in 10 rectangles of 9 cm x 7.5 cm. The frame is positioned over the crop and the mark 1 is allocated to each rectangle when the green area covers more than half the rectangle and 0 if the opposite is true. The total marks divided by the total number of rectangles is the coverage level; two measurements were carried out on each block, i.e. eight measurements per treatment twice a week.

Marks are given on the shape of the tubers (second growth, malformation, cracks) and the number of tubers per plant is counted a few days after the haulm killing operation on 15 plants in each basic plot.

Finally, the yield is measured at harvest on 2 rows of 9 m for sizes below 30 mm, 30-55 mm and above 55 mm. The dry matter content is measured with a feculometer on the 30-55 mm size tubers. The presence of common scab is noted on 60 tubers per basic plot with a size of 30-55 mm. The potatoes are irrigated with drip systems positioned at the top of the ridge on each row. The irrigation of each plot is controlled by meters. The main growing operations were as follows:

	2002 trial	2004 trial
Planting date	08/04	13/04
Fertiliser, date, quantity/ha	0-8-30 + 8 MgO 1 t/ha 07/03	0-8-30 + 8 MgO 1 t/ha 12/03
Fertiliser, date, quantity/ha	N39 165 U 22/04	N39 130 U 27/04
Herbicides, product, quantity, date	Defi 5 l/ha + Sencoral 0.5 kg/ha 25/04	Defi 5 l/ha + Sencoral 0.5 kg/ha 05/05
Fungicides	Weekly protection	Weekly protection
1st haulm killing : product, quantity/ha, date	Reglone 2 l/ha 25/07	Reglone 2 l/ha 20/07
2nd haulm killing : product, quantity/ha, date	Reglone 2 l/ha 30/07	Reglone 2 l/ha 25/07
Harvest date	27/08	07/09

Results

In 2002, the potatoes were planted on 9th April (Table 1) and the weather was relatively dry between planting and emergence (Figure 1).

Soil water potential values were therefore high: the total tension figures at 30 and 60 cm depth reached 50 to 60 kPa and two irrigations were carried out on treatment T1 between the emergence and the tuber initiation stage. Treatments T2 and T3 were not irrigated during this period; instead they were irrigated at the tuber initiation stage. In all, T1 was irrigated 7 times, i.e. a total of 122 mm, while T2 was irrigated 3 times, with a total of 58 mm and T3 twice, with a total of 36 mm. The tension values measured in the soil were consistent with the irrigation timetable: the total tension at 30 and 60 cm depth was all the lower when the treatment was irrigated. The yield in 35-55 mm marketable size was significantly higher for treatment T1 compared to the yield from the other irrigated treatments, T2 and T3, and the non-irrigated treatment T0 (Table 2). The increased yield was in excess of 5 t/ha. The number of tubers per plant for treatment T1 was also significantly higher than that of treatments T2, T3 and T0 (Figure 2).

It appears, therefore, that carrying out early irrigation between the emergence and tuber initiation stages and maintaining low tension throughout the cycle (total tension values at

Table 1. Timetable of irrigation, rainfall and potato stages, Charlotte 2002.

	Irrigation in mm				rainfall in mm
	T1 low tension	T2 middle tension	T3 high tension	T0 non irrigated	
		Planting 9th April Emergence 19th May			
21/5	14				
22/5 to 26/5					18.5
28/5	14				
2/6		tuber initiation			
3/6	14	14	14		
3/6 to11/6					43.5
17/6		full canopy			
17/6	14				
20/6					24
24/6	22				
26/6					
28/6	22	22	22		
1/7 to 12/7					53
13/7		start of senescence			
19/7	22				
22/7		22			
25/7		haulm killing			
Total mm	122	58	36	0	139

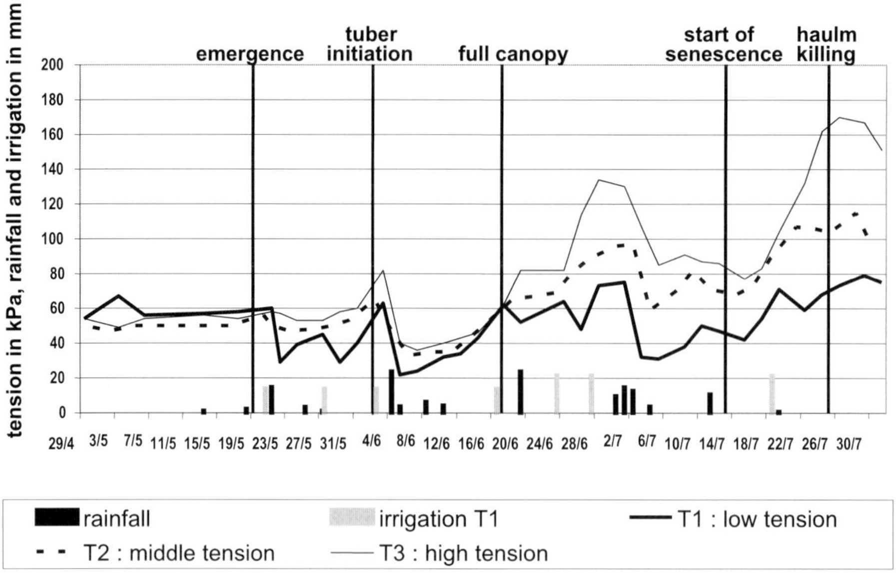

Figure 1. Total tension values at 30 cm and 60 cm depth, Charlotte 2002.

Table 2. Yield and dry matter content, Charlotte 2002.

	Yield t/ha			DM %
	<35 mm	35-55 mm	>55 mm	
T0 non irrigated	2.29	43.72 B	8.75	20.46
T1 low tension	2.59	49.29 A	9.48	19.24
T2 middle tension	1.77	44.20 B	11.46	19.54
T3 high tension	1.99	44.25 B	9.50	19.90
		Sed=2.31		
		P=0.02		

30 and 60 cm depth of between 40 and 60 kPa until full canopy and between 30 and 80 kPa until haulm killing) resulted in more tubers per plant and tubers of a larger size.

In 2004, there was more rain between emergence and tuber initiation, but the tension values remained high and close to those observed in 2002 (Figure 3).

Irrigation was therefore triggered during this period on treatments T1 and T2 (Table 3). The irrigation of treatment T3 was triggered at the end of June. In all, T1 had 104 mm, T2 76 mm and T3 34 mm. The tension values were all the higher when there was little irrigation (Figure 3). The yields were low (-19.3 t/ha for the highest yield of marketable size) and in these conditions there was no significant difference between the yields for the different water regimes (Table 4).

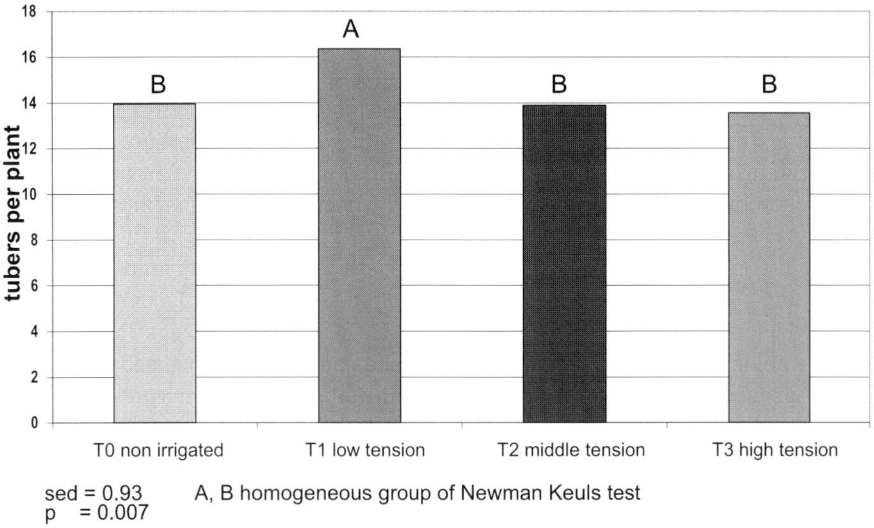

Figure 2. Number of tubers per plant, Charlotte 2002.

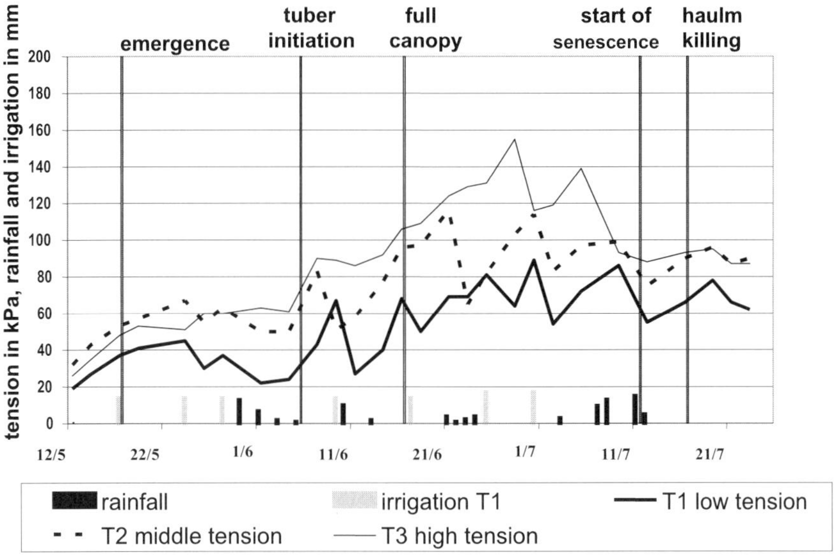

Figure 3. Total tension values at 30 cm and 60 cm depth, Charlotte 2004.

The number of tubers per plant was low (12.45 tubers per plant maximum in 2004 compared with 16.35 in 2002) and identical for the irrigated treatments T1, T2 and T3 (Figure 4). Early irrigation between emergence and tuber initiation and maintaining low tension throughout the cycle did not result in more tubers per plant or a higher yield. The level of presence of

Table 3. Timetable of irrigation, rainfall and potato stages, Charlotte 2004.

	Irrigation in mm				rainfall in mm
	T1 low tension	T2 middle tension	T3 high tension	T0 non irrigated	
		Planting 13th April Emergence 18th May			
17/5	14				
24/5	14	14			
28/5	14	14			
30/5					13
1/6					7
6/6		tuber initiation			
8/6		14			
9/6	14				
10/6					10
17/6	14				
20/6 to 23/6					11.5
21/6		full canopy			
23/6		17			
25/6	17				
28/6			17		
30/6	17	17			
3/7					3
5/7			17		
7/7 to 12/7					42.5
		Haulm killing 15th July then 17th July			
Total mm	104	76	34		87

Table 4. Yield and dry matter content, Charlotte 2004.

	Yield t/ha			DM %
	<35 mm	35-55 mm	>55 mm	
T0 non irrigated	1.21	23.28	4.10	19.75
T1 low tension	0.53	29.95	3.30	19.17
T2 middle tension	0.37	28.91	2.30	19.17
T3 high tension	0.44	28.10	3.47	19.07
		Sed=4.71 P=0.09		

common scab shows (Figure 5) that the frequency of attack was reduced for treatments T1 and T2 compared to the non-irrigated treatment T0 and the less-irrigated T3.

The beneficial effect of early irrigation on common scab has already been demonstrated by Goffart and Ryckmans (1996).

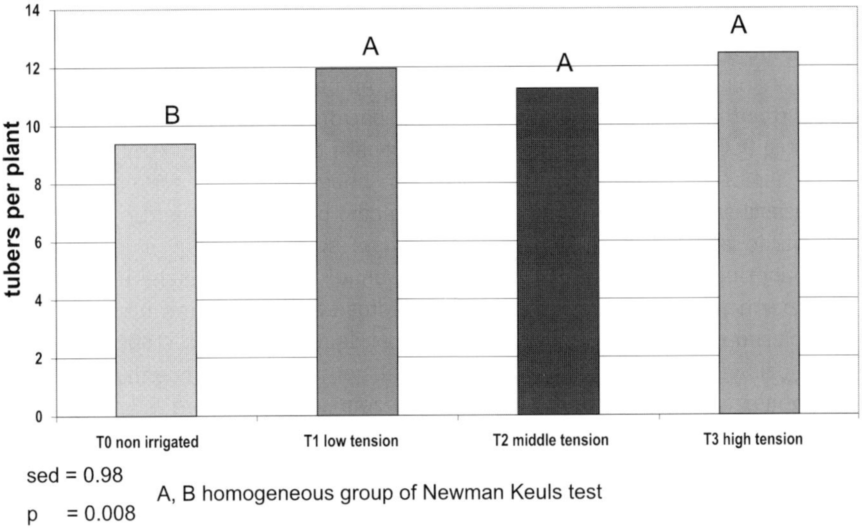

Figure 4. Number of tubers per plant, Charlotte 2004.

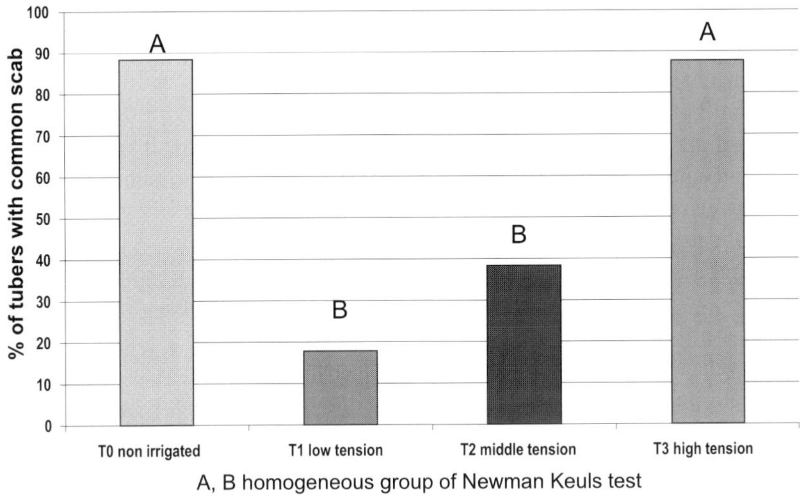

Figure 5. Common scab: Frequency of attack as a %.

Discussion

In 2004, observations clearly showed that potato growth was limited. The number of primary stems per plant was between 3.25 and 3.5 for T1 and T3, while in 2002 it was between 5.15 and 5.30. The level of coverage by the vegetation reached only 80% for the T2 treatment in 2004, while in 2002 it was 100%. Finally, the weights of fresh sprouts of 50 parent tubers kept in a damp atmosphere at 21 °C in the dark for 3 weeks, 6 weeks and 9 weeks were considerably lower in 2004 than in 2002.

These last observations suggest that the seed tuber quality was defective in 2004. Maybe the low growth made it impossible to demonstrate the beneficial effect of low soil water potential values on the number of tubers. In 2002, on the other hand, irrigating early and maintaining low tension resulted in a high number of tubers per plant and a high yield.

We presume that the high number of tubers per plant in T1 was due to the irrigation carried out between emergence and tuber initiation which helped to maintain low tension in the soil. This fact is consistent with what we find in literature (Haverkort *et al.*, 1990; Walworth and Carling, 2002). However, the fact of maintaining low tension in T1 throughout the cycle was probably beneficial.

Moreover, we measured very high air temperatures on 16th, 17th and 18th June (Figure 6); it was therefore possible that the irrigation carried out on 17th June on T1 (and not on T2 and T3) also had a beneficial effect on maintaining a high number of tubers.

New trials are currently under way to find out more about the effects of irrigation on the establishment of the number of tubers per plant, following the developments during tuber initiation.

Conclusion

These trials showed that to produce high quality, firm-fleshed potatoes, it is advantageous to maintain low soil-moisture tension values from emergence to haulm killing. This results in a

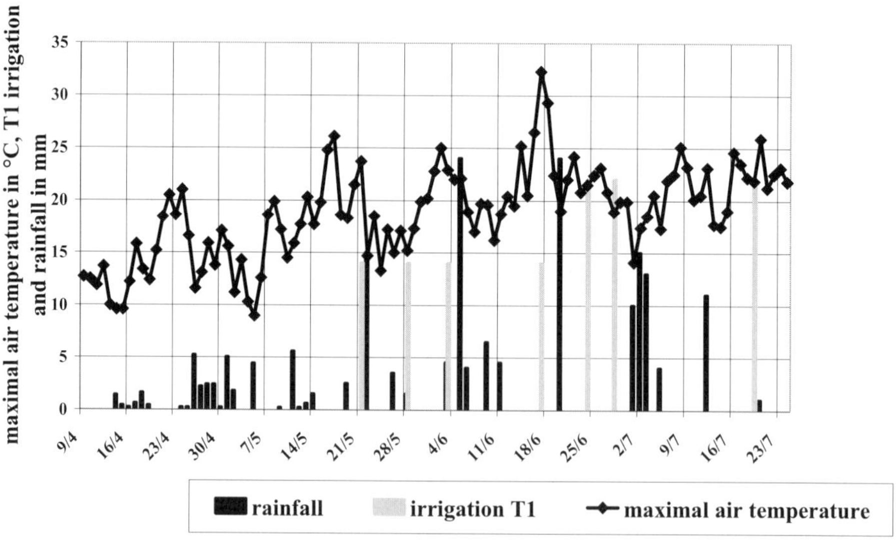

Figure 6. Maximal air temperature and irrigation T1, Charlotte 2002.

high number of tubers per plant, provided that other limiting factors apart from the water supply do not disrupt the results. This also helps to protect the tubers against the development of common scab. These recommendations have been set out in the IRRINOV® method user's guide, the method of irrigation scheduling circulated in partnership with McCain, the GITEP and the Chambre d'Agriculture du Pas-de-Calais. However, trials are continuing to find out more about the effect of early irrigation on the number of tubers per plant and to give more precise rules on the irrigation patterns.

References

Deumier J.M. and S.Vallade (2002). 3.5 mm/j en Bintje : les pommes de terre s'irriguent aussi (3.5 mm/d for Bintje : potatoes are irrigated too). Perspectives Agricoles n°280, 5 pages.

Goffart J.P. and D. Ryckmans (1996). Problématique de l'irrigation en culture de pomme de terre : intérêts dans la lutte contre la gale commune et effets sur la culture (The irrigation problem in potato growing : advantages in combating common scab and effects on the crop). La pomme de terre de consommation (Centre de Recherche de Gembloux), 8 pages.

Haverkort A.J., M. Van de Waart and K.B.A Bodlander (1990). The effect of early drought stress on numbers of tubers and stolons of potato in controlled and field conditions. Potato Research, 33, 89-96.

Walworth J.L. and D.E. Carling (2002). Tuber initiation and development in irrigated and not irrigated potatoes. Amer. J. of Potato Res., 79, 387-395.

Guide de l'utilisateur et carnet de terrain : pilotez l'irrigation des pommes de terre avec la méthode IRRINOV®, modules variétés Bintje, Russet-Burbank et Santana pour la production de frites et Charlotte, Amandine, Chérie, Franceline, Ratte et Exquisa pour les variétés à chair ferme en sols de limon du Santerre et du Pas-de-Calais (2005), 34 p.

(User's guide and field log : controlling potato irrigation with the IRRINOV ® method, modules Bintje, Russet-Burbank and Santana cultivar for the production of French fries, and Charlotte, Amandine, Chérie, Franceline, Ratte and Exquisa for the firm-fleshed cultivars in the silt soil of Santerre and Pas-de-Calais (2005), 34 p.)

Curriculum vitae – J.M. Deumier

J.M. Deumier studied agronomy at the Institut National Agronomique de Paris-Grignon, from which he received his engineering degree in 1977. After completing his military service, he joined the Chambre d'Agriculture de la Haute-Garonne where he was in charge of farming irrigation and drainage projects. He joined ARVALIS - Institut du végétal in 1985, where he has been in charge of research about irrigation of cereals, corn and peas since appointment. His responsibilities expanded in 1995, since when he has also been responsible for researching irrigation of potatoes. His job is to set up irrigation research programmes company and to propose experimental programmes for each experimental centre. He turns experimental results into practical advice for farmers, and he puts in place tools for improving irrigated crop system management. Between 1990 and 1998 ARVALIS developed three programmes that model irrigation plans with regard to key variables: available water resources, equipment and manpower:

- LORA which enables farmers to achieve the optimal cropping plan on an irrigable area;
- IRMA, an irrigation simulator which allows farmers to test and to improve a provisional irrigation program;
- IRRELIA for scheduling irrigation according to a strategy.

Since 1998 the institute has developed:

- IRRIPARC® which is software designed for simulating irrigation water distribution which depends on gun settings and wind speed and direction. This software put in place with Cemagref, allows users to find the best settings for each type of gun in order to improve water distribution.
- IRRINOV®, an irrigation management method to irrigate peas, cereals, maize and potatoes. This method includes decision-making rules for starting, resuming and ending irrigation. The indicators take into account soil water status (tensiometering), climatic data (rainfall) and crop (development stage and growing conditions).

Curriculum vitae – François-Xavier Broutin

François-Xavier Broutin studied at ISA University College in Agriculture, Food and Environmental Sciences (Lille, North of France), from which he received his Ingenieur degree (master level) in 2004. After several training periods in potato agronomy, he worked on the international promotion of the 4th International Technical and Commercial Potato Exhibition that was held in Villers-Saint-Christophe (Aisne, France) at the beginning of September 2004, and in which nearly 10 000 persons (among them 3 000 international) took part. He joined ARVALIS - Institut du vegetal / ITPT at the end of 2004, working now on potatoes as a regional Ingenieur for the North-East of France. He is also in charge of the experimental station of Villers-Saint-Christophe, following the experimental programmes elaborated by the specialists in different thematic such as irrigation, new varieties, and fertilization.

Curriculum vitae – Denis Gaucher

Denis Gaucher studied the environmental and agronomic sciences at the Ecole Nationale Supérieure Agronomique et des Industries Alimentaires (ENSAIA) de Nancy (France), from which he received the agronomic ingenior degree in 1989. In 1991, after a one year period in Africa (Guinea) as hevea and rice researcher, he joined the Agronomic Department of Institut Supérieur Agricole de Beauvais (ISAB) as potato researcher and agrometeorology teacher. In 1996 he joined ARVALIS-Institut du vegetal (ITCF until 2002) as a researcher on potato diseases. The main themes of research are potato skin parasites (black scurf, silver scurf, black dot, common scab) and the fight against late blight. He was initiator of the project Mildi-LIS®, a new online decision support system for treatment against potato late blight.

Company profile – ARVALIS - Institut du végétal / ITPT (Institut Technique de la Pomme de Terre)

ARVALIS - Institut du végétal / ITPT (Institut Technique de la Pomme de Terre) is a technical institute involved in applied research. It is founded and managed by the professionals of the French Potato Sector.

Its goal is to give to farmers and the different actors of the whole French Potato Sector technical informations improving the competitivity for the production and the quality for the products while integrating the environmental impacts. Orientations, programmes and financements are decided by the Administrative Council in which are represented CNIPT

(Interprofessional Organization for the "Fresh" raw potato Market), GIPT (Interprofessional Organization for the processing potato industry) and FNPPPT (Seed Potato Growers Organization). The Council is chaired by a potato producer: Mr Arnaud DELACOUR.

The applied research programmes of ITPT and the results diffusion are done by the engineers and technicians of ARVALIS - Institut du vegetal/ITPT (Applied Research Institute for Cereals, Maize, forages and pulses) on the base of a pluriannual partnership contract signed in 1995. The orientations of ITPT rely on a Scientific and Technical Council, chaired by Mr Daniel ELLISECHE (INRA) and composed by numerous personalities coming from the research side and the private enterprises. This Council plays an important role for prospective vigilance, proposals and exchanges in order to orientate and evaluate the activities.

An operational board and technical commissions meet the engineers of the Institute and the three professional organizations (CNIPT, FNPPPT, GIPT) to follow, discuss and promote the results. The main financial resources of ITPT come from professional fees, ONIFLHOR subsidies and services selling.

The R & D studies of ITPT are carried by engineers and technicians over a large number of topics into ARVALIS - Institut du vegetal since 1995. Experiments are conducted either in ARVALIS facilities on two main research centres: Boigneville (91) and Villers-Saint-Christophe (02) which own specific equipments to well develop potato works or in different parts of FRANCE most of the time in partnership with a large number of organizations or companies also involved in the potato area.

Presentation of a Decision-Support System (DSS) for nitrogen management in potato production to improve the use of resources

J.P. Goffart, M. Olivier and J.P. Destain
Walloon Agricultural Research Centre (CRA-W), Crop Production Department, 4 rue du Bordia, 5030 Gembloux, Belgium

Abstract

A simple Decision-Support System (DSS) has been developed at our Research Centre aimed at achieving better match, at the field level, between a potato crop's nitrogen requirements and the mineral N supply from soil and fertilizers. The system requires, initially, the assessment of a total N-recommendation based on the predictive balance-sheet method (software Azobil, INRA, Laon, France). Before planting, 70% of the recommended N amount is applied to the crop. Subsequently, between 25 and 55 days after crop emergence, the need for supplemental N is monitored through a fast and non-invasive tool, a chlorophyll meter. In-season measurements of leaf chlorophyll-concentration are made directly on the foliage in the field. Based on specific measurements in the field and in a small window in the field without any applied N fertilizer, the potato grower can decide whether to apply the remaining 30%. When required, the strategy enables the grower to save on N fertilizer (annually in about 35% of cases) and reduce the risk of excessive N fertilization and its negative impact on tuber quality and the environment, without a negative effect on tuber yield. It therefore also results in an improvement in the use of resources, particularly nitrogen. The DSS strategy is economically feasible and easy to operate.

Keywords: nitrogen, field specific N-advice, N-splitting, crop N-status, N-efficiency

Introduction

There is a clear need to provide guidance to the potato producer on the rational management of the nitrogen fertilization of the crop. The objective, at a field-specific level, is to ensure a better match between N supply and crop N requirement, resulting in expected yield and tuber quality being achieved and a reduction in undesirable losses of N to the environment. Such an approach should also improve the use of resources, such as nutrients, land, energy and money.

With regard to applying the whole recommended amount of N fertilizer at planting, it is generally agreed that the establishment of a field-specific N recommendation for potato at planting time can never be accurate as it is not possible to predict with precision the total crop N requirements and soil mineral N supply that will occur during the growing season. These variables are influenced by several predictable factors as well as by unpredictable ones, such as weather conditions, chemical and physical soil properties, type and evolution of organic matter previously incorporated in the soil, cultural practices, variety earliness and crop duration.

The Decision-Support System (DSS) developed by our department to help to solve this problem is based on an approach defined by Vos and MacKerron (2000). It combines splitting the N fertilizer applied with estimating the crop N requirement during the growing season and thus, indirectly, the soil mineral N supply.

This paper outlines the basic strategy underlying the DSS and its practical use, and describes demonstrated or expected improvements in the use of resources. The strategy has been described in detail by Goffart and Olivier (2004).

Description of the DSS

The strategy involves five steps, as shown in Figure 1. Step 1 aims to determine, at a specific field level, the total N recommendation for the crop before planting and is based on the predictive balance-sheet method. The software used for this is AZOBIL, developed at the

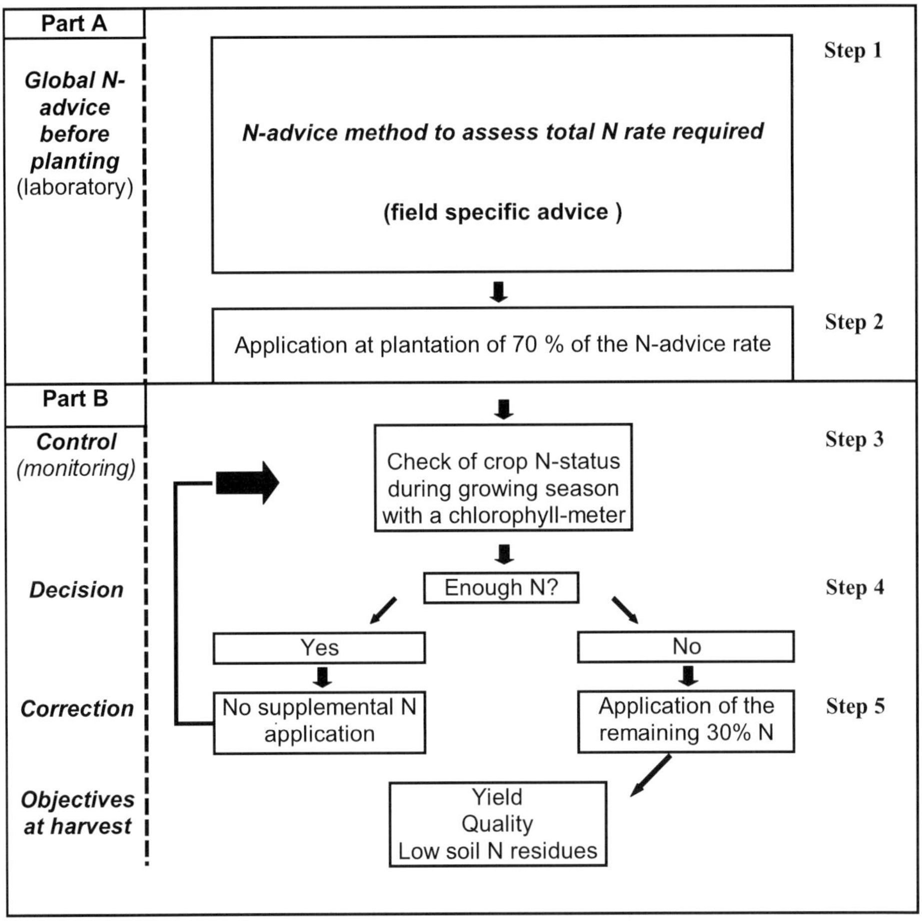

Figure 1. Steps in the strategy for managing the nitrogen fertilization of a potato crop.

Institut national de la recherche agronomique (INRA, Laon, France) by Machet *et al.* (1990) and recently upgraded by Chambenoit *et al.* (2004) specifically for the potato crop. In step 2, N is applied at planting at 70% of the recommended rate. Part A of the strategy (steps 1 and 2) are implemented before planting. The next three steps (part B of the strategy) occur during the growing season. In step 3, the crop is monitored to assess its nitrogen status, that is, whether it has sufficient nitrogen N or not, based on a threshold value for a measured parameter. According to our research results and due to the very close relationship with the leaf N concentration (Vos and Bom, 1993) and the crop N-uptake (Goffart *et al.*, 2002), the leaf chlorophyll concentration was chosen for monitoring.

A hand-held chlorophyll meter (model HNt 'Hydro N-tester', marketed by Yara International ASA, Oslo, Norway) enables fast and non-invasive measurements on individual leaves in the field to be made. Depending on the difference between HNt measurements in the fertilized part of the field and HNt measurements in a reference 'zero N' plot in the field, relative rather than absolute values of the meter are used for the next step. In step 4, the results of the measurements determines the decision whether or not to apply the remaining 30% of the field-specific recommended N according to the crop's estimated need for it. Finally, in step 5, when supplemental N is required, the grower has to decide on the type of N fertilizer and mode of application to ensure the greatest efficiency of the supplementary application.

Specific threshold values of HNt measurements are used for potato cultivars or groups of potato cultivars. To date, values have been established for the most popular potato cultivars cropped in Belgium (including Bintje, Agria, Astérix, Saturna, Nicola, Charlotte and Franceline).

Potential of the DSS for improving the use of resources

Use of nutrients

Two characteristics of the DSS lead to an improvement in the use of nitrogen.
The first one is the application of the balance-sheet method to establish the N recommendation at field level. The various elements of the balance sheet taken into account in the AZOBIL model are presented in Figure 2.

As inputs into the soil/plant system, all the N sources are considered. They are: the measured and readily available soil mineral N at planting (Ri); the estimated net soil mineral N in the growing season (Mn) that results from the sum of four components (net mineralization from soil organic matter, mineral N supply from previous crop residues, mineral N supply from previously applied organic manures (such as farmyard manure and slurry) and catch crops, and mineral N supply from previous grass crop); and the recommended N rate applied as mineral N fertilizer (X). The inputs are counterbalanced by the outputs that include the estimated crop N requirements (B), eventual potential losses through leaching during the period between soil analysis and fertilizer N application (L), and soil mineral N residues at harvest (Rf) that are expected to be as low as possible.

In this complete approach to N supply and N requirement, taking account of the sources of mineral N available to the crop, other than N fertilizer, improves the soil mineral N recovery

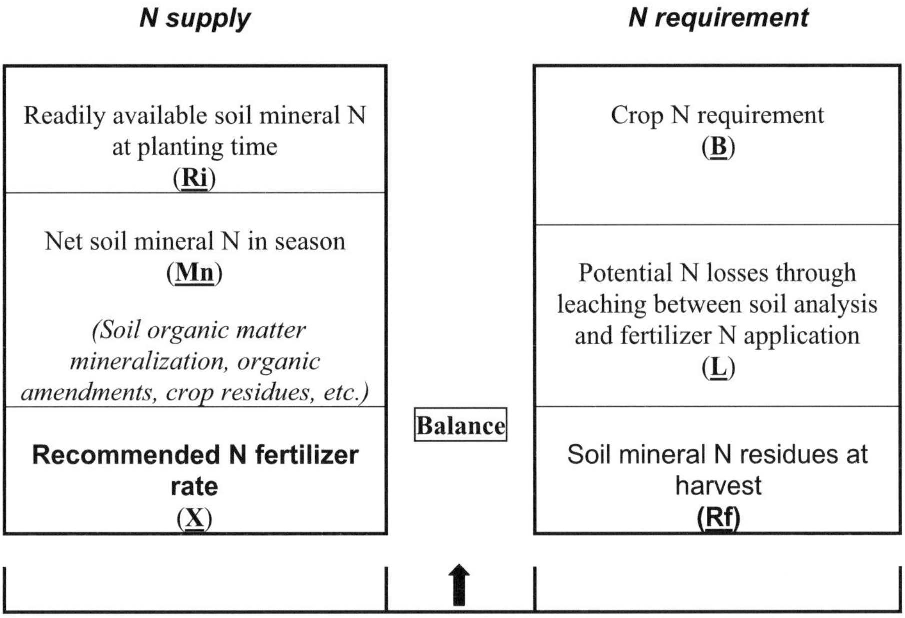

Figure 2. Outline of the balance-sheet method at field level, developed using AZOBIL software (version model 1.2; Machet et al.*, 1995).*

by the crop. It therefore prevents the application of an excessive N fertilizer rate with a lower N efficiency (lower N recovery coefficient).

The second characteristic is the application at planting of 70% of the recommended N amount. Depending on the HNt measurements and the decision on the need for supplemental N, the DSS makes it possible for the grower either to save 30% of the recommended amount of N fertilizer (that would normally have been applied at planting time) or to apply this remaining 30%.

In the first case, in relation to sites with little or no response to supplemental N application, the saving of 30% of the recommended rate leads to a total tuber yield and N uptake similar to that with the application of the whole recommended amount. Consequently, fairly high N fertilizer recovery coefficients are generally deduced. Table 1 illustrates this for some field trial sites in Belgium.

In the second case, the application of the remaining 30% when the crop requirement is determined also usually leads to a better recovery of total rate of N fertilizer, and there have been similar or higher yields of large tubers (from 1 to 6 tons ha^{-1}) (Goffart *et al.*, 2002). For field trials with the cultivar Bintje, Table 2 shows the comparison of yields for similar recommended N rates, either applied in one application at planting (with CAN) or as a split application, 70% at planting (CAN) and 30% during the growing season (urea solution sprayed on the foliage). Similar effects were observed with CAN as supplemental N in sufficiently wet conditions.

Table 1. Comparison of total tuber yield, crop N uptake (at harvest) and N fertilizer recovery coefficient for the recommended N rate and 70% of this rate applied as calcium ammonium nitrate (CAN) at planting in sites where no supplemental N was required following HNt measurements (results from field trials conducted in Belgium in 1999 and 2000) for the cv. Bintje.

N rate	Tuber yield (ton ha^{-1})	N uptake (kg N ha^{-1})	N recovery (%)
Site 1			
70% recommended rate	51.6	177	42
Recommended rate	52.6	184	34
Site 2			
70% recommended rate	83.7	230	51
Recommended rate	83.4	245	44
Site 3			
70% recommended rate	68.4	200	78
Recommended rate	68.9	209	60
Site 4			
70% recommended rate	44.5	157	53
Recommended rate	45.4	167	44

Table 2. Effect of the splitting of the total recommended N rate on tuber yield characteristics for the cv. Bintje (results from field trials conducted in Belgium in 1997 and 1998) in sites where supplemental N application was required following HNt measurements.

N fertilizer rate	Total tuber yield (tons ha^{-1})	Proportion of large tubers (%)
1997		
0	50	55
70%	68	68
100% (150 kg N ha^{-1})	71	69
70%+30%[1]	72	73
1998		
0	44	42
70%	53	59
100% (176 kg N ha^{-1})	56	61
70%+30%[1]	62	66

[1]N-fertilizer applied as CAN at planting (70 or 100%), and as a urea solution sprayed on the canopy (30%) during the growing season.

It was also demonstrated in field trials using an N^{15} isotope as tracer of the N-fertilizer (Destain *et al.*, 1993). Table 3 shows that when the supplemental 30% of the recommended N rate is applied as urea sprayed on the canopy, N^{15}- fertilizer recovery coefficients are equal to or higher than the whole recommended N rate applied at planting.

Table 3. Effect of the splitting of the total recommended N rate (following HNt measurements) on N^{15} fertilizer recovery for the cv. Bintje (results of field trials in 1997 and 1998) in sites where supplemental N application was required following HNt measurements.

N^{15}-fertilizer supply	N^{15}-fertilizer supply at planting	N^{15}-fertilizer recovery (%)[2] in growing season
1997		
100% CAN[1]		77%
70% CAN	30% urea solution sprayed	75%
1998		
100% CAN		47%
70% CAN	30% urea solution sprayed	55%

[1] CAN: calcium ammonium nitrate.
[2] Ratio between amount of N^{15} measured in the plant and amount of N^{15} applied as fertilizer.

Other expected improvements in the use of resources

Where supplemental N is not required and not applied following HNt measurements (which represents about 35% of the field cases), the observed tuber yield was similar to that reached with the recommended N rate. Lower amounts of N fertilizer used also means less energy spent on producing fertilizer. Increase in tuber quality was also observed: there was a higher dry-matter content (about 1%) compared to the recommended N rate. This could benefit industrial processes for chips or crisps (improvement in use of oil and energy).

In terms of biomass production, the saving of N is also important as it contributes to reduced potato canopy development and consequently to the reduced occurrence of foliar diseases such as late blight during wet seasons. The expected advantage is an improvement in the use of pesticides (mainly fungicides).

Finally, the saving of N fertilizer is important for reducing the risks to the environment (surface and groundwater pollution by nitrate) and for contributing to the protection of one of the planet's most important resources: water.

Where supplemental N (the remaining 30%) is required, the large tuber yield increase generally observed (from 1 to 6 tons ha^{-1}), compared to the yield when the total recommended N amount was applied in one application at planting, indicates the potential for improved yields in a same area. We could therefore expect improvements in land use. In addition, with the higher proportion of large tubers, there would be a higher recovery of potatoes for processing, with a resultant improvement in the use of energy.

Is the strategy economically feasible?

The price of the AZOBIL software (or its recently upgraded version AZOFERT) is about 1,250 Euros; the software is licensed to and marketed by INRA. An advisory laboratory or consultant usually uses it. The cost of the soil analyses for mineral nitrogen and organic matter content is about 25 euros per field; the grower meets these costs.

The HNt chlorophyll meter is commercially available and costs about 1,100 euros (excluding VAT). The cost could restrict its purchase by individual growers and thus hamper widespread adoption of the method. However, this cost can be reduced by sharing it by a small syndicate of growers and by using the meter on other crops such as winter wheat. Service laboratories can also make the measurements.

Taking the costs into account, and the case where one producer has to meet the full HNt cost alone, it is estimated that the cropped potato area required to compensate for the cost of using the method over a 5-year period is not significant (Goffart *et al.*, 2002).

Where supplemental N is not required and not applied following HNt measurements (the case in about 35% of the studied situations), it has been demonstrated, based on the mean cost of N fertilizers (CAN or urea applied on the foliage) and the mean selling price for tubers, that 9-18 ha of potato crop are sufficient to compensate for the measurement costs by saving on N fertilizer (30% of the recommended rate) for an AZOBIL-based recommended rate of 200 to 100 kg N ha^{-1}.

Where supplemental N is required according to HNT measurements (65-70% of cases), as mentioned above a small yield increase has generally been observed due to good timing of the supplementary application of N. Here, the potato crop area required to compensate for the cost of soil analysis and measurements compared to the cost of the whole amount of N being applied in one application at planting ranges between 1.3 and 7.2 ha.
Obviously, smaller cropped areas will be required to compensate for the costs of the method when the HNt is bought by a small syndicate of growers or is used for other crops. It can then be argued that the strategy is economically feasible at least for medium to large potato-producing farms.

Conclusion and prospects

Through a better assessment of the growing season N requirements of the crop and, indirectly, of the soil mineral N supply, the DSS that has been developed clearly gives the grower the opportunity to save N fertilizer when justified (mean value of 30% of the cases). In this way, it also prevents the negative impact of excessive N on tuber quality and on the environment, and makes a considerable contribution to improving the use of resources, particularly for nitrogen.
Further improvements in the DSS include:

- Improving the accuracy of the AZOBIL-based recommended N rates (partly done with the upgraded version, AZOFERT);
- Modulating the supplemental rate of 30% nitrogen by assessing more accurately the required supplemental N using HNt measurements;
- Confirming the validation of the DSS for irrigated conditions and for other important cultivars;
- Integrating this strategy for N management into a more global DSS developed for potato;
- Building into the strategy the use of tools other than HNt for assessing crop nitrogen status, such as crop reflectance (e.g., Cropscan and GPN), that is being developed in several countries.

Acknowledgements

This project was funded by the Belgian Federal Ministry of Agriculture from 1997 to 2000 for research activities, and the Ministry of the Walloon Region from 2001 to 2005 for development activities.

References

Chambenoit, C. , F. Laurent, J.M. Machet and H. Boizard (2004). Development of a decision support system for nitrogen management on potatoes. In: D.K.L. MacKerron and A.J. Haverkort (eds) Decision support systems in potato production: bringing models to practice. Wageningen Academic Publishers, Wageningen, The Netherlands, pp. 54-67.

Destain, J.P., E. François, J. Guiot, J.P. Goffart, J.P. Vandergeten and B. Bodson (1993). Fate of nitrogen fertilizer applied on two main crops, winter wheat (Triticum aestivum) and sugar beet (Beta vulgaris) in the loam region of Belgium. Plant and Soil, 155/156, 367-370.

Goffart, J.P., M. Olivier, J.P. Destain and F. Frankinet (2002). Stratégie de gestion de la fertilisation azotée de la pomme de terre de consommation (Management strategy of nitrogen fertilization of the potato). Centre Wallon de Recherches Agronomiques (CRA-W) (eds), Gembloux, Belgium, pp. 118. (web site : www.cra.wallonie.be).

Goffart, J.P. and M. Olivier (2004). Management of N-fertilization of the potato crop using total N-advice software and in-season chlorophyll meter measurements. In: D.K.L. MacKerron and A.J. Haverkort (eds) Decision support systems in potato production: bringing models to practice. Wageningen Academic Publishers, Wageningen, The Netherlands, pp. 68-83.

Machet, J.M., P. Dubrulle and P. Louis (1990). AZOBIL: a computer program for fertilizer N recommendations based on a predictive balance sheet method. Proceedings of the 1st Congress of the European Society of Agronomy.

Machet, J.M. , P. Dubrulle and P. Louis (1995). AZOBIL: Manuel d'utilisation agronomique du logiciel. Station d'analyses agricoles de Laon et Institut National de la Recherche Agronomique (INRA) (eds). Laon, France, pp. 33.

Vos, J. and M. Bom (1993). Hand-held chlorophyll meter: a promising tool to assess the nitrogen status of potato foliage. Potato Research, 36, 301-308.

Vos, J. and D.K.L. MacKerron (2000). Basic concepts of the management of supply of nitrogen and water in potato production. In: A.J. Haverkort and D.K.L. MacKerron (eds) Management of nitrogen and water in potato production. Wageningen Pers, Wageningen, The Netherlands, pp. 15-33.

Curriculum vitae – Jean-Pierre Goffart

Dr.ir. Jean-Pierre Goffart is a research agronomist and crop physiologist specialized in plant nutrition, and particularly nitrogen nutrition of potato and vegetables, at the Walloon Agricultural Research Centre (CRA-W) of Gembloux in Belgium. He started to study Agricultural Sciences at the Catholic University of Louvain-la-Neuve (UCL) in Belgium, from which he received the MSc degree (ingenieur) in Agricultural Sciences in 1984. After a one year period for Monsanto Company as biologist (in pesticides research), he joined the Research Unit of Plant Pathology of UCL, from which he received in 1992 a PhD degree (for a work on epidemiology and control of Rhizomania disease in sugar beet). In 1990 he had already joined the Crop Husbandry Research Station of the CRA-W where he started working on different arable crops (potato, cereals, sugar beet, maize, catch crops) mainly in field trials (applied research) and especially on nitrogen management and related topics (N-fertilizers choice and use, varietal assessment, soil mineral nitrogen, crop N requirement, reduction in N emissions to the environment). Since 1996 his research interests focus on the development of cultural practices or methods and decision support systems to improve N management of the ware potato crop and of vegetables crop (spinach, lettuce, bean, endives, ...). He particularly develops the practical use of relevant tools to monitor in season crop N-status used to decide on the need for supplemental nitrogen to be applied. From 1997 to 2000, he was partner of the European Project (concerted action) "EUROPP" initiated by Dr. D.K.L. MacKerron on management of water and nitrogen in potato crop. In 1996, he was appointed as Project leader at the Crop Production Department of CRA-W. He is also active in the set up of extension services and development activities to the producers and the whole potato chain partners. Since 2004, he is President of FIWAP, the Walloon potato chain association in Belgium.

Company profile – The Walloon Agricultural Research Centre (CRA-W) in Gembloux (Belgium)

The Walloon Agricultural Research Centre in Gembloux (CRA-W), founded in 1872, is a top-level scientific institution depending of the Ministry of the Walloon Region in Belgium. It is a multidisciplinary research facility, organised into seven departments, standing on a 300 hectares site with laboratories, greenhouses, orchards, experimental fields and offices, employing over 400 people including about 100 scientists. CRA-W has fundamental as well as applied research activities and services in biotechnology (molecular biology, propagation techniques), in crop production (soil and fertilization, crop husbandry, plant accessions and recommended cultivars), biological control and plant genetic resources (plant diseases and applied zoology, plant improvement), pesticides research (chemistry, physical chemistry and biological activity of plant protection products), agricultural engineering (mechanisation, energy and industrial use of biomass), animal production and nutrition, quality of agricultural products (quality and technological value of plant and animal products, spectrometry for the assessment of quality), biometry, data management, agrometeorology, farming systems, grasslands, fodder crops. Funding of the research and development projects are mainly coming from the Walloon and Federal Governments and from the participation of several scientists to European Research Projects. CRA-W operates activities all-over the Walloon Region area or in Belgium together with other national or international, public or private institutions or organisations (Universities, Research Institutes, official extensive services, growers and farmers associations, traders, industry, consumers associations, ...). CRA-W specific activities in potato research and development mainly includes: fertilization, varietal assessment, late blight control, classical and new methods for tuber quality analysis, organic farming, storage, aphids control, seeds research, viral/fungal/bacterial diseases.

NemaDecide: a decision support system for the management of potato cyst nematodes

T.H. Been[1], C.H. Schomaker[1] and L.P.G. Molendijk[2]
[1]Plant Research International, PO Box 16, 6700 AA Wageningen, The Netherlands
[2]Aplied Plant Research, Edelhertweg 1, 8219 PH Lelystad, The Netherlands

Abstract

A consortium of eight agri-business companies and research organizations are currently developing *NemaDecide*, a support system (DSS) that, when completed, will contain all relevant quantitative knowledge about plant parasitic root nematodes (cyst nematodes, root knot nematodes and *Pratylenchus* spp.) of some major crops, including potatoes, with emphasis on rotations. The primary task of *NemaDecide* is to keep nematodes at low, economically acceptable density levels.

The first phase of the project, the development of a simple module for the management of potato cyst nematodes (*Globodera rostochiensis and G. pallida*), has been completed. The interface has been developed in close cooperation with extension officers. It was tested in 2004 with growers and is now used to provide growers with advice. The potato cyst nematode module is provided with an internet connection to soil sampling data, enabling providers to collect sampling results easily. The development of a second, more sophisticated module, including other root nematodes and GEO-information based visualization of farmer fields was started in 2004 and is being tested in 2005.

All the results from 50 years of Dutch quantitative nematological research have been structured into stochastic models and integrated in a software package. The quantitative information system enables growers to estimate risks of yield losses, to determine population development, to estimate the probability of detection of nematode foci by soil sampling, to calculate the cost/benefit of control measures and to provide adequate advice for growers to optimize financial returns. Growers can compare cropping scenarios and ask 'what if' questions. The system can be used for agricultural practices, administrative duties required by government, certification schemes and education institutions.

The decision support system is tested constantly on farms with different groups of users. Based on comments from growers and extension officers, the system has been modified and extended to incorporate new functions. Gaps in the knowledge, based on uncertainties in the advice, are pinpointed in cooperation with the agro sector, user groups and extension services. Thus, the most problematic unanswered questions now receive priority in research programmes, which are now better focused on practical problems.

Keywords: stochastic models, theory building, prediction, risk analysis, extension, software

Introduction

Plant parasitic nematodes are important causes of crop losses in temperate zones. This especially applies for *G. rostochiensis and G. pallida*, *Heterodera schachtii*, *Meloidodyne* species, *Pratylenchus penetrans* and *Trichodorid* species. In the past, control of these pests was relatively straightforward using crop rotation schemes and a heavy input of nematicides, based on information on mean multiplication rates under hosts and decline rates under non-hosts. Nowadays, chemical control is increasingly restricted by governments and, as prices of agricultural products are under pressure, farmers need to be more critical about the cost/benefit ratios of control measures.

In the Netherlands, resistant potato cultivars can control *G. rostochiensis* completely. Cultivars with a high level of resistance to *G. pallida* are also available in starch potatoes, but the control of potato cyst nematodes still causes problems in these areas, probably because of the common practice of cropping starch potatoes with insufficient levels of resistance in short rotations, usually one potato crop every two years. In cropping areas with ware and seed potatoes, rotations are longer, but potato cultivars with high level of resistance against *G. pallida* are scarce. As the resistance of potato cultivars against *G. pallida* is partial and covers the whole resistance spectrum from 0 to >95% and farmers want to utilize this resistance in ware and seed potatoes optimally, a knowledge based system for nematode management is needed. Such a system must be based on accurate estimates of the nematode situation in a specified field based on adequate sampling techniques, information on the degree of resistance of potato cultivars and on the effectiveness of (chemical) control measures. The control of other, more polyphagous, nematodes requires the addition of the effect of cropping sequence and green manure crops on nematode numbers. To put this knowledge system into practice requires a high level of skill from farmers and extension officers. Thus, it was decided to develop a decision support system, called *NemaDecide*, for nematodes in close cooperation between governments, agribusiness, breeders, soil sampling laboratories, farmers and nematologists.

To begin with *NemaDecide* focuses on rotations with potatoes. To enhance the acceptation and use of *NemaDecide*, it will minimize data input by farmers by trying to establish webservices to sampling agencies. Ideally, farmers only have to ask their questions to the decision support system and get a straight answer. Therefore, the availability of data, whether it be the GEO-information or the monitoring results, should be standard. In relation to the rapid development of Farm Management Systems, which can function as the data-providers to DSS-systems, it was decided to design *NemaDecide* on a modular basis by developing the scientific engine *NemaMod* and the Geo-tool *SampView* as auxiliary components which can be used by these Farm Management Systems in the future.

Outline

NemaDecide is a quantitative, stochastic DSS. Figure 1 presents a general outline of the data and models used by *NemaDecide*. The following sections will address these models, the origin of the data to parameterize them and the additional external information required by the DSS. A brief outline of the software is presented and, finally, some examples of the functionality and use of *NemaDecide* are provided.

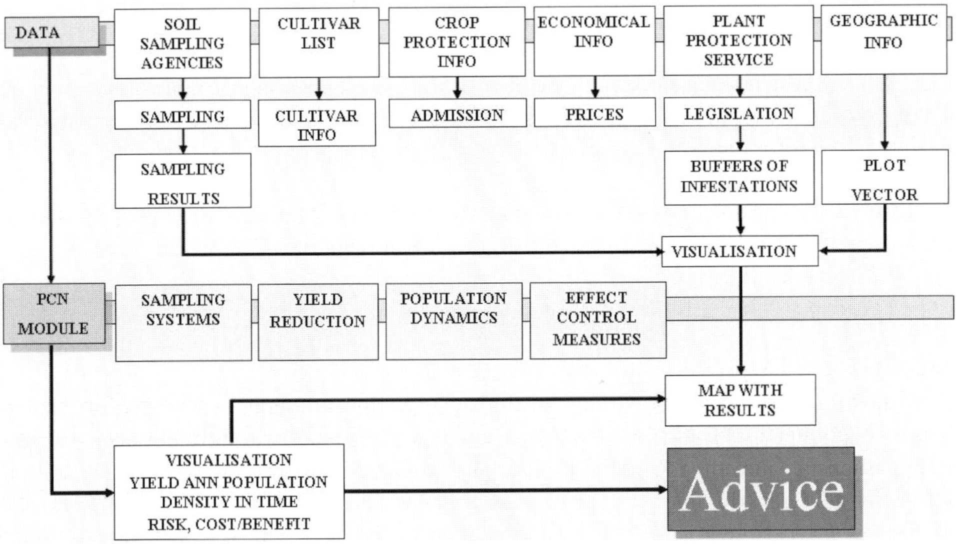

Figure 1. Layout of NemaDecide structure.

The final decision support system will consist of the following parts:

- An ActiveX-component *NemaMod* incorporating all nematological knowledge; 'the engine'.
- An ActiveX-component *SampView* incorporating sampling logic and geometry algorithms to divide fields into agronomic areas (sampling units, cropping areas, etc.).
- A visualisation component.
- Databases.
- A user interface developed in cooperation with extension officers and farmers.

The most important part of *NemaDecide* is *NemaMod*, an ActiveX-component that provides the functionality of the DSS. It contains all the models and nematological information required to execute the required calculations. The databases, in which parameters of the different models for each nematode/host combination and data are stored, is external. It can be updated or extended with new nematode/host combinations without the necessity for changing the *NemaMod* component or *NemaDecide* itself. This feature makes the engine independent of the current state of knowledge concerning parameter values and their stochasticity. New results can be incorporated in the component by updating the database. This not only extends the use of the component in actual applications, but also provides the possibility to adapt parameter sets for different countries, if required.

Models used

The submodels used in *NemaDecide* are mathematical analogues of nematological theories developed over a time span of more than fifty years. The following subjects are covered:

1. Plant growth and tolerance.
2. Population dynamics, plant resistance and nematode virulence.
3. Spatial distribution patterns and sampling methods.

4. The effect of control measures on (1) and (2).

Parameters in these models are stochastic quantities, have a clear nematological meaning and can be estimated directly from data sets. The models have adequate predictive value and contribute to theory building.

Plant growth: The relation between initial population density (P) and plant weight (Y) is derived from a growth model for the first mechanism of growth reduction (Seinhorst, 1986b):

$$r_P/r_0 = k+(1-k)\cdot z^{P-T} \quad \text{for } P > T \qquad (1)$$
$$r_P/r_0 = 1 \quad \text{for } P \leq T$$

One of the most important assumptions in the model is that for plants with nematode density P and without nematodes, of the same total weight (and, therefore, of different age), the ratio r_p/r_0 is constant during the growing period. The variables r_0 and r_P are the growth rates for plants without nematodes and for plants at nematode density P, respectively. The parameter T, the tolerance limit, is the maximum density below which the nematodes do not reduce plant weight. The parameters z and k are constants smaller than 1. The parameter k means that nematodes cannot stop plant growth completely and that, even at very large nematode densities, some growth still takes place: $r_P/r_0 = k$. The value of the parameter k is independent of nematode density and time after planting but varies between experiments.
The fundamental model of the relation between nematode density P and plant weight Y was derived from cross sections orthogonal to the time axis through growth curves of plants for ranges of densities P/T and different values of k. These cross sections are in close agreement with equation (2).

$$Y=Y_{max}\cdot\{m+(1-m)\cdot z^{P-T}\} \quad \text{for } P > T \qquad (2)$$
$$Y=Y_{max} \quad \text{for } P \leq T$$

The parameter m is the minimum relative plant weight and usually is slightly larger than k, the parameter z is a constant < 1 with the same or a slightly lower value than in equation (1) and the parameter T is the tolerance limit with the same value as in equation (1).

Yield can be derived directly from plant weight, Y, using relevant quality and financial data. Equation (2) was validated by fitting it to data from 36 plant/nematode combinations out of 29 experiments about the relation between Pi of 14 tylenchid nematode species and the relative dry plant weight (Y) of 27 plant species/cultivars several months after planting (Seinhorst, 1998). For cultivars susceptible to Seinhorst's second mechanism of growth reduction, equation (2) was expanded to a double exponential function. The phenomenon of "early senescence" of potato plants at tuber formation is not included in *NemaDecide* as it appeared to occur only in exceptional cases.

Tolerance: Tolerance of potato cultivars was quantified by expressing it in values of the tolerance limit T and the minimum yield m. The parameter T manifests itself at small nematode densities, m at larger ones. Yet, one needs a whole range of pre-plant nematodes densities (P) to estimate either one of the parameters. The value of the tolerance limit, T, seems unaffected by differences in external conditions and could, therefore, be estimated from both pot experiments in the glasshouse and field experiments. The only requirement of glasshouse tests, to prevent biased estimates of T, was that large enough pots had to be used to prevent the plants from becoming pot bound. The minimum yield, m, is more sensitive to external conditions than T. Therefore, differences in tolerance in plant cultivars were established in one field experiment under the same conditions. In addition, a sufficient number of values of m were estimated in field experiments to establish a distribution function of m.

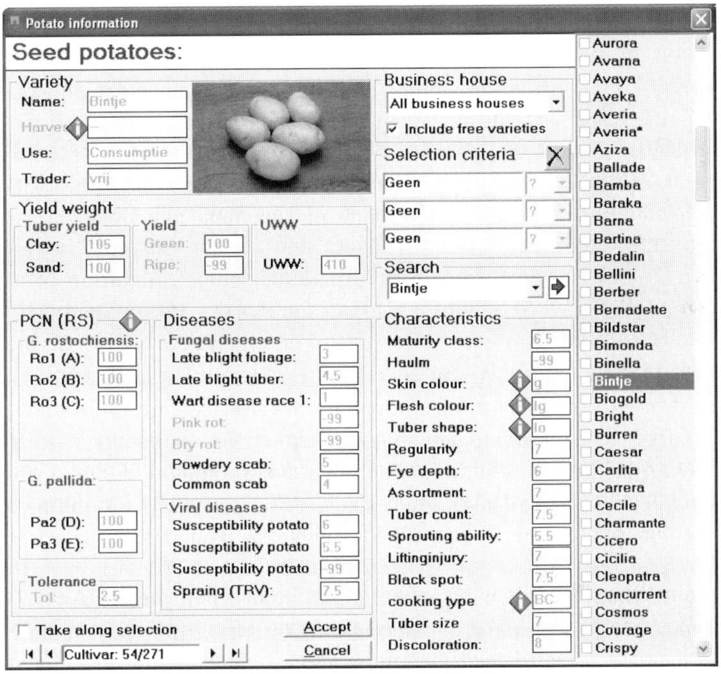

Figure 2. Screenshot of the potato cultivar information window. To the left: Resistance to PCN pathotypes. Middle left: Fungal and viral diseases. Middle right: cultivar characteristics. Selections can be made on Business house and three criteria. One can search for a cultivar or pinpoint only some preferred cultivars (to the right). If the checkbox is checked only selected cultivars will be available in the calculations.

Population dynamics: For nematodes species with one generation per year, which become sedentary after invasion, the fundamental population dynamic model (Seinhorst, 1986a) is:

$$Pf = M \cdot (1 - e^{-a \cdot \frac{Pi}{M}}) \qquad (3)$$

in which

Pi Initial nematode densities (before planting) as juveniles/g soil;
Pf Final nematode densities (after harvest) as juveniles/g soil;
a the maximum rate of reproduction;
M maximum population density.

In this exponential model it is assumed that the offspring is proportional to the part of the root system that is exploited for food and that $Pf = a \cdot Pi$ if $Pi \rightarrow 0$ and $Pf = M$ if $Pi \rightarrow \infty$. Further, this model is based on the same principles as equation (1). Equation (3) only applies to the soil area containing roots. To predict population dynamics under field conditions, equation (3) was expanded in three ways. First, we accounted for the reduction in plant size caused by the nematodes. Second, with root invading nematodes species, some of the nematodes cannot take part in plant infestation if, for example, the food source is reduced and/or the nematodes are not in the vicinity of the roots. In the soil area without roots, the nematodes slowly decrease independently of *Pi*. Therefore, equation (3) was expanded with an expression for the population dynamics in the unrooted soil. Third, account has been taken of the reduction in numbers of juveniles per parent nematode caused by intraspecific competition.

Resistance: To predict population dynamics of nematodes on resistant cultivars a stable measure for resistance is needed. Therefore, the concept of relative susceptibility (r.s.) was used. Relative susceptibility is defined as the ratio of the maximum multiplication rate *a* of a nematode population on the resistant cultivar and on the susceptible reference cultivar ($a_{resistant}/a_{susceptible}$) or the equivalent ratio of the maximum population density *M* on these cultivars ($M_{resistant}/M_{susceptible}$) (Phillips, 1984, Seinhorst, 1984; Seinhorst and Oostrom, 1984). These ratios present two equal measures of partial resistance or r.s., provided that the tested cultivar and the susceptible reference are grown under the same conditions in the same experiment.

Control measures: *Nematistatics and contact nematicides*: Treatment of plants with nematistatics delays nematode penetration into the roots and results in a certain fraction of the root system escaping nematode attack and, thus, remaining healthy. As a result, the minimum yield *m* is increased by the fraction of the root system untouched by nematodes. Nematode penetration is postponed until the chemical is no longer effective or when the roots grow into soil layers where the nematicide is not present. Experiments on root feeding nematodes confirm that nematistatics increase *m* and sometimes influence Y_{max} but hardly affect *T*. The parameter *T* is only affected when nematodes die because of long lasting effects of nematistatics. Dependent on dosage, nematistatics also influence the parameters *a* and *M* of the population dynamic relation. The effect of contact nematicides, which can be fumigant or non-fumigant, was modelled as an increase in the tolerance limit *T* and a decrease in the maximum multiplication rate *a*.

Distribution patterns: Crop losses and financial returns of nematode control can only be predicted within the limits of estimation of nematode densities. The distribution of nematodes in fields generally is far from random, even in small areas. Therefore, sampling practices and sample size must be adapted to expected small scale and medium scale irregularities in the distribution of nematodes. The small scale distribution is the result of root distribution in the rooted area and of multiplication on the roots. The medium scale distribution refers to the size and the shape of an infestation focus which is the result of redistribution of nematodes by farming practices and the population dynamics of the nematodes. In the case of full field infestations the medium scale distribution refers to an area of about 25 square metres. This information can be used to design sampling methods with predictable, adequate sampling error, and to evaluate current sampling methods. Both the small (1 m^2) scale and medium scale (30-500 m^2) distribution patterns of potato cyst nematodes have been extensively studied. Forty fields with infestation foci in five different cropping areas and 10 fields with more uniform distributions in the North-East part of the Netherlands were mapped to model and parameterize the spatial distribution of PCN (Schomaker and Been, 1999; Been and Schomaker, 2000). The models are used in *NemaDecide* to estimate the variation of *Pi*-values and enable the cost/benefits calculations of the various sampling methods

Parameters: The parameters used in the submodels are entered in *NemaDecide* as stochastic quantities, meaning that their distribution functions, mean and variance are known, unless they are true constants, as seems to be the case with the tolerance limit *T*. This information facilitates risk analysis, which is important for predictions on individual fields and for detection of sources of uncertainty that need further investigation. Field data collected by Seinhorst and Den Ouden, over more than forty years, Molendijk and Been & Schomaker were used to parameterize the growth model and population dynamic model. Data to estimate the tolerance parameters of starch potatoes were kindly made available by Prof. Dr. Lauenstein, Landwirtschaftskammer Weser-Ems, Germany. Data collected by Seinhorst and Den Ouden, Molendijk and data made available by the plant protection industry were used to estimate the effect of nematistatics and contact nematicides on all relevant parameters. Since 1999, all new Dutch potato cultivars have been tested according to a new protocol to estimate relative susceptibility. The protocol is safeguarded by the Plant Protection Service and this information is available for all new cultivars. The current cultivar list was screened for cultivars with possible resistance, based on the presence of resistance genes in their ancestors, and the relative susceptibility of promising cultivars was estimated.

Data required

1. *Sampling results*: The most important information required as input for *NemaDecide* is the average initial population density *Pi* in farmer's fields. This information is provided by about 10 different commercial sampling agencies in the Netherlands. Currently, about thirty different sampling methods for nematodes are in use, varying from rather extensive (200 cc. from 60 cores/ha) to intensive (13 kg from 330 cores/ha). The ten sampling methods for PCN are included in the system, together with their uncertainties. NAK AGRO is part of the *NemaDecide* consortium and a web service has been developed to enable extension officers to download sampling results from this sampling agency to *NemaDecide* when farmer permission is obtained.
2. *Cultivar list*: In the cultivar list, properties of potato cultivars are made available to farmers to enable selection of those cultivars which are most beneficial with respect to resistance, tolerance and other properties. Figure 2 gives an overview of the presented information.
   - Resistance: At the moment more than 270 potato cultivars are present in the potato cultivar database. The r.s. to the pathotypes 2 and 3 of *G. pallida* of 50% of these cultivars was tested and entered into the cultivar list. This information exceeds the information about partial resistance as required by EPPO.
   - Tolerance: Tolerance is the ability of potato cultivars to prevent yield reduction by nematodes. This quality is expressed in parameter values *T* and *m* and in susceptibility to the second mechanism of growth reduction.

- Cultivar properties: Except for information on nematodes several other properties of the potato cultivars, like susceptibility to other diseases or physical properties of potatoes such as yield, skin colour, etc., are entered into the database. A substantial amount of these data can be found in the 'List of Cultivars of field crops' which includes 'The Recommended List of Cultivars' and the National List', annually provided by the Directorate of Agricultural Research. However, the majority of data were provided by the breeders involved in the development of *NemaDecide*. All other breeders kindly provided the required information for *NemaDecide* after the project was started.

3. *Chemical control*: To obtain objective decision rules for chemical control, the effectiveness of nematicides was expressed in terms of model parameters. The contact nematicides DD, Metamsodium, and the nematistatics Temik, Mocap, Vydate and Nemathorin are entered in the system. The nematistatics were entered in full field/full dosage; full field/halve dosage and row application/quarter dosage.
4. *Financial Information*: Data on the cost of sampling methods, chemical control etc., required for calculating the cost and benefits of sampling methods and control measures, were provided by the companies involved. Marketing information of farmers (expected maximum yields, market prices of products) can be added manually.
5. *Legislation*: Government rules concerning the management of quarantine nematodes are incorporated in the system. Pesticides allowed in the various crops and buffers added to infested areas are implemented in the system
6. *Geographical data*: In the late 1990s, the Dutch government developed the so-called 'Basisregistratie Percelen', a GEO-referenced database containing the polygon of every farmer's field in The Netherlands (750.000) and each crop area within these fields. In *NemaDecide* sampling results and the geo-information from the government are, together with the *SampView* component, integrated to enable visualization.

Outline of the software

NemaDecide is split into two main simulation sections: one for infestation foci (Figure 3) and one for full field infestations (Figure 5). Seed and ware potato growers use different sampling methods than starch potato growers. The EPPO method has a different name in the two areas. On the basis of the sampling method requested by the farmers, *NemaDecide* differentiates between the two simulation environments.

Focus model

In seed and ware potatoes PCN sampling is carried out in strips. Ideally these strips are 300 meters long and 6 meters wide when dedicated methods are used or 300 m by 18 m when the EPPO method is applied. The software first analyzes the sampling data and decides whether the infested strips found belong to one or more separate infestations. These infestations will then be presented in a colour code (Figure 3). Next, an infestation can be selected for advice. The programme then tries to recalculate the actual size and shape of the selected infestation focus. If too many cysts are found in the infested strips the programme issues a warning that reconstruction is impossible and that a switch should be made to the full field simulation. If the data indicate the existence of an infestation focus, this size of the focus is estimated and displayed. The farmer now can choose various cropping scenarios, apply nematicides or nematistatics, calculate the effect of volunteer plants and see the effect displayed both in

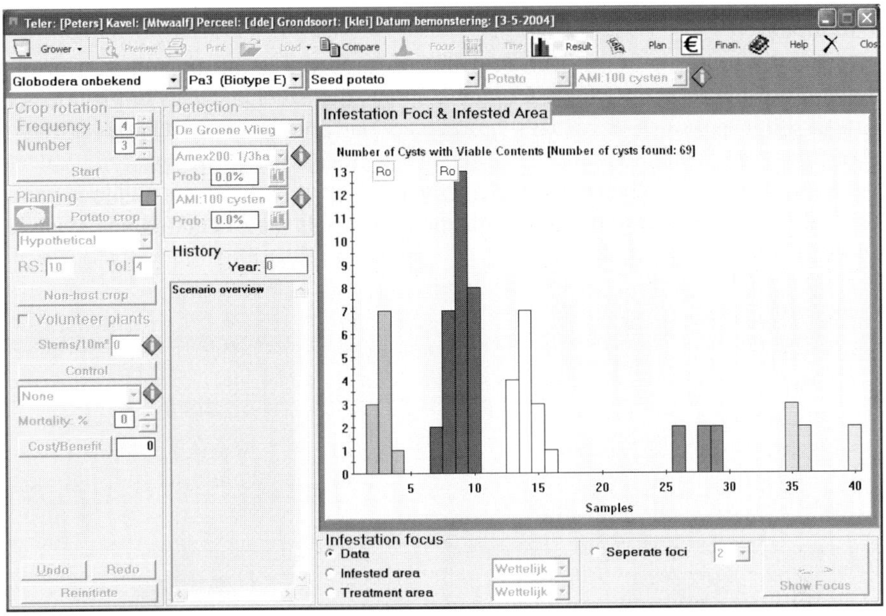

Figure 3. Screenshot of the focus model interface, displaying the sampling results of the sampling agency. Separate infestations are recognized and identified by different colours.

the 3D graph and in the financial information frame. To keep the farmer informed about the consequences of his actions for detection, the programme consistently calculates the detection probability of the focus, using the EPPO sampling method and a chosen dedicated sampling method. Farmers can investigate which crop scenario will lead towards small cost/benefit ratios and a low detection probability by statutory soil sampling methods. The long-term effect of partial resistant cultivars on the population density of populations of *G. pallida* can be predicted within the limits that are determined by the uncertainties in the entered parameters and variables. Also long-term risks and financial losses caused by inadequate sampling methods can be calculated and visualised. Obviously, an initial population density estimated by an intensive sampling method is more precise than a *Pi*-value estimated by an extensive sampling method. In *NemaDecide* the *Pi* will be entered as a stochastic quantity, its uncertainty being determined by the sampling procedure used. This enables cost/benefit calculations on sampling methods and helps farmers to choose an optimal procedure for their purpose. To do this, we estimated the variance of all the sampling methods offered by NAK-AGRO, a sampling agency that contributes to the development of *NemaDecide*. In Figure 4 the difference in focus size is displayed when 2 cysts are detected with the EPPO sampling method and with the AMI100 sampling method. The AMI100 method was developed for seed potatoes in a 1:4 rotation. It is clear that the AMI100 method detects significantly smaller infestation foci than would be detected by the EPPO method. These scenario studies clarify that investment of some extra money in accurate sampling methods has a positive long-term financial effect.

Full field model

Figure 5 displays a screenshot of the full field model interface. The functionality of the interface is kept similar to the focus interface. A graph displays the average values of yield loss (bars) in the years when potatoes have been grown and the average population densities (line) throughout the years. When the models are set to full stochasticity, the result of a calculation is a frequency distribution. Long-term effects of partial resistant cultivars and long-term risks and financial losses caused by inadequate sampling methods are calculated and visualised as in the focus model.

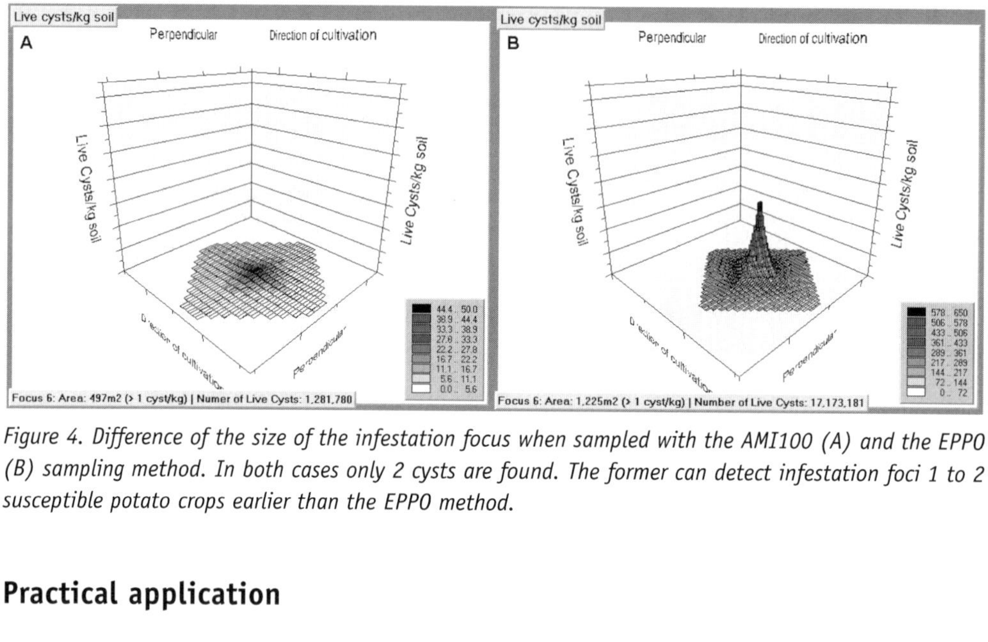

Figure 4. Difference of the size of the infestation focus when sampled with the AMI100 (A) and the EPPO (B) sampling method. In both cases only 2 cysts are found. The former can detect infestation foci 1 to 2 susceptible potato crops earlier than the EPPO method.

Practical application

The following cases illustrate how *NemaDecide* is used to explain and discuss the general principles of *Globodera pallida* control with extension officers and farmers. In these cases pathotype Pa3 of *G. pallida*, is presumed to be present with an initial population density of 5 juveniles/g soil.

Case 1: The use of resistant cultivars in a 1:2 rotation: Farmers often presume that it is impossible to control *G. pallida* without chemical control in short rotations. But when the r.s. of cultivars is 2%, as is the case with cultivar Seresta, the yield reduction during the first year varies between 2.3 and 6.4%, with an average of 4.4%. In the next years, yield reduction is negligible if cultivars are grown with the same degree of resistance. After two year, the population density has declined below the tolerance limit.

Case 2: The use of a cultivar with r.s.=30% (formally susceptible). On the basis of the status of a cultivar within EPPO regulations, farmers are made to believe that a cultivar like Santé (r.s.=30%) is not useful to manage PCN. This is true in 1:2 rotations, where yield reduction after some years stabilizes at an average of 24.4%, with a maximum of 50% (Figure 6A).

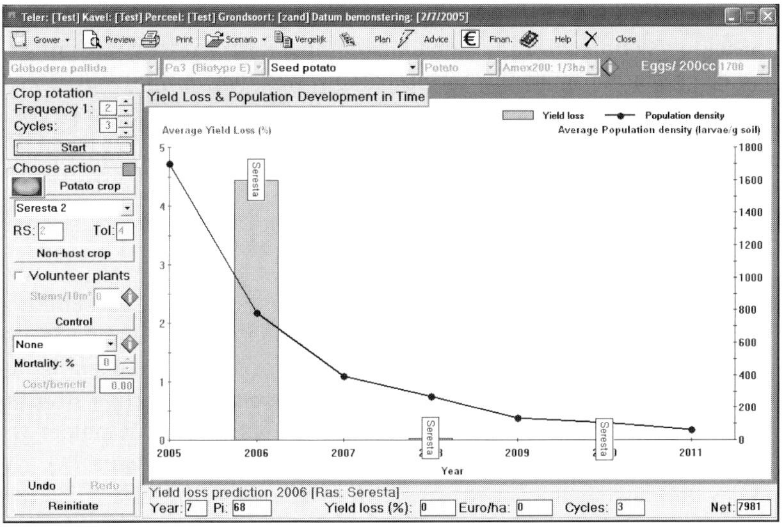

Figure 5. Screenshot of full field infestation. Case 1: The resistant starch potato cultivar Seresta (r.s. = 2% for Pa3) grown in a 1:2 crop.

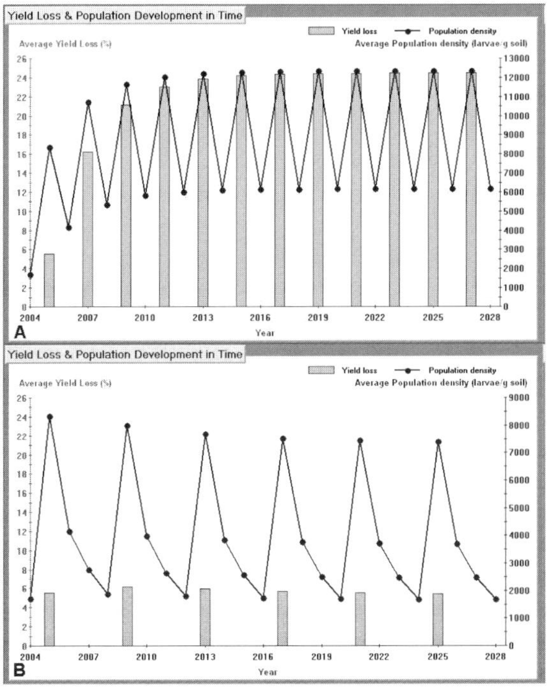

Figure 6 A and B. Case 2: Population density and percentage yield loss when the resistant variety Santé (r.s.= 30% ; susceptible according to EPPO-standards) within a 1:2 rotation (A) and in a 1:4 rotation (B).

However, when Santé is cropped in a 1:4 rotation, average yield loss is limited to 5.4%, while the risk of a yield reduction of >10% is only 5.5% (Figure 6B). In longer rotations with ware potatoes this degree of resistance causes acceptable yield losses. Presently 31 cultivars are available with an r.s. of 30% or smaller for Pa3. *NemaDecide* visualizes the concept of r.s., resulting in an equilibrium population density based on the level of resistance and helps farmers to familiarize themselves with the relation between cropping frequency and partial resistance to choose levels of resistance that are sufficient for their specific needs.

Case 3: Alternating susceptible and highly resistant cultivars in 1:4 rotations. Another common misconception is that infestations can be managed by alternating susceptible and highly resistant cultivars. As shown in the *NemaDecide* simulation, the population increase on the susceptible cultivar (Bintje) causes an average crop loss of 15% (95% quantile: 35%) in the next resistant crop (Innovator, r.s.=1%). Because of the population decline after the resistant cultivar, the susceptible cultivar can be grown almost without damage. Therefore, the resistant cultivar is believed to be less tolerant than the susceptible cultivar (Figure 7A). This example helps the farmer to understand that resistance does not mean that the crops cannot suffer damage and that the use of fully susceptible cultivars should be avoided, even at small densities. Even the row application of a nematistatic under optimal conditions, a sandy soil, can not prevent an average yield loss of 7%. The net benefit of the use of the nematicide is in this case €165,- (Figure 7B).

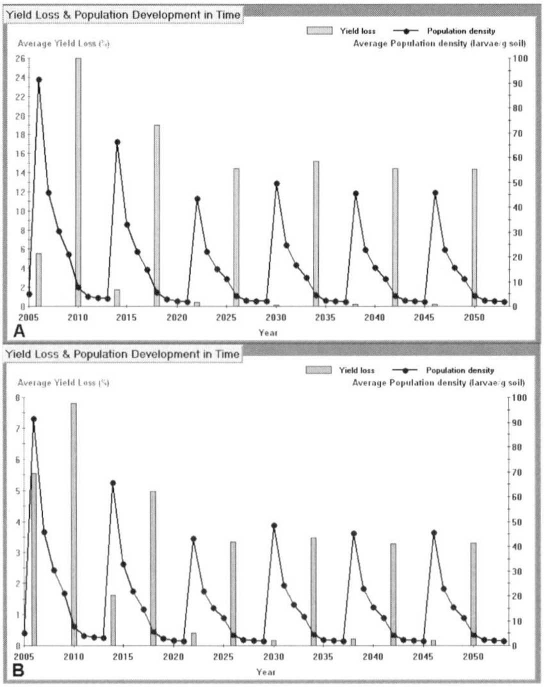

Figure 7 A and B. Case 3: Population density and average percentage yield loss in a 1:4 rotation where Bintje (r.s.=100%) and Innovator (r.s.=1%) are alternated. (A) without and (B) with a row application of a nematostatic applied in the resistant cultivar.

Case 4: The effect of soil fumigation on the detection of G. pallida. In this example Bintje is grown every four years. We are dealing with an infestation focus. In one scenario a soil fumigation is applied in 2008 and in an alternative scenario there is no fumigation. It is shown that even though the soil fumigation results in a mortality rate of 70%, the detection probability is not reduced substantially. This is caused by the fact that after fumigation, there is hardly any reduction in the number of cysts containing live juveniles. Of course, the infestation levels of numbers of juveniles per unit of soil are decreased (Figure 8).

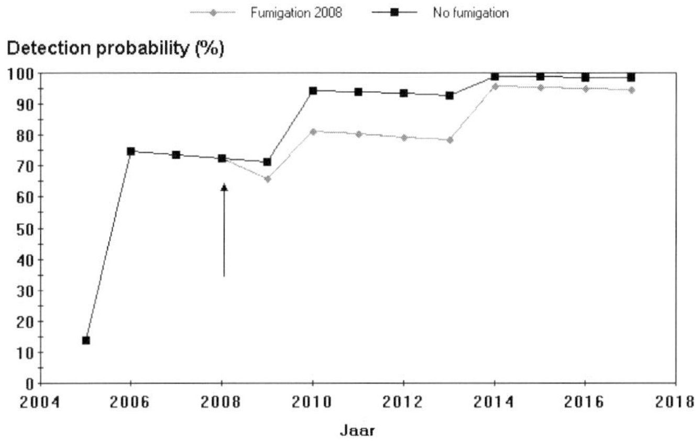

Figure 8. Case 4: Soil fumigation in 2008 (↑) with a mortality rate of 70% hardly decreases the detection probability of a focus.

References

Been, T.H. and C.H. Schomaker (2000). Development and evaluation of sampling methods for fields with infestation foci of potato cyst nematode (*Globodera rostochiensis* and *G. pallida*). Phytopathology, 90, 647-656

Phillips, M.S. (1984). The effect of initial population density on the reproduction of *Globodera pallida* on partially resistant potato clones derived from *Solamum vernei*. Nematologica, 30, 57- 65.

Schomaker, C.H. and T.H. Been (1999). A model for infestation foci of potato cyst nematodes *Globodera rostochiensis* and *G. pallida*-. Phytopathology, 89, 583-590.

Seinhorst, J.W. (1984). Relation between population density of potato cyst nematodes and measured degrees of susceptibility (resistance) of resistant potato cultivars and between this density and cyst content in the new generation. Nematologica, 30, 66-76.

Seinhorst, J.W. and A. Oostrom (1984). Comparison of multiplication rates of three pathotypes of potato cyst nematodes on various susceptible and resistant cultivars. Mededelingen Faculteit voor Landbouwwetenschappen, Rijksuniversiteit Gent, 49/2b, 605-611.

Seinhorst, J.W. (1986a). The development of individuals and populations of cyst nematodes on plants. In: Lamberti, F. and C.E. Taylor (Eds), Cyst nematodes. New York & London, Plenum Press: 101-117.

Seinhorst, J.W. (1986b). The effect of nematode attack on the growth and yield of crop plants. In: Lamberti, F. and C.E. Taylor (Eds), Cyst nematodes. New York & London, Plenum Press: 191-210.

Seinhorst, J.W. (1998). The common relation between population density and plant weight in pot and micro plot experiments with various nematode plant combinations. Fundamental and Applied Nematology, 21, 459-468.

Production and storage

Technology developments in potato yield and quality management

V.T.J.M. Achten
Wageningen University and Research Centre, Agrotechnology and Food Innovations, P.O. Box 17, 6700 AA, Wageningen, The Netherlands, vincent.achten@wur.nl

Abstract

Potato quality demands are pushing today's potato farmers to new technologies and a more management intensive production. New technology and developments in management support systems are presented in this article. Soil characteristics have a large influence on potato quality. Therefore, location specific soil information is important for quality management. Although several soil sampling techniques are available at the moment, there is a need for more cost-effective systems for dense sampling of fields. Variation of potato planting distances is a way to achieve a more uniform yield. Site specific systems for varying planting distances show positive results but need to be optimised and validated. Centimetre precise GPS guidance systems allow potato ridges to be formed exactly above the seed potatoes. This ensures optimal growth and reduces the risk of green colouring. Plant nutrition is a key instrument for managing potato quality. Site specific crop fertilization proved to reduce fertilizer input while maintaining quality and yield. Opportunities for further optimisation exist. Spraying systems for site specific weed control are under development. Sensor based potato haulm killing is also feasible and is evaluated in practice. Yield mapping provides important feedback on site specific actions during crop growth. A prototype system for potatoes, using a line scan camera, produces estimates of the total yield, the amount of dirt tare and the potato size distribution. Information management systems will play a key role in achieving an integral approach in quality management. Standardisation of data interchange is necessary for integration of various management tools and for user friendliness. Potato quality management should be regarded as an iterative learning process in which the farmer has the most influence. Technology just provide the means to achieve optimal potato quality management.

Keywords: quality management, precision agriculture, system integration, information management

Introduction

Agricultural production is subject to fundamental changes on a worldwide scale. A shift has been made from simply maximising gross yield to maximising economic returns whilst reducing the load on the environment and preserving food quality. The quality of the product is of the utmost importance. For the potato processing industry, a high quality batch of potatoes results in lower processing costs and a higher quality of the end-product. The potato farmer is required to grow a high-quality and uniform product to ensure a better market position. Therefore the pursuit of quality is pushing today's potato growers towards a more management intensive crop production. Technology can play a major role in assisting the farmer in this new management role. Since the early 1980s information technology is penetrating agriculture; terms like sensors, on-board computers, GPS, yield mapping and precision farming are becoming more and more adopted among farmers. This paper discusses a number of new and emerging technologies that can improve potato quality production and -management at farm level. A typical one-year potato crop cycle is followed to present the various technologies and, subsequently, the farm management system (Figure 1).

The cyclic arrow in Figure 1 represents the iterative character of farm management with a central role of the farmer, assisted by an integral farm management system. New and emerging technologies are drawn on the outside, with dotted arrows depicting the information flow to and from the management system.

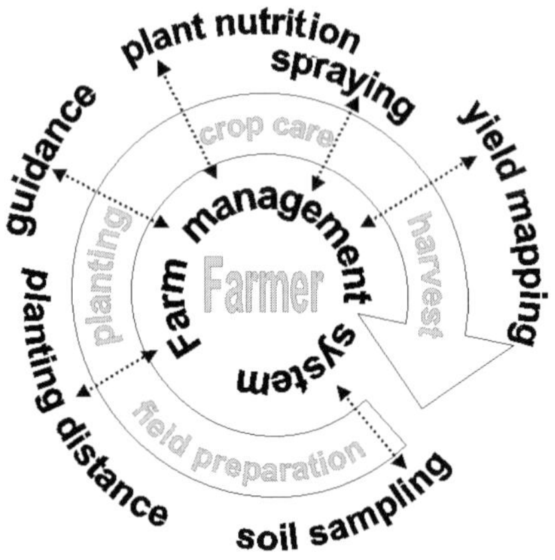

Figure 1. New and emerging technologies in the potato cropping cycle.

Soil sampling

Soil characteristics have a large influence on the quality of the yield. Therefore, knowledge of the variation of soil characteristics in the field offers opportunities to better manage potato quality. Soil properties like clay content, nitrogen content, pH and organic matter content can be input to many management decisions like choice of potato variety, time and quantity of spring fertilisation, irrigation intervals, etc.

The most reliable technique to determine soil characteristics is probably soil sampling in the field and (chemical) analysis in a laboratory. Ideally, a large number of samples has to be taken in one field to account for variations within the field. In practice only one composite sample is analysed representing a fairly large area as soil sampling and analysis is time consuming and expensive.

Alternative methods have been investigated to obtain soil properties in a fast and inexpensive way. Measurements on soil conductivity (Veris, 2005), draft of soil tillage implements (Van

Bergeijk and Goense, 1996), electromagnetic induction (Jaynes and Heikens, 2001) and soil reflection (Shibusawa *et al.*, 1999) all provide a means of determining soil properties indirectly. All these sampling systems are based on the assumption that various agronomically useful soil characteristics such as clay content and nitrogen content are correlated with the measured soil characteristics (force, EC, EMI, reflection).

Some systems use soil sampling and laboratory analysis as a reference. The measured variation in the soil is then used to 'interpolate' agronomically useful characteristics, obtained by sparse soil sampling. In the Netherlands, a new promising system has been introduced that uses natural radioactive irradiation of the soil to determine soil characteristics. The system is used in conjunction with sparse soil sampling. The advantage is that the sensor is sensing above the soil surface. Therefore, there is no soil disturbance and the system can also be used when there is a crop in the field (e.g. grasslands). The accuracy of the system is still to be determined.

Although there are several systems on the market nowadays, there is still a need for the development of cost-efficient and reliable soil mapping technology. This calls for further improvement, assessment of combinations of current technologies and development of new soil mapping systems.

Planting distance

Obtaining a uniform grade of potatoes from a field with spatially non-uniform soil characteristics is difficult. Usually potato planting distance is uniform throughout the field. In the parts of the field with a heavy soil potato tuber numbers tend to decrease while their size of the individual tubers tends to increase (Van Der Zaag, 1992). The idea is to plant potatoes more densely in the areas with a heavy soil and more sparsely in the lighter soil areas. Generally, a smaller planting distance will result in more, but smaller tubers; exactly what is needed in the areas with the heavy soil type. In the areas with a lighter soil type the larger planting distance will result in less, but bigger tubers.

In the Netherlands, a group of farmers is experimenting with variable planting distances. A belt-planter is equipped with a system that is able to vary the planting distance based on a virtual planting map: a map with geo-referenced planting distances. In the field, the planter determines its position in this virtual map and adapts its planting distance on the go. Planting maps are mainly derived from soil sample maps. Grain yield maps of the preceding year are also taken into account an estimate of site specific yield potential. Currently, planting distances are varied in a 20% range around the nominal planting distances: 10% denser on the heaviest parts of the field and 10% sparser on the lightest part of the field. This 20% band is based on an initial guess and should be optimised in the future.

Exact numbers of the impact of planting distance variation on yield quality cannot be presented after only one year of trials. According to the project group (Spinof, 2005), the experiments showed promising results on all fields in the first year. Therefore variation of planting may be an important component in an overall quality management system.

Guidance

In the past, guidance systems based on LASER or dug-in wires provided accurate positioning of farm equipment but were troublesome and inconvenient in use. With the introduction of Real Time Kinematic (RTK) GPS (Trimble, 2005), centimetre precision guidance in agriculture became a workable option. Nowadays, all major tractor manufacturers offer GPS guidance systems on their equipment. The accuracy of these systems is depending on which type of GPS is used but only the RTK systems provide centimetre accuracy.

Most systems automatically steer the tractor. This means that the tractor is kept on a predetermined course. However, this does not mean that the implement is following the same course. Due to slopes or variations in the soil the implement tends to drift. To be sure the implement is following the correct course GPS controlled hydraulic side-shifts have been developed (Achten *et al.*, 2001). These implement guidance devices, mounted in the three-point hitch, are able to move the implement perpendicular to the driving direction. With these implement guidance systems it is possible to steer potato ridgers exactly over the planted rows. Because the seed potatoes can be positioned exactly in the middle of the ridge the risk of potatoes colouring green due to exposure to sunlight is reduced. The overall quality of the produce is improved and the work load of the tractor driver is reduced.

Guidance systems likely will become commonplace in agriculture, like navigation systems in cars nowadays. The investment involved with (especially RTK based) guidance systems is decreasing. Multi-antenna guidance systems could be used to steer both tractor and implement accurately at the same time. Guidance systems are the first step towards fully autonomous (robotic) field operations.

Plant nutrition

An important instrument to manage potato quality is plant nutrition. Nitrogen (N) is an essential component of chlorophyll, cell walls, amino acids and enzymes. Most crops therefore respond strongly to N-fertilisation in both quality and yield, in comparison to other nutrients (Smit, 1994). In the past, an excess of nitrogen was supplied to prevent N deficits. Nowadays, with strict environmental legislation, controlling N-fertilisation is a precision work. Supplying the crop with the right amount of nutrients at the right time is crucial for optimal crop growth and a minimal environmental load. For an optimal N-efficiency, site specific crop fertilisation should be implemented in spatially non-uniform fields.

In 2003 a project was started to introduce site-specific N-fertilisation in the Netherlands. The focus of the project was on integration of components and development of a commercial service for approximately 1200 ha. Site specific fertilisation was done before the growing season and during the growing season (at crop closure).

Before the growing season soil samples (lab-analysis) were taken to get an estimate of the N-supply and mineralization. The sampling results were analysed using a Geographic Information System (GIS). The goal was to supply sufficient nutrients for the growing season and to adjust according to spatial differences in nutrient supplies. Site specific application maps were created and nutrients were applied using a GPS-controlled fertiliser spreader.

During the growing season crop development and nitrogen demand was monitored using leaf stem analysis (Lokhorst *et al.*, 2003). Spectral crop reflection was measured using the YARA N-Sensor, a sensor that can be mounted on top of a tractor cabin (YARA, 2005). This sensor measures an area of approximately 30 m^2 on both sides of the tractor. The crop reflection was geo-referenced using a GPS receiver and stored in a GIS.

The leaf stem analysis determined the moment and amount of N-fertiliser to be applied. The spatial distribution of the fertiliser was determined based on N-Sensor readings. The fertiliser was applied site specifically with a GPS controlled fertiliser spreader.

The soil analysis showed that soil fertility varied a lot within fields. Before the growing season only the pH needed to be corrected in the trial fields. Soil sampling after the site specific spreading of lime showed that the differences in pH levelled out. With the aid of the N-Sensor differences in nitrogen status could be detected during the growing season. There are indications of differences in the relation between crop reflection and N-status between potato varieties. The top dressing of fertiliser based on leaf stem analysis and N-Sensor measurements resulted in a 10 to 40 kg/ha reduction of fertiliser input whereas total yield and yield quality remained the same. At the end of the project a commercial service of over 1200 ha was realised. YARA recently developed new software for their sensor with which a potato crop can be scanned and fertilised in one operation. This system should be tested.

A relation between N-deficiency and crop reflection is assumed implicitly. However, a deviant crop reflection is not always caused by differences in N-availability, it can also be caused by e.g. spatial differences in the soil (like moisture content and soil compaction). Therefore multiple factors such as soil characteristics, precipitation (soil moisture content) and temperature should be taken into account. This integral approach could lead to an even more precise control of plant nutrition the future.

Spraying

Potato quality should not only include the quality of the end product. Quality should also be associated with the side-effects of production. Is a perfect potato, grown under environmentally polluting circumstances, really a quality potato?

Reducing inputs of crop protection chemicals during potato production can therefore add to the quality of the produce. Systems are under development that are able to reduce the amount of herbicide for weed control by using cameras and advanced image processing software (University of Bonn, 2003). Management support systems have been developed to control late blight (Opticrop, 2005). These systems use inputs like crop development and weather predictions to tell the farmer when to spray and what spraying agent to use.

A recent development is sensor based potato haulm killing. This system is based on the concept of Minimum Lethal Herbicide Dosage (MLHD): applying the smallest lethal dose of herbicide.

The idea behind the potato haulm killing application was to determine the dose of haulm killing spraying agent based on spectral reflections of the crop. Initial trials at field level showed that a reduction in agrochemical use of 30% was achievable whilst maintaining efficacy (Kempenaar *et al.*, 2004). A handheld CropScan multi-spectral radiometer (CropScan, 2005) was used in these experiments.

Currently a system is being developed that is able of controlling the amount of spray liquid on the go. To obtain spectral crop reflection the YARA N-Sensor (normally used for fertilisation) is used. In 2004, field experiments showed that the N-Sensor can also be used for the potato haulm killing application. Spectral crop information, obtained by the N-sensor, is directed to a Personal Digital Assistant (PDA) that houses algorithms for determining the herbicide dose (MLHD). The PDA sends the calculated dose to a sprayer with a chemical injection system. The PDA also acts as a user interface and datalogger. A GPS can be connected to the system for site specific data logging.

Because spray liquid is applied site specifically, an even larger reduction agrochemical use is expected compared to the 30% that was achieved earlier. The focus in the first year will be on system integration and measurement of deposition and efficacy. In the coming years the system will be optimised.

Yield mapping

Potato quality management is an iterative process like most control systems. Feedback is needed to see what the results of specific actions are and to optimise them. Therefore, measurement of the output is required. Yield mapping is a well-known method to determine the output of the potato production process site specifically.

In grain production, yield mapping systems are available for some time now. Yield mapping systems for potato production are seen rarely. Most potato yield mapping systems are based on weighing the product flow in the potato harvester by means of weighing cells that are mounted below a conveyer belt. A computer uses the measured weight, the speed of the conveyer belt, the forward speed of the harvester and GPS information to create a yield map. These yield maps can give a estimate of the site specific yield (kg/ha) but do not give any information about potato quality and measurement results can be influenced by the presence of soil, clods or crop residues. Under wet harvesting conditions yield measurement may be inaccurate due to the large amounts of soil present in the produce.

In the Netherlands, a prototype yield mapping system has been developed based on machine vision (Hofstee and Molema, 2002). A line-scan camera is placed above a conveyer belt in the potato harvester to generate two-dimensional (2D) images of the product flow. Advanced image processing techniques are able to singularise potatoes and determine the 2D shape of each individual potato on-line. Soil sticking to individual potatoes can also be 'removed' and an estimate of the amount of dirt tare can be calculated. Potato volume (3D information) is calculated from the 2D potato shapes by using variety specific models (Hofstee and Molema, 2003). These models were created by hand-measurement of potato size, -shape and -volume. Regression analysis was used to determine the relation between potato dimensions and potato volume. An average volume prediction error of only 0.27% was achieved in the varieties Bintje and Agria.

Tests with the camera system and a moving belt were carried out under laboratory and field conditions. Volume estimation errors ranged from only 1.5 to 2.6% at batch level. These results show that a vision based yield mapping system has potential and is ready for implementation in potato harvesters. The resulting data on yield quantity and yield quality (such as potato shape and size distribution) provides feedback to evaluate (site specific) actions during production and is a useful management instrument, therefore.

Information about dirt tare could be used to optimise harvester settings automatically. For example, if a too high dirt tare amount is measured the sieve intensity could be increased to decrease the amount of tare.

Quality information is also useful for the potato processing industry. The processing of potatoes can be performed more efficiently when quality information of batches of potatoes is available. This may lead to higher financial and environmental profits for the processing industry and could lead to a new system of pricing for the farmers in which quality production is rewarded.

Management system

The farm management system is becoming the key component in potato quality management. The management system's core component is a database in which all relevant information is stored. For site-specific farming purposes, an important approach to improve potato quality, the farm management system should also include a GIS to store geo-referenced information. The grid size of the geo-referenced information does not necessarily have to be at sub-field level; it can also concern a whole field, such as information about agents used when spraying a field uniformly.

This central database provides information for various management support systems and also stores information that is generated by these systems. This means that information is flowing into two directions and that information is reusable by other management support systems. To ensure this reusability and to maintain data integrity, database information should be defined in a indisputable way. It should not be possible to have two different definitions of e.g. yield in the same database.

Several initiatives for standardisation such as Computer Integrated Agriculture (Esprit, 1996) and Pre-Agro (Pre Agro, 2003) were taken but none led to an internationally accepted overall standard. A first (small) step has been made in the standardisation of data interchange between farm machinery and management systems (ISO, 2005).

To bring management information systems to the next level, international cooperation is needed to create a broadly accepted framework for data definition. Standardisation will promote reusability of information and integration of components and, most importantly, user friendliness.

Farm management systems are getting larger, their functionality increases and operating them is getting a complex task. Software houses face a difficult task in creating management systems that can be used by non-specialists and that still offer sufficient functionality.

Another approach is to create large, all-round systems intended for use by so-called 'service providers'. Service providers are companies that perform (site specific) management for multiple farmers. The service provider is an intermediate between farmer and management system and acts as a consultant, the farmer takes the decisions. The big advantage is that complex management systems can be used in practice. Implementation of new models, systems or technologies will be easier and a central storage of information offers opportunities for benchmarking: farmers are able to learn from each other.

Discussion and conclusions

Technology can assist the farmer in the production of high quality crop. New technologies are emerging in almost every aspect of potato production. An important shift has been made in the size of the management units. In the past, operations were carried out at field level. Modern technology is enabling management on much smaller areas, usually determined by the width of an implement. In the future, control within the width of the implement (sections) is possible and may shift to management units the size of individual plants.

The effect of each individual tool on potato quality is difficult to assess and may even be working adversely on the effect of another tool. Therefore an integral approach is needed. Integration of different management tools prevents systems 'contradicting' each other and enables the evaluation of the side effects of interventions in the potato production process. An important aspect in system integration is standardisation. Standardisation will ensure interchangeable information and easy integration of components. Above all, standardisation is necessary to ensure user friendliness and acceptation.

To perform (site specific) quality management sufficient and reliable information is needed. It is hard to make decisions about this year's fertilisation rate based on last year's yield map only. Better decisions can be made on information form multiple sources: yield maps, soil sampling, crop monitoring etc. Potato quality management is an iterative learning process and information should be gathered and formalised in the central database of the management system. Spectacular results should not be expected in the starting year(s). It should be seen as an investment; those who persevere will reap the benefits.

References

Achten, V.T.J.M., A.T. de Groot and D. Goense (2001). Implement guidance based on RTK-DGPS. VDI tagung 2001, Hannover, Germany.

Bergeijk, J. Van and D. Goense (1996). Soil tillage resistance as a tool to map soil type differences. In: Proceedings of the 3rd International Conference on Precision Agriculture, Minneapolis, MN, 605-616.

CropScan (2005). CropScan home page. www.cropscan.com.

Esprit (1996). Computer-Integrated Agriculture. http://www.cordis.lu/esprit/src/results/res_area/iim/iim5.htm.

Hofstee, J.W. and G.J. Molema (2002). Machine vision based yield mapping of potatoes. ASAE Annual International Meeting, Chicago, USA, 10 pp.
Hofstee, J.W. and G.J. Molema (2003). Volume estimation of potatoes partly covered with dirt tare. ASAE Annual International Meeting, Las Vegas, USA, 12 pp.
ISO (2005). Tractors, machinery for agriculture and forestry - Serial control and communications data network, Part 1, General Standard. ISO/ WD 11783-1.
Jaynes, D. and K. Heikens (2001). Electromagnetic induction for precision farming. www.nstl.gov/research/onepage/1pageemi.html.
Kempenaar, C., R.M. Groeneveld and D. Uenk (2004). An innovative dosing system for potato haulm killing herbicides. XIIeme colloque international sur la biologie des mauvaises herbes, Dijon, France, 11 pp.
Lokhorst, C., P. Dekker, K. Grashoff, T. Guiking and S. van 't Riet (2003). Perspectieven geleide bemesting in de open teelten (Perspectives of controlled fertilisation in arable farming). Nota 2003-51, IMAG, Wageningen, The Nederlands, 43 pp.
Opticrop (2005). Prophy. www.opticrop.nl.
Pre Agro (2003). Verbundprojekt Pre Agro (National Project Pre Agro). http://www.preagro.de.
Shibusawa, S., M. Z. Li, K. Sakai, A. Sasao and H. Sato (1999). Spectrophotometer for real-time underground soil sensing. ASAE Paper No. 99-3030.
Smit, A.L. (1994). Stikstofbenutting (Nitrogen use). In: Haverkort, A.J., K.B. Zwart P.C. Struik en P.H.M. Dekker (eds.), Themadag Stikstofstromen in de vollegrondsgroenteteelt. PAGV themaboekje 18, PAGV, Lelystad, 111 pp.
Spinof (2005). Stichting Precisie Landbouw in Noord Friesland (Precision Farming Foundation in North Friesland). www.gpslandbouw.nl.
University of Bonn (2003), Herbicide Sprayer sees where it needs to spray (press release). www.uni-bonn.de/en/News/25_2003_druck.html.
Trimble (2005). High-Accuracy RTK/DGPS. www.trimble.com
Veris technologies (2005). Sensors and Controls for Soil-Specific Agriculture. www.veristech.com.
YARA (2005). N-Sensor. www.sensoroffice.com/hp_home2.
Zaag, D.E. van der (1992). Potatoes and their cultivation in the Netherlands. NIVAA, The Hague, 47 pp.

Curriculum vitae - Vincent T.J.M. Achten

Vincent T.J.M. Achten was born on November 13th 1975 in a small town in the south-east of the Netherlands. At an early age he started working in agriculture as a seasonal worker. Later on he was employed by several farm contractors doing machine maintenance and field work. His interest in agricultural engineering grew and he decided to study agricultural engineering at Wageningen University. He specialised in robotics, system measurement and -control. His internship was done at the Institute of Environmental and Agricultural Engineering (IMAG) in Wageningen, The Netherlands. During this internship he was involved with the development of a robotic cucumber harvester. After this internship he received his MSc degree (2000) and he was employed by IMAG as a researcher in the field system- and software development. In 2004 IMAG became part of Agrotechnology and Food Innovations as a result of a merger. IMAG became a separate business unit and was renamed Agrisystems and Environment. In his function as a researcher he became involved in national and international standardisation in the field agricultural electronics and data interchange between tractors and implements (ISO11783). Currently, he is involved in several national and international projects concerning measurement systems, automation, autonomous vehicles and precision agriculture.

Company profile – Agrotechnology and Food Innovations (A&F)

Agrotechnology and Food Innovations (A&F) is a research institute. The research of A&F covers the whole production chain; 'from the field to the mouth ' is our slogan. Research is performed in the field of greenhouses, stables, agricultural production, on-field technology, design and production, transport and retail. We work on consumer-driven product design, new logistical concepts, environmentally friendly production, innovative production technologies, efficient processing and on better or newer products for both food and non-food applications. Research is based on scientific insights and technological know-how. Together with our customers (both commercial and governmental) we realise concrete, innovative solutions in all parts of the agricultural production chain. A&F is organised in four Business Units (Agrisystems & Environment, Quality in Chains, Biobased products and Food Quality) and has an annual turnover of approximately EUR 35 millions. A&F is part of Wageningen University and Research Centre and employs approximately 400 people. More information can be found on our company website: www.agrotechnologyandfood.nl.

Comparing the effects of chemical haulm desiccation and natural haulm senescence in potato by the use of two different skin set methods

Eldrid Lein Molteberg
Norwegian Crop Research Institute, Apelsvoll Research Centre, N-2849 Kapp, Norway

Abstract

Norway has a fairly short growing season, and it is important to optimize growing techniques to avoid immature tubers and undesirable skinning. The traditional killing of haulm with diquat shortens the growing season by around two weeks, and it has been questioned whether this really gives the best effect on skin set. In this study, the skin set after natural senescence and chemical haulm desiccation is studied. Two different skin set methods are compared. The results indicate that a somewhat better skin set is generally obtained when the haulm is desiccated two weeks prior to harvest than with natural senescence. However, the greatest differences in skin set were found between sites and between varieties. Between sites, skin set rate was affected by the length of the growing season and by haulm greenness close to harvest, and it was also related to dry matter content of the potato. Both the Torquometer method and a well defined washing method could be used to rank skin set, but the Torquometer gave more reproducible results and was less dependent on tuber shape.

Keywords: skin set, haulm desiccation, maturity

Introduction

Killing of the haulm using the desiccant diquat (Reglone) a fortnight before harvesting is the normal procedure in Norwegian potato cropping. This protects the tubers against late blight (*Phytophthora infestans*) and hastens maturity. However, the practical effect of desiccation on skin set in Norwegian potato cropping has been questioned, as this shortens the already limited growing season. The aim of this study was to compare the skin set after natural senescence and after chemical haulm desiccation, measured by two different methods.

Material and methods

The experiment was performed in 2004, with the variety 'Folva' grown at three different sites about 200 km apart (South-East Norway). The sites are considered to differ in both soil quality and climate. 'Folva' was grown in three replicates at each site, using a split-plot design. The potatoes were harvested two weeks after haulm desiccation, or on naturally wilted haulm. Seven days prior to harvest, the three plots varied in greenness from 10 to 88%, as compared to full growth values.

The skin set was measured 7 days after harvest, both by a tough washing method and by the use of a Torquometer. With the washing method, 20 medium size tubers (each ca 100 g) were placed in a rotating container with a continuous water flow and treated at a high and defined speed for one minute. The value indicates the remaining percentage of skin. The Torquometer measures skin set of individual tubers (Lulai, 2002). A rubber end is held against a flat area of the tuber at constant pressure. The method measures the rotational force applied (in mNm) before the skin slips. The given values are means of 30 repetitions (10 tubers) per sample.

Results and discussion

There were significant differences in yield, tuber size and dry matter content between both sites and haulm treatments. Skin set differed far more between sites than between haulm treatments, as it also did between varieties (not shown). For haulm treatment, the differences were small, and only significant for the Torquometer method. In a related experiment with 5 different varieties, the effect of haulm killing was found to be larger (0.44 mNm). Skin set rate could not be related to any single factor, but between sites it seemed to be somewhat related to the length of the growing season, to haulm greenness close to harvest and to the dry matter content of the potato.

Table 1. Significant differences among sites and of treatment prior to harvest (P<0.01).

	Days between planting & desiccation	N-tester, 7 d prior to harvest, %*	Yield> 40mm, t/ha	Dry matter content, %	Mean tuber weight, g	Torquo-meter, mNm	Remaining skin after "washing", %
Site							
Rygge	86	94	45.6	18.2	110	2.9	15
Solør	88	62	42.9	19.8	96	3.7	76
Apelsvoll	97	23	40.7	21.6	76	3.9	87
Treatment							
"Green" haulm	91	60	45.8	20.1	98	3.4	not
Desiccated h.	106	0	40.2	19.5	89	3.6	sign.

*in % of maximum value (primo July)

Conclusion

Desiccation of the haulm two weeks prior to harvest resulted in a somewhat better skin set than natural senescence of the haulm. However, skin set differed much more between sites and varieties. The length of the growing season and greenness of the haulm close to harvest were both important factors for the skin set rate. The Torquometer method was found to be a fairly objective and reproducible method for measurement of skin set on single tubers. Within the same variety (same shape of the tubers) the washing method generally ranked the samples in the same order, but was less sensitive than the Torquometer.

Acknowledgements

This project was funded by the Fund for Research Tax on Agricultural Products, and the work was performed in cooperation with the Plant Protection Centre of the Norwegian Crop Research Institute, Norwegian Food Research Institute, and the Agricultural Extension Services of SørØst/Rygge and Solør-Odal.

References

Lulai, E. C. (2002). The roles of phellem (skin) tensile-related fractures and phellogen shear-related fractures in susceptibility to tuber-skinning injury and skin-set development. American Journal of Potato Research, 79, 241-248.

Curriculum vitae – Eldrid Lein Molteberg

Eldrid Lein Molteberg finished her MSc studies at the Agricultural University of Norway, Department of Food Science, in 1989. She worked in the cereal group at the Norwegian Food Research Institute (Matforsk) from 1989 to 1997. A PhD degree on lipid quality of oat products for human consumption was completed in 1996. From 1997 she has been working at the Norwegian Crop Research Institute, Apelsvoll Research Centre. She has been coordinator of the potato research in the Norwegian Crop Research Institute since 1998. The main focus of her work has been on quality, and often related to problems with immature tubers. This has included work on fertilization, physiological quality of seed potatoes, haulm desiccation and quality parameters related to frying quality, skin set and storage ability.

Company profile – Norwegian Crop Research Institute

The Norwegian Crop Research Institute is the leading institute for applied research on crop production in Norway. It has research units all over the country, of which Apelsvoll Research Centre at Kapp serves as administration centre for the regional units in Eastern Norway.
The Norwegian Crop Research Institute participates in most of the plant production chain, from variety development, plant protection, agronomy, soil science to harvesting and storage of products. It covers horticulture as well as agriculture, and also includes work on grassland, green areas, managed landscapes and ecological agriculture, as well as on traditional and new agricultural products within cereals, vegetables, herbs, fruits, berries and potatoes.
The Research centre of Apelsvoll is situated around 120 km north of Oslo, and has a climate typical of the Norwegian inland. In addition to potatoes, the centre is responsible for research within vegetables, cereals and oilseed rape, and also for the official variety testing in Norway. The Centre also plays a major role in research concerning organic farming and precision farming. Subjects concerning plant protection are covered in cooperation with colleagues at the Plant Protection Centre of the Norwegian Crop Research Institute. More information may be found on www.planteforsk.no.

Volunteer potatoes

Melvyn F. Askew
Visiting Professor Harper Adams University College and Head of Agricultural and Rural Strategy Group, Central Science Laboratory, York, YO41 1LZ, UK

Abstract

The volunteer potato is fundamentally a weed potato developed from viable tuber or tuber portions remaining in the field post harvest, or from true potato seed (TPS). There are many misconceptions about its propensity to survive and multiply, which have led to it becoming a pernicious weed in some crops. Additionally, volunteer potatoes act as host to potato pests and pathogens, thereby causing added risk and cost to producers and necessitating additional pesticide use.

Multiplication and vigour of volunteer potato plants can be managed through rotation and cropping, allied to herbicide use. Without well planned and focused control measures, volunteer potato populations are unlikely to decline in a rotation, except where very severe winters occur.

Keywords: volunteers, weed, ground-keepers

Introduction

The European or White Potato (*Solanum tuberosum*) is the fourth largest tonnage food crop in the world. Production varies somewhat, but approximates to 330 million tonnes per annum. Production area approximates to 20 million hectares. This compares with wheat, the world's largest area crop at approximately 215 million hectares per annum, and 600 million tonnes per annum production (FAO, 2005).

Market specifications for potato vary considerably, with some markets (e.g. French fries) requiring production of a specific potato variety, to a company blue-print, whilst generally lower value 25 kg bags of ware potatoes may have no closer specification than size and a requirement to name the variety itself.

In mass consumption markets (e.g. Europe) potatoes for ware or processing sectors are usually produced from certified/classified seed tubers or sometimes from home-saved "seed" tubers, once-grown from this latter material. It is possible to produce potatoes from botanically true potato seed (TPS), although this has some disadvantages in mass production markets. TPS is more commonly used in Africa and Asia. Clearly, TPS is only an option where cultivars set viable seed. In UK, for example, several popular cultivars set TPS.

It is not surprising therefore that volunteer potatoes arise widely in intensive potato rotations in particular. Currently, control of volunteer potatoes in a particular potato crop is not generally practicable, although should genetic modification technologies become acceptable and developed, then they may offer some opportunities to do so. Possible opportunities in this instance could be specific herbicide resistance or even 'terminator genes'. If G. M. technologies were used to enhance frost tolerance, then winter frosts would have no impact and the geographical area infested with volunteer potatoes would expand. Degree of control in other crops is variable, but is particularly challenging in horticultural species.

Incidence of volunteer potatoes from TPS

TPS can retain a high degree of viability for at least 10 years and some seed will germinate after 20 years storage (Barket and Johnson, 1980). This capability is well proven in gene banks where TPS is a means of storing material. (Jackson - Pers. Comm.); field experience suggests that TPS can remain viable for at least 7 years (Lawson, 1983). Since TPS can survive in the field for long, it is important that long term cultivar popularity for cultivation and not solely current cultivar interest be assessed in order to assess TPS/volunteer potato risks.

Some cultivars such as the outclassed Dutch cultivar Bintje tend to produce flower buds that abscise before flowering, whereas other cultivars flower and set berries profusely, thus producing enormous amounts of viable TPS. Details of studies on some cultivars in this latter group was undertaken between 1977 and 1981 at the Scottish Crop Research Institute, Invergowrie, near Dundee (Lawson, 1983) and is shown in Table 1. Maximum seed potential reported was 249 million per hectare in the cultivar Pentland Ivory, approximately 6000 times a normal seed tuber population. Production in Désirée was high too. Désirée and similar cultivars have been of major interest in the Iberian Peninsula and parts of North America; hence the TPS potential must be similar in those areas.

Table 1. Berry and seed production in some potato cultivars.

Year	No of plants recorded/ cultivar	Cultivar	Fresh weight berries (kg ha^{-1})	No of berries (10^3 ha^{-1})	No of berries $plant^{-1}$	Mean weight $berry^{-1}$ (g)	No of seeds $berry^{-1}$	No of 'Seeds/ha x 10^6' (estim.)
1979	75	Pentland Ivory	5343	820	17.9	6.5	304	249
		Pentland Dell	2248	311	6.7	7.2	261	81
		Maris Piper	1340	240	5.2	5.6	186	45
		Record	816	116	2.5	7.1	NR	29
1981	44	Bonte Désirée	1736	221	6.3	7.8	301	67
		Pentland Ivory	1384	172	4.9	8.0	401	69
		Pentland Dell	664	129	3.7	5.2	326	42
		Cara	456	71	2.1	6.4	391	28
		Maris Piper	156	43	1.2	3.6	NR	11

*NR = Not Recorded.

Volunteer potatoes from potato tubers or tuber portions

The conventional harvesting procedure for potatoes leaves small and, in some instances, discarded large tubers in the field. Lutman (1977) reported that up to 370,000 tubers per hectare could remain and Lumkes (1974) 460,000 tubers per hectare. These data are indicative of a general high level of field leavings but these populations could vary.

The vast majority of potatoes are harvested by machine and machinery design and mode of use undoubtedly have effects upon levels of tubers left upon the field. In practice, no harvester removes every tuber. Hence, there is an automatic "inoculum" of propagative material into the volunteer potato system following each potato crop.

Factors affecting volunteer potatoes

Frost

There is commonly held but erroneous belief that a frost will kill volunteer potatoes. However, it has been shown that 50 accumulated Celsius degree frost hours below -2 °C are required to kill tubers (Lumkes and Sijtsma, 1972). Logic would suggest therefore, that should these conditions prevail, then surface borne tubers would be killed. Regrettably, in real-time agricultural practice, tubers are spread throughout the soil profile.

Lumkes and Beukema (1973) reported only 67% of tubers left after harvest to be more than 30 cm deep, post ploughing, which is the usual primary operation post potato harvest. Frosts rarely penetrate to this depth at all in most cool-temperature (or warmer) agriculture (Barric - Pers. Comm).

Cultivations

Mechanical damage is generally accepted as being deleterious to potato tubers. However, for obvious reasons, mechanical damage must be minimised as the potato crop is harvested. Undoubtedly, cuts and bruises allow ingress of disease, although no critical evidence on disease or impact upon volunteer potatoes has been found.

Any process, which surely crushes tubers, is likely to destroy any significant growth potential. Hence, tuber crushing systems may be beneficial in reducing populations. However, these are slow, require high levels of energy, tend to destroy soil structure and permeability to air and water, and cannot be used when stones occur significantly. Additionally, crushed stone causes damage to machinery and excessive wear on tyres.

Crop competition

The volunteer potato crop like any other competes for light, water and nutrients. Hence growth of potato volunteers is affected by crops in which the volunteer occurs. Perombelon (1975) found a multiplication rate in volunteer potato of 1.7 to 2.1 in winter wheat and Lumkes (1974) 2.6 in winter wheat and 3.9 in sugar beet. Clearly, crop competition reduces multiplication rate of volunteer potatoes.

Moreover, rigorous, well-established crops (relative to volunteer potato emergence and subsequent canopy development) appear to have a greater impact on volunteer potato plants than spring established row crops. This seems to be confirmed by casual observation in vegetable and pea crops also.

However, whilst crop competition does contribute to reducing rate of volunteer multiplication, it cannot succeed alone in eradicating the problem.

Adverse effects of volunteer potatoes

In this section it is intended to highlight adverse effects only, rather than discuss them in depth.

Volunteer potatoes have the capability to create several detrimental effects:
i) Disease carry over to other potato crops;
ii) Pest carry over in foliage or tubers;
iii) Yield and quality problems in potato and other crops, especially vegetables;
iv) Additional cost.

i) Fungal diseases
There are many fungal diseases of potatoes. Those diseases which are most likely to be affected by the presence of volunteer potatoes are:
- Late Blight (*Phytophthora infestans*);
- Powdery Scab (*Spongospora subterranea*);
- Silver Scurf (*Helminthosporium solani*);
- Verticillium Wilt (*Verticillium albo-atrum* and *Verticillium dahliae*);
- Skin Spot (*Polyscytalum pustulans*);
- Black Scurf (*Rhizoctonia solani*);
- Sclerotinia (*Sclerotinia sclerotiorum*);
- Wart Disease (*Synchytrium endobioticum*);
- Violet Root Rot (*Helicobasidium purpureum*);
- Gangrene (*Phoma exigua*);
- Black Dot (*Colletotricum coccodes*);
- Dry Rot (*Fusarium coeruleum, Fusarium* spp.);
- Pink Rot (*Phytophthora erythroseptica*).

The author considers that of these, Late Blight and Wart Disease would be key problems.

Virus diseases
The two key viruses of potato crops are leaf-roll virus (PLRV) and severe mosaic virus (PVY). Both are very important in seed growing areas and isolation of seed crops from virus sources and aphid vectors is a major part of virus control strategies.

Volunteer potatoes in following crops are a likely source of virus, particularly if they are from ware crops where no action was taken to prevent aphid ingress and subsequent virus spread. Volunteer potatoes in following crops are potential sources of virus for seed crops which may be grown nearby.

Bacterial diseases

The main bacterial diseases of potatoes are listed below. Not all are frequently occurring or indigenous to all countries but if such diseases were to become established their survival may be aided by volunteers in subsequent crops:

- Common Scab (*Streptomyces scabies*)
- Blackleg/Bacterial Soft Rots (*Erwinia* spp.)
- Brown Rot (*Pseudomonas solanacearum)*
- Ring Rot (*Corynebacterium sepedonicum*)

Brown Rot and Ring Rot are the diseases where volunteer potatoes would impact most.

ii) Pests

Potatoes are attacked by a wide range of vertebrate and invertebrate pest species. In terms of impact of or/on volunteers outcomes may be different. Potatoes are attacked by a range of slug species, but especially *Derocerus pulmonata.* Should these slugs damage volunteer potatoes, the impact would then be positive from the agricultural perspective. Whilst data to support the following comment are lacking it seems likely that volunteer potatoes will have no significant impact overall on slug populations.

Of those species where a negative agricultural impact may occur, Potato Cyst Nematode (*Globodera* spp.), and perhaps, Colorado Beetle (*Leptinotarsa decemlineata*) are likely to be most important.

Populations of the two common species of PCN - *Globodera rostochiensis* and *Globodera pallida* normally decline over time in the absence of a suitable host. Rates of decline are different for the two species, *Globodera pallida* being slower than *Globodera rostochiensis* (Whitehead, 1992). The presence of volunteer potatoes would reduce the rate of decline overall and where mixed species populations of *Globodera* occurred, might, according to response of the potato genotype, marginally alter the balance of *Globodera* spp. It is to be noted that the initial PCN population would have a major impact upon magnitude of effect of volunteer potatoes.

iii/iv) Yield, quality and cost impacts

Any plant which grows close to another, has the potential to compete for water and light and thereby reduce yield. In some commercial production value is often related to a particular size grade, such, that strong competition would cause reduced size of produce, or, other issues remaining unchanged, a longer period to harvest.

Where cultivars of volunteer potatoes, which set fruits (so-called "berries") occur in some food crops there is an additional cost of field roguing, or, if "berries" occur in the harvested produce, total crop rejection. Alkaloids occurring in "berries" are toxic. Knott (1993) reported 20% of vining pea crops to be infested with volunteer potatoes. A little less than 1% of the total crop was rejected for this reason.

Anecdote suggests that cost of removing potato berries from broad beans is £15-20 per tonne sorted. Similar impacts occur on several other crops. Additionally, volunteer potatoes can impact upon formal crop certification procedures in many species.

Certification requirements usually include rotational, disease, crop species and crop genotype standards. Where volunteer potatoes occur in other species, their effect from a certification aspect alone is likely to be small albeit that the pressure of volunteer potatoes could increase production costs. However, where volunteer potatoes occur in potato crops grown for classification/certification for use as seed tubers, two much more important issues arise. Firstly, volunteer potatoes will be the same cultivar as that which is being grown; if they are a different cultivar they may be relatively easy to identify and remove, albeit at a cost, but count as 'off types'. Additionally such volunteers may be infected with diseases, for example, viruses, and could therefore reduce the standard of the whole crop.

Discussion and conclusions

The volunteer potato is a persisting problem and its existence is not heavily threatened at present. Clearly, data exists to show some impacts that will reduce its vigour (frost, vigorous crop competition, pests or diseases) though other farm operations favour its increase (more intensive potato production; ploughing post-harvest; production of less/non-competitive crops). Logic would suggest that as climate warms, or if cold tolerant potatoes are fully commercialised then volunteer potatoes and their adverse effects will increase at least in current cool or warm temperature areas.

Ever increasing quality specifications upon food crops, including potatoes will increase need for and cost of control of volunteer potatoes. However and simultaneously, there is an increased emphasis on use of pesticides and considerable monitoring for residues. Additionally, many herbicides are being removed as registration procedures become even more strict.

The development of modelled population dynamics for volunteer potatoes and the possibility to use these in real-time scenarios, would be helpful in the development of integrated control systems for volunteer potatoes. Such broad spectrum approaches would appear to be the only practicable way to manage this weed.

Acknowledgements

The author wishes to acknowledge the assistance of Mr Jonathan Stein and Miss Zara Court in sifting some data for this paper and Miss Court for its typing.

References

Barker, W.G. and G.R. Johnston (1980). The longevity of seeds of the common potato, *Solanum tuberosum* . *American Potato Journal,* 57, 601-607.

FAO (2005). Website of Food and Agriculture Organisation of United Nations. http://www.faostat.fao.org/ [Accessed Monday 27th June 2005].

Knott, C.M. (1993), Volunteers in legumes for processing. *Aspects of Applied Biology,* 35, Volunteer Crops as Weeds.

Lawson, H.M. (1983). True potato seeds as arable weeds. *Potato Research,* 26, 237-246.

Lumkes, L.M. and H.P. Beukema (1972) Informed Paper.

Lumkes, L.M. and R. Sijtsma (1972) Mogelijkheden aardappelen als onkruid in volggewassen te voorkomen en/of te bestrijden. *Landbouw en Plantenziekten,* 1, 17-36.

Lumkes, L.M. (1974). Research on control of volunteer potatoes in The Netherlands. Proc. 12th British Weed Control Conference, pp. 1031-1040.
Lutman, P.J.W. (1977). Investigations into some aspects of the biology of potatoes as weeds. *Weed Research,* 17, 123-132.
Perombelon, M.C.M. (1975). Observations on the survival of potato groundkeepers in Scotland. *Potato Research,* 18, 205-215.
Whitehead, A.G. (1992). Emergence of juvenile potato cyst nematodes *Globodera rostochiensis* and *Globodera pallida* and the control of *Globodera pallida. Annals of Applied Biology,* 120, 471-486.

Curriculum vitae – Melvyn F. Askew

Melvyn Askew took his first degree in the School of Agriculture at University of Newcastle upon Tyne and upon completion, joined the former National Agricultural Advisory Service in England. (Later this became ADAS). Whilst working in the East Midlands area of Britain, he undertook his second degree by research, at University of Nottingham.

His career with ADAS took on a more strategic perspective when he became the National Specialist for Non Cereal Crops. In 1994, he became Head of Arable Crops Development Centre for ADAS, then Head of Alternative Crops and Biotechnology. Shortly afterwards he was invited to transfer to newly built Central Science Laboratory of the then Ministry of Agriculture, Fisheries and Food, near York. He remains there today as Head of Agriculture and Rural Strategy.

He has considerable interests in international collaboration and has held Section Chair and Council posts in European Association for Potato Research. Additionally he has led the c. 31 nation IENICA project on non-food products from plants, which has been funded by EC. He is also a member of GCIRC (based in Paris), IIRB (based in Brussels), Institute of Biology (based in London). He has travelled and lectured widely and has recently been appointed as Visiting Professor at Harper Adams University College in England.

Company profile – Central Science Laboratory(CSL)

CSL is one of three laboratories owned by Department for Environment, Food and Rural Affairs. All have Executive Agency status and whilst owned by Defra, are funded from the marketplace. CSL earns approximately £45 million per annum.
CSL is the only Defra laboratory underpinning agriculture, food and terrestrial environment. The 600+ staff are brigaded into two Directorates, namely Agriculture/Environment and Food.

Agriculture and Rural Strategy Group works within the Agriculture and Environment Directorate for CSL and has interest in adding value to the rural economy, introducing new technologies and catalysing and assessing impacts of change in the rural economy. It has been one of the main drivers in producing the CSL portfolio of potato business.

Present state and future prospects of potato storage technology

A. Veerman and R. Wustman
Applied Plant Research, Wageningen University & Research Centre (WUR), P.O. Box 430, 8200 AK Lelystad, The Netherlands, Arjan.veerman@wur.nl, Romke.wustman@wur.nl

Abstract

This paper documents achievements and future prospects in potato storage in NW and Central Europe. Major achievements are the substantial reduction in long-term storage losses; from 10-15% down to 6% in about 25 years. Quality of the stored product has improved due to better storage facilities and store management. Recent developments are highlighted and discussed. These development are: increased product quality awareness, introduction of horizontal ventilation in box storage, effect of elevated CO_2 levels on fry colour index, prediction of sugar levels during storage, improved control of silver scurf through quicker drying and application of imazalil, increased storage losses caused by mechanical damage and promising effects of the natural sprout inhibitor carvone. Future prospects are listed in four categories: disease control and sprout inhibition, hardware, software and management.
Improvements will require more users' knowledge and information on potato rather than on hardware. We foresee a need for a stronger development of the interaction between product (potatoes as well as others) and storage technology in order to make post-harvest technology a full-grown discipline.

Introduction

Potato storage technology in N.W. and Central Europe has progressed considerably in terms of hardware during the last five decades and since two decades in software development. Long-term storage losses in well-equipped buildings have been reduced to about 6%, coming down from 10-15% 25 years ago. Optimization of storage technology played a key role in significantly improving potato quality due to improvements of storage buildings and their equipment, the widely applied use of sprout inhibition compounds (mainly IPC-CIPC), storage disease control and of increased knowledge and skills on storage management. Small and large scale storage complexes have been commissioned and are operating successfully in many parts of Europe and in other parts of the world. These developments have contributed significantly to year-round availability of various types of potatoes: table potatoes, potatoes for processing (food and starch), and seed potatoes. This paper will provide a more detailed picture of the present state of technology and will deal with future prospects in potato storage technology.

Present state

Market demands have led to specific temperature regimes for various outlets. This applies to consumer potatoes and to seed potatoes. The basic requirements in potato storage are control of:

- Temperature (ranges in °C):
  - Seed 3-4
  - Table 4-5
  - French fries 5-7
  - Chips (crisps) 6-9
  - Starch 4
- Air composition
  - Avoidance of elevated carbon dioxide levels
- Ventilation rate and air distribution
  - Up to 100 m^3 $air.m^{-3}$ $potatoes.hour^{-1}$ (≈ 150 m^3 $air.t^{-1}$ $potatoes.hour^{-1}$)
  - Even air flow throughout the potato pile or boxes
- Sprout growth
  - Propham and chloro-propham (IPC and CIPC) are commonly used in Europe to suppress sprout growth
- Diseases
  - Dry rot (*Fusarium* spp.)
  - Silver scurf (*Helminthosporium solani*)
  - Soft rot (*Erwinia* spp.)

The above-mentioned requirements are executed in insulated buildings, insulation being required to keep out both cold in winter and heat in spring, to stabilize the product temperature and thereby to maintain the desired product quality. The buildings are equipped with fans, air distribution systems and flaps to let ambient air in and out. Ambient air is used for drying and cooling in temperate climates. Mechanical cooling is applied in warm climates (Hesen, 1986). During the last two decades refrigeration has increasingly been introduced in storage of table stock and seed potatoes as a supportive or even main source of cooling.

This technology is operating successfully when storage management is applied in a proper way, i.e. when the right decisions are executed at the right time. The human component is crucial in maintaining the right control. During the last decades the human component has partly been substituted by computer hard and software. With help of sensors software can continuously and more accurately assess the suitability of ambient air needed for drying and cooling. Similarly, it is able to monitor the air composition within stores; especially the carbon dioxide content. However, the human expertise and skills remain decisive in storage management since they determine the criteria by which the software instructs the hardware. The storage period consists of a number of phases, each phase having its own specific requirements with respect to temperature and relative humidity of the ventilation air. The yearly different properties of potatoes and storage conditions determine the optimum settings of hard and software in each storage phase, the settings thus never being a standard recipe.

General developments in potato storage

Potato buyers are increasingly demanding: consumers and processors require high external and internal quality. No damage and bruising and the absence of diseases are demanded. The trend is to optimize storage management at cultivar and even at lot level. Specific market demands result in specific storage regimes. A clear example is the introduction of PCN resistant potato cultivars for the French fry industries in The Netherlands during the 1980s and 1990s. The commonly applied storage temperature for cultivar Bintje needed to be differentiated for new cultivars (Veerman *et al.*, 2003). This development clearly demonstrates the necessity for increased input (research and knowledge transfer) to optimize storage management for newly developed cultivars. Improvements require more users' knowledge and information on potato rather than on hardware. There is need for a stronger development of the interaction between product (potatoes as well as others) and storage technology and to make post-harvest technology a full-grown discipline.

Recent developments

This section will present recent (i.e. the last two decades) findings in potato storage research in The Netherlands.

Horizontal ventilation

Forced horizontal ventilation was introduced in The Netherlands during the 1980s as an economically attractive bulk storage system for potatoes (Wustman, 1987). The principle of horizontal ventilation is at present successfully applied in box storage of seed potatoes (Figure 1).

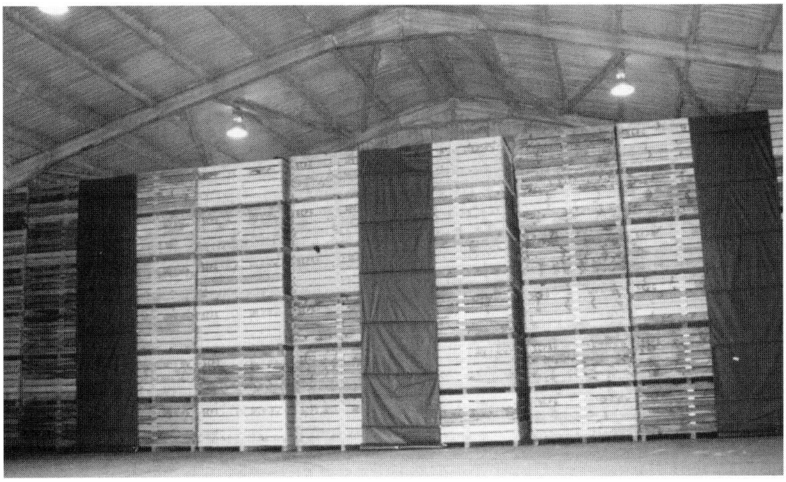

Figure 1. Application of forced horizontal ventilation in a box store.

Effect of carbon dioxide on fry colour

Higher levels of carbon dioxide give poorer fry colour. This phenomenon was found during the late 1980s particularly in newly constructed potato stores and in older stores adapted through foam (PUR) insulation. Applied Plant Research (Wageningen UR) conducted investigations during the 1990s and came to some interesting conclusions. Significantly darker fry colours were obtained after storage at an average carbon dioxide level of 0.5% during the entire storage season (Veerman, 1996). The short line fragments in Figure 2 also illustrate that over time carbon dioxide related increase of reducing sugars follows the same pattern of reconditionability

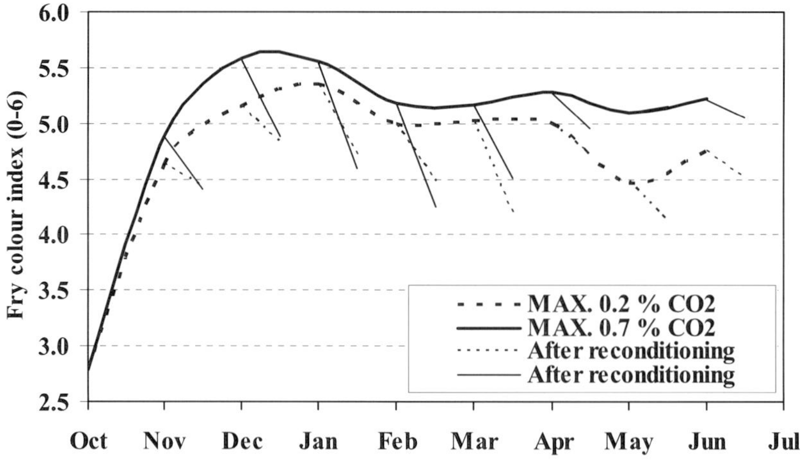

Figure 2. Elevated levels of carbon dioxide increased fry colour index.

as low temperature related sweetening. It is important to note that reconditioning was in absence of CO_2. Presence of CO_2 during reconditioning was able to strongly increase fry colour index (data not shown).

Monitoring carbon dioxide levels in farmers' stores proved that levels may be high, in some stores up to levels that are dangerous for humans: levels of 5% were found after store loading. But even during mid-storage season at pile temperatures of 6-7 °C, levels exceeding 3% may be found (Figure 3). Apart from differences in respiration rates between potato lots, especially the insulation of the building and tightness of doors and hatches determine the accumulation of carbon dioxide.

From the research results and the monitoring of stores in practice, it was concluded that store specific monitoring of carbon dioxide is the only way to define a flushing regime with ambient air. At full hatch opening, flushing the store does in most stores not take more than a few minutes per day to keep average values below 0.5%.

Tuber respiration rate appeared to be increased by the application of sprout suppressants (both CIPC and Carvone, data not shown). The application increased the carbon dioxide levels in the store which could be ruled out to be caused by the fuel consumption of the application technique itself. The increased respiration rate and carbon dioxide levels gradually went down to the levels before application in about 7 to 10 days time. During this period the flushing regime had to be intensified to meet the threshold value that had been set.

Prediction of sugar levels during storage

Agrotechnology & Food Innovations (Wageningen UR) has developed a model to predict the reducing sugar levels in stored potatoes for the processing industry (Hertog *et al.*, 1997). The model has been designed with the goal to predict fry colour, depending on cultivar, soil, cultural practices (e.g. nitrogen level, Figure 4) and seasonal effects. Parameters correctly reflecting

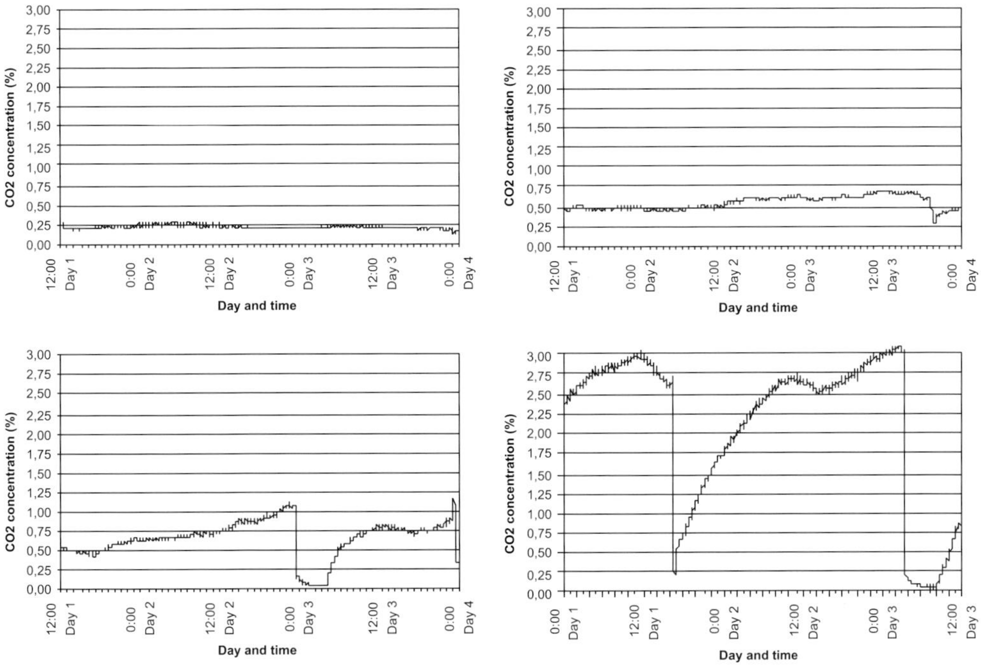

Figure 3. CO_2 levels depend on crop and store properties resulting in levels to differ strongly between stores.

Figure 4. Fry colour gets darker with increasing nitrogen level (from left to right: 0, 100, 200, 300 and 400 kg/ha, the four rows being four replicates (Veerman et al.*, 2003).*

and thus predicting the effect of these factors are yet to be found. At this time they are investigated through a genomics approach.

Silver scurf

Helmintosporium solani causes silver scurf in potatoes and the infection rate may strongly increase during storage. The fungus increases water loss and can have a negative impact on growth vigour of seed potatoes. These are sufficient reasons for Dutch potato companies to fund Applied Plant Research work on better control. Mother tubers are the main source for infection of daughter tubers. This was clearly reflected in research on mother tuber treatment before planting. With some fungicides a consistent reduction of over 90% of the infection level of daughter tubers at harvest could be realized, which means a better start of the storage season.

The first weeks in storage are a critical period with regard to the development of silver scurf. At harvesting and store loading spores become dispersed throughout the crop. The crop being mostly wet or moist at store loading and having a temperature between 10 and 20 °C is creating opportunities for new silver scurf infections. For three decades it has therefore been advised in The Netherlands to dry the crop within one week after harvesting. However, one of our experiments demonstrated that the infection opportunities for silver scurf shortly after store loading were already fully utilized within 5 days. To prevent silver scurf infections during the initial days of storage drying should therefore be completed at least within 5 days. Further research is necessary to more accurately assess the critical duration of the drying period.

Imazalil is an effective fungicide to control silver scurf during the storage season (Figure 5). However, it not only protects tuber surface from silver scurf infection, it also strongly reduces sporulation during the entire storage season, resulting in substantially lower disease pressure (Figure 5). Applied Plant Research has demonstrated that the efficacy of imazalil remains stable during the entire storage period. Increasing rates of silver scurf infection are more likely to be caused by other factors than decreasing efficacy of imazalil. These factors may be: insufficient coverage of the tuber surface while applying imazalil at store loading or high fractions of the tuber surface being shielded from the treatment by adhering soil. After removal of the soil unprotected tuber surface becomes available for infection.

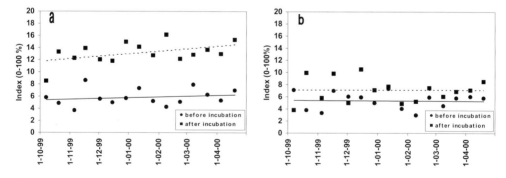

Figure 5. Development of the silver scurf index during storage season 1999/2000 without (a) and with (b) a pre-storage treatment with imazalil (Diabolo SL, Certis Europe). Development of the fungus at low storage temperature (4 °C) only became apparent after incubation for 10 days at 15 °C.

Mechanical damage increases storage losses

Some 30% of potatoes produced in The Netherlands are processed into potato starch. Dutch and German farmers grow specifically bred cultivars with high dry matter contents. Due to their high dry matter content, starch potatoes are prone to external but especially internal damage. Assessment of internal damage is done through calculating a damage index after tuber peeling and damage assessment. The index ranges from 0 (no damage) to 50 (100% severely damaged tubers). Tubers losing water from the internal damaged tissue show a decrease of underwaterweight (specific gravity) during storage. As payment of starch potato growers is based on specific gravity, the financial return when delivering starch potatoes from storage is strongly reduced by internal harvesting damage. Experiments by Applied Plant Research showed an average 0.67% increase in storage loss for each unit increase of the internal damage index. Figure 6 illustrates the moisture loss from the damaged tissue, while actual dry matter and starch contents are only slightly affected. This illustrates that the increasing water loss from the internally damaged tissue compromises the relation between specific gravity (under water weight) and dry matter and starch content.

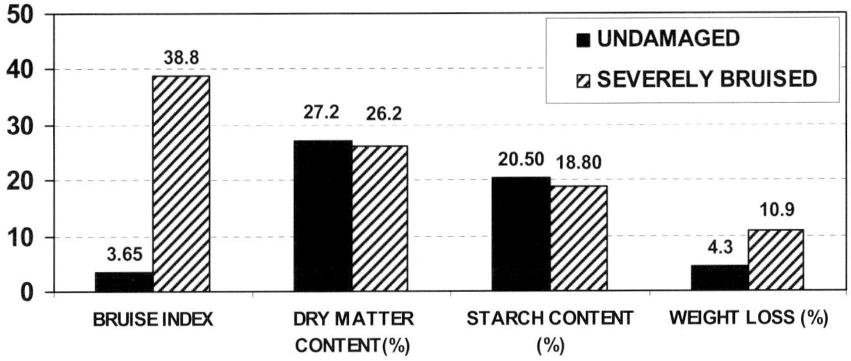

Figure 6. Effect of mechanical damage on weight loss of starch potato cultivar Seresta during storage (Oct-March).

Sprout inhibition

Potato growers in the Andean region in Latin America have traditionally been using naturally occurring compounds to control sprout growth and diseases and pests in stored potatoes (Oosterhaven, 1995). During the 1980s extensive Dutch research developed carvone (Talent®, Luxan BV) into a registered compound for effective sprout control in ware and starch potatoes. At present the admission in The Netherlands of carvone for sprout growth regulation in seed potatoes is awaited. Next to sprout growth control the chemical has some control on the spread of major storage diseases: dry rot (*Fusarium* spp.) and silver scurf (*Helminthosporium solani*). The effect of carvone on Rhizoctonia (*Rhizoctonia solani*) is currently investigated at Applied Plant Research.

Future prospects

In our opinion potato markets will never cease to ask for quality improvements and product diversification. This will be pushing the development not only of breeding and growing potatoes but of their storage as well. We foresee developments in four areas of potato storage:

Disease control and sprout inhibition
- Quicker drying to accomplish improved control of storage diseases like silver scurf and black dot;
- New (synthetic and natural) compounds for sprout inhibition and storage disease control for both ware and seed potatoes contributing to more sustainable storage: lower residue levels in stored produce.

Hardware
- New application techniques for disease control and sprout inhibition compounds;
- Increased use of box storage for high quality markets;
- Use of air humidification in temperate climates;
- Sensory techniques for monitoring and predicting sugar levels;
- Sensory techniques for monitoring disease levels.

Software
- Physiological knowledge specifically for cultivars and lots within cultivars, with respect to e.g. sugar building, dormancy, physiological age and seed vigour;
- Specific storage regimes for different lots of potatoes for various markets;
- Specific storage regimes to optimize growth vigour of seed potatoes;
- Interaction between product characteristics and the use of technology will be key issue in software development leading to new "Techknowledgy".

Management
- Further optimization of storage management at store managers' level. ICT is offering new possibilities not only for developing new potato storage software, but also for a faster integration of hard and software in daily practice.
- The former not only offers possibilities in a technical sense, but also with respect to integrating new developments into the knowledge and skills of store managers, thus realizing transfer of knowledge more quickly than in the past.

References

Hertog, M.L.A.T.M., L.M.M. Tijskens and P.S. Hak, 1997. The effects of temperature and senescence on the accumulation of reducing sugars during storage of potato (Solanum tuberosum L.) tubers: a mathematical model. Postharvest Biology and Technology. Vol. 10, P. 67-79.

Hesen J.C., 1986. Storage management influences potato quality in the Dutch Polders regions. P. 551-560. In: Engineering for potatoes. Editor: B.F. Cargill. Michigan State University (MSU) and American Society of Agricultural Engineers (ASAE). ASAE, St. Joseph, Michigan, USA. 644 pp.

Oosterhaven, J., 1995, Different aspects of s-carvone, a natural potato sprout growth inhibitor, PhD Thesis, Landbouwuniversiteit Wageningen, 152 pp.

Veerman A., 1996. The effect of increased CO_2 levels on fry colour during storage. P. 419-420. In: Abstracts of Conference Papers, Posters and Demonstrations. 13th Triennial Conference of the European Association for Potato Research (EAPR). Veldhoven, The Netherlands. 14-19 July 1996. 714 pp.

Veerman A., P.C. Struik and C.D. van Loon, 2003. An analysis of the effects of cultivar, nitrogen, potassium, location and year on yield and quality of ware potatoes in The Netherlands. P. 44. In: Abstracts of Papers and Posters. 15th Triennial Conference of the European Association for Potato Research (EAPR). Hamburg, Germany. 14-19 July 2003. 354 pp.

Wustman R., 1987. Horizontal ventilation in potato stores. P. 339-340. In: Abstracts of Conference Papers and Posters. 10th Triennial Conference of the European Association for Potato Research (EAPR). Aalborg, Denmark. 26-31 July 1987, 449 pp.

Curriculum vitae – Arjan Veerman

Arjan Veerman studied arable crop science at the Agricultural University, Wageningen, The Netherlands, and graduated in 1988. In the same year he started his professional career at the Research Station for Arable Crops and Field Grown Vegetables (PAGV, presently known as Applied Plant Research) at Lelystad, The Netherlands. He obtained his doctorate at Wageningen University in 2001. His PhD thesis dealt with variation in potato tuber quality between and within ware potato lots. His present research activities include work on potato storage disease control. His present main field of work is the programme leadership of an extensive yield and quality improvement programme in starch potatoes in The Netherlands. Besides he has a major responsibility in potato research projects within Applied Plant Research. Furthermore he is active as advisor in a number of committees, one of these being the Advisory Committee on Seed Potatoes of the Netherlands Inspection Service (NAK).

Curriculum vitae – Romke Wustman

Romke Wustman studied arable crop science at the Agricultural University, Wageningen, The Netherlands, and graduated in 1976. In the same year he started to work for the International Potato Center (CIP) at Lima, Peru, and was stationed at Islamabad, Pakistan, working in Pakistan and Afghanistan. Subsequently he worked in potato projects in Bangladesh. He started to work for the Research Institute for Storage and Processing of Arable Crops (IBVL, presently Agrotechnology & Food Innovations) at Wageningen in 1984 with work in The Netherlands and assignments abroad. In 1990 he moved to the Research Station for Arable Crops and Field Grown Vegetables (PAGV, presently Applied Plant Research) at Lelystad, The Netherlands, and was in charge of the Netherlands potato varietal assessment programme. From 1998 till 2003 he analysed for the Netherlands starch potato processing company AVEBE the starch potato production at growers' level in The Netherlands and in Germany. His most recent work includes projects and activities within The Netherlands and abroad (Central Europe and S.E. Asia).

Company profile – Applied Plant Research

Applied Plant Research (Praktijkonderzoek Plant en Omgeving B.V. (PPO)) is a modern enterprise which, in close cooperation with its clients, is working towards practical and sustainable solutions for agricultural crops, horticultural crops, mushrooms, trees and fruit crops. By means of combining forces and the integration with the Wageningen University & Research Centre (WUR), Applied Plant Research is flexible with respect to research demands and is offering a broad scope of approaches. Our research facilities are located throughout the Netherlands, facilitating possibilities to conduct research under various situations. Within the organisation, practical skills are available at crop-, company-, and sector level as well as disciplinary level. PPO has four sectors for research of and demonstrations in:

- Arable crops, multifunctional agriculture and field grown vegetables (Lelystad);
- Bulbs and trees (Boskoop);
- Bees, fruit trees and mushrooms (Randwijk);
- Glasshouse horticulture (Naaldwijk).

Each sector has at least one research location. Besides specific knowledge of the crops (cultivation, quality control, integrated control of pests and diseases) and marketing, there is ample knowledge available on practical farm management.

Applied Plant Research boasts a broad network in the sector. Applied Plant Research holds a strong position within the scientific world, achieving that newly developed insights (foreign or national) can quickly be fitted into the research, where practical applicability plays the major role. Applied Plant Research has obtained a central position within the primary sector and is very well known for its well attended demonstration days. Also the Applied Plant Research crop manuals, the contribution to and the organisation of courses and cooperation with farmer study groups enhance its strong position. Increasingly, Applied Plant Research has become active abroad, including Central and Eastern Europe.

Amongst its clients Applied Plant Research has the Ministry of Agriculture, Nature and Food Quality, the Commodity Board for Arable crops and the Commodity Board for Horticultural crops, Dutch Organisation for Agriculture and Horticulture co-operations and enterprises within the agricultural sector such as producers of propagation material, crop protection products, agricultural machinery etc. Large internationals such as Unilever and AVEBE have also been included in our extensive database of clients.

Link between knowledge, experience and capacity within Applied Plant Research
Applied Plant Research boasts extensive knowledge in the areas:

- Practical solutions oriented research;
- Operating research farms;
- Knowledge exchange by means of demonstration days;
- Crop manuals of most crops cultivated in areas with a moderate climate;
- Courses and workshops;
- Project management.

The courses are presented internationally as well as nationally. We are very well up to date concerning all aspects of the potato, vegetables and fruit chain.

Applied Plant Research Business Unit AGV has been mandated as programme coordinator for the EU Access Programme, in which five new member states participate concerning the "market access of fruits and vegetables". Applied Plant Research AGV coordinates the activities deployed by these five new member states and those of the LEI (Agricultural Economical Institute) and the IAC (International Agricultural Centre).

Applied Plant Research AGV also is involved in the establishment and execution of the DUCATT project in Ukraine and has initiated activities in Rumania.

Comparison of different transport and store-filling methods

T. Horlacher and R. Peters
KTBL-Potato Research Station, Dethlingen 14, 29633 Munster, Germany

Abstract

Compared to bulk storage, for farmers the use of pallet boxes for the transport and storage of potatoes implies quality advantages. However, from a comparison of the three operational transport and store-filling methods 'bulk store/farm', 'box store/farm' and 'box store/field' it could be learned that the 'bulk store/farm' method provided the lowest damage to tubers while the damage rate was the highest for the 'box store/field' method. Nevertheless, a significant reduction of the amount of damage to the potatoes in connection with the 'box store/field' method could be obtained by a shift of the box size from the usual 1-ton box to a box of 1.5 or 3 tons as well as the use of a box filling chute at the discharge end of the hopper of the potato harvester. In connection with the method 'box store/farm', four different basic box-filling systems are available, i.e. the 'dipping elevator', the 'tipping box', the 'vertical elevator' and the 'cascade elevator'. On the whole, the tuber damage caused by these machines is relatively low and therefore, a system-based differentiation is not possible. Considering the aspect of higher economic efficiency, the 'bulk store/farm' method is likewise to be preferred in comparison with the other two box methods. On account of the comparatively high machine costs in connection with the 'bulk store/farm' and 'box store/farm' methods there is, however, no marked difference between the operational costs of all three methods.

Keywords: store-filling methods, box-filling systems, tuber distribution, tuber damage, operational costs

Introduction

As the quality requirements have been experiencing a constant rise, the transport and storing of potatoes in pallet boxes also gained in importance. The pallet boxes are either directly filled in the field by the harvester or by a box filler after having been transported in bulk to the farm. When using these solutions, the farmers are aiming at a reduced mechanical load on the tubers - compared to their bulk storing - which should be reflected by a lower damage level, a reduced risk of rotting and a better suitability for storing and, thus, provide a higher marketing quality. However, judging from practical experiences, such box-filling methods imply a lower digging or store-filling performance as well as higher additional costs.

Materials and methods

On the occasion of a first experimental trial (2001), the three transport and store-filling methods 'bulk store/farm', 'box store/farm' and 'box store/field' were compared with each other on a farm with a cultivated area of 100 hectares of potatoes (Figure 1). For both box methods,

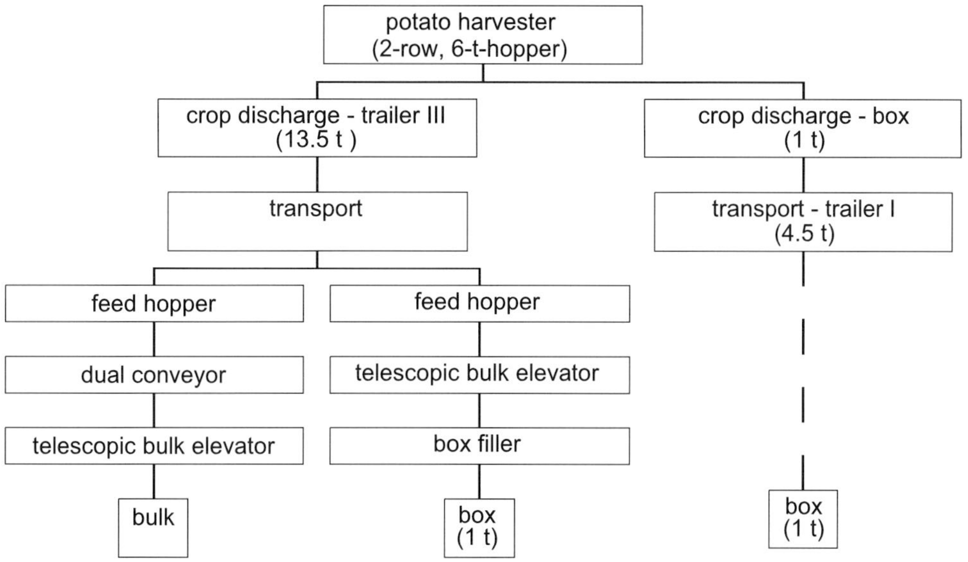

Figure 1. Transport and store-filling methods (trial 1).

usual pallet boxes made of wood with an inside dimension of 1.15 x 1.30 x 1.05 m (W x D x H) and a capacity of 1 ton of potatoes were used. The respective studies included the required working time, the mechanical loads, the damage to tubers as well as the operational costs. The calculation of the operational costs of the three methods comprised the costs of harvest, transport and store filling with new machines (Figure 1).

In order to capture the mechanical load on the potatoes during the three transport and store-filling methods, the instrumental sphere IS 100 was used. The IS 100 registered the number of impacts as well as their impact force and - after multiplying the two mean values registered at the respective data points - the damage index resulted thereof. The samples for bruise detection - four repetitions of 12 kg potatoes each - were taken from the raw material out of the hopper of the potato harvester and from three layers of the bulk and five layers of the pallet box respectively. The tubers were pealed by hand after four weeks of storage.

On the occasion of a second trial (2002/03) studies of the 'box-store/field' method referred to the influence of the size of the boxes on the required working time, the damage to tubers as well as the distribution of potatoes and loose soil in the boxes (Table 1). The walls and the bottom of all pallet boxes were made of wood but the two bigger boxes had a frame of steel. The bruise samples - four repetitions of 12 kg potatoes each - were taken from six layers out of the three different pallet boxes and pealed by hand after four weeks of storage. In order to capture the distribution of potatoes and loose soil the pallet boxes of 1.5 tons were divided into the three sections 'Front', 'Middle' and 'Back' and were then emptied in layers. In this respect, the 'Front' section refers to the part of the pallet box or the moving floor hopper, respectively, which face the tractor.

Table 1. Conducted investigations (x) at the different box sizes (trial 2).

Box	1 ton	1.5 tons	3 tons (1)	3 tons (2)
Inside dimension (m) (W x D x H)	1.35 x 1.15 x 1.05	2.18 x 1.06 x 0.95	2.35 x 1.95 x 1.00	2.35 x 1.45 x 1.36
Working time	X	X		X
Damage	X	X	X	
Soil distribution		X		
Tuber distribution		X		

The calculation of the operational costs for the 'box store/field' method include the costs of harvest - 2-row, bunker type - , transport - farm-field-distance 8 km, trailer for six 1 ton, four 1.5 tons and two 3 tons boxes respectively - , store filling - fork lift - and six month storage - 23 x 45 m building, insulation, ventilation system, pallet boxes, energy.

Both studies (trials 1 & 2) as well as the calculations on the basis of the same referred to the stage of procedure from the filling of the trailers and pallet boxes, respectively, in the field to the end of the storing chain in the potato store.

A third trial (2003/04) in connection with the 'box store/farm' method included all four basic box-filling systems available on the market. The central investigation parameters of the four box-filling systems 'dipping elevator', 'tipping box', 'vertical elevator' and 'cascade elevator' referred to the distribution of the tubers in boxes of 1 ton and 1.5 tons as well as the tuber damage caused by the machines in connection with their respective maximum operational capacity (Table 2).

Standardized potato lots of an oval and a long-shaped tuber variety with a mix containing 10% of the size 28 - 35 mm, 60% of the size 35 - 50 mm and 30% of the size > 50 mm were used as base material. In order to ascertain the risks of damage inherent in the four stationary box filler systems hand harvested, white coloured tubers of medium size were placed on a conveyor belt shortly before the tubers reached the box filler. After the filling of the pallet boxes the boxes were divided into three vertical sections and then emptied in seven layers. The distribution of potatoes was classified and the dyed tubers were collected. These samples for bruise control were pealed by hand after four weeks of storage.

Table 2. Operation parameters of the four box-filling systems (trial 3).

Box-filling system	Filling belt width (m)	Filling belt speed (m/s)	Capacity (t/h)	Pallet box size (max. W x D x H) (m)
Dipping elevator	0.67	0.50	30	2.25 x 1.25 x 1.25
Tipping box	0.70	0.65	40	2.40 x 1.30 x 1.50
Vertical elevator	0.60	0.5	30	2.24 x 1.25 x 1.25
Cascade elevator	0.57	0.4	13	1.40 x 1.20 x 2.00

Results

Mechanical loads

The lowest damage index measured by the instrumental sphere IS 100 of about 190 load units (g) referred to the 'bulk store/farm' method and the highest with 300 - 350 load units (g) to the 'box store/field' method (Lüdemann *et al.*, 2002). The high mechanical load on the potato tubers in case of their direct filling into pallet boxes in the field was due to their fall from a height of 2.15 m from the discharge end of the moving floor hopper onto the bottom of the box. Although this drop was interrupted by a fall breaker in the chute of the hopper, there still remained a height of fall of 1.25 m to the bottom of the boxes of 1 ton used in this trial. A further reduction of this height was not possible as the box filling chute was already lowered to the rim of the pallet box.

Tuber damage

After deducting the damage caused by the harvester to the tubers of the ware potato variety 'Cilena' (damage susceptibility low/medium), the 'bulk store/farm' method caused 6.8 damages to 100 tubers each (Lüdemann *et al.*, 2002). For the 'box store/farm' method this figure increased to 10.6 damages to 100 tubers each while 18.8 damages to 100 tubers each were registered for the filling of the boxes in the field (Figure 2). The differences between all methods were significant.

After changing over to a bigger box dimension the 'box store/field' method in connection with the highly susceptible starch variety 'Ponto' showed a reduction of the damage to tubers from 143 damages to 100 tubers each in case of boxes of 1 ton to 51 damages to 100 tubers each in case of boxes of 1.5 tons (Horlacher, 2004a) (Figure 3). This is mainly due to the potatoes' lower height of fall at the beginning of the filling process. The box filling chute of the moving floor hopper could almost be lowered to the bottom of the box of 1.5 tons thus reducing the damage level of the tubers in this area by half. When using boxes of 3 tons, a further decline

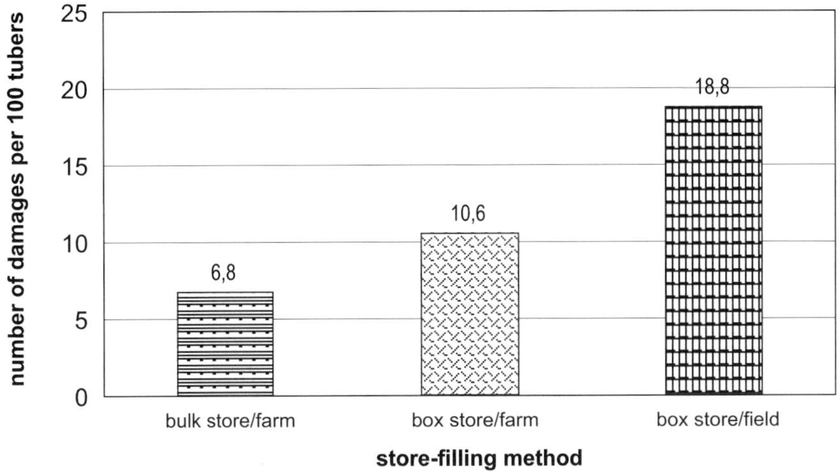

Figure 2. Damage level for the three transport and store-filling methods (trial 1).

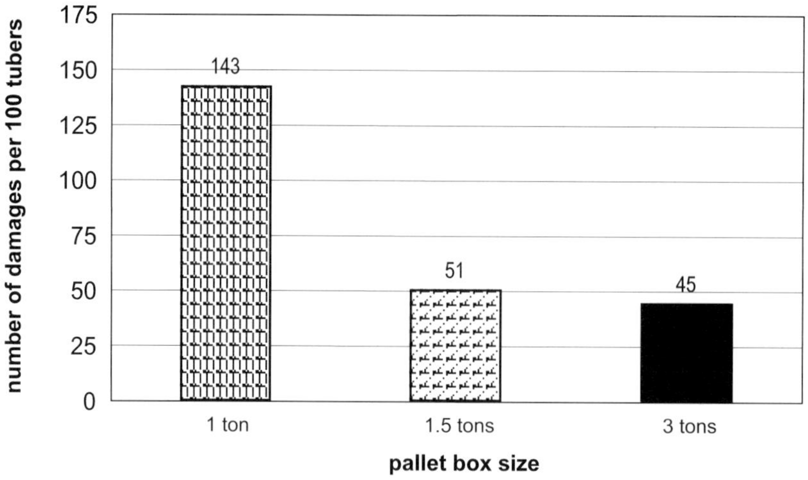

Figure 3. Damage level for the 'box store/field' method with different box sizes (trial 2).

to 45 damages to 100 tubers each could be registered. In addition to the very low height of fall, this is also due to the advantageous ratio that the box volume bears to the wall surface as tubers in the areas at the box edges show a higher damage level than those in the middle.

After passing the box fillers, an average increase occurred in the damage level by 60 - 90 damages to 100 tubers each (Horlacher, 2004b) (Figure 4). In this respect, higher differentiations occurred in the boxes of 1 ton than in the boxes of 1.5 tons which were, however, not supported by statistics in all corresponding cases. Relatively high damages were shown in the lowest layer of the box of 1 ton filled with the 'dipping elevator' system in case

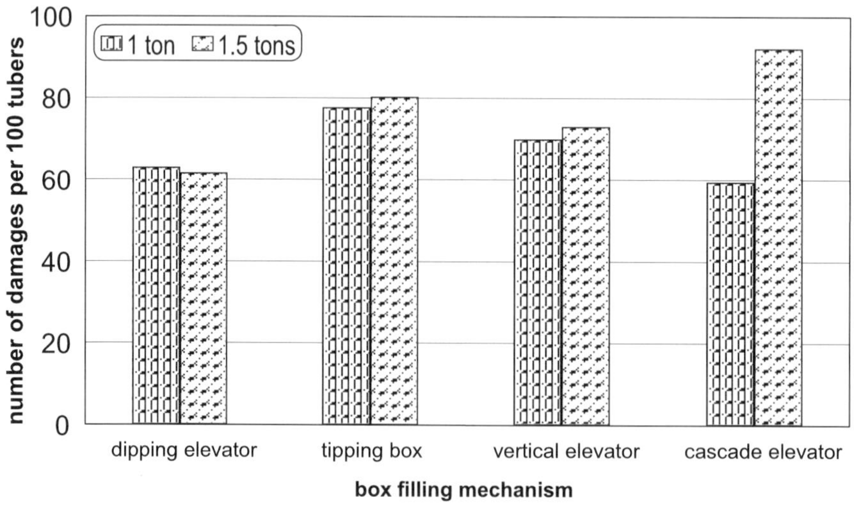

Figure 4. Damage level of tubers as a function of the filling system of the box filler.

of a speed of 0.7 m/s of the filling belt. By means of a reduction of the speed to 0.4 m/s and a respective adjustment of the brake flaps, the high sloping of the filling belt could by defused so that the tuber damage in both types of boxes was halved. For all systems, it is of importance to use an adjustment which provides an especially gentle handling of the tubers at the end of the grading chain since at that stage the box fillers are usually run with low filling capacities and a mutual protection of the potatoes is not possible on the other side.

Distribution of potatoes and loose soil in the pallet boxes

The loose soil is mainly concentrated in the middle section of the pallet boxes while about the same concentration of loose soil could be registered in the front and back parts (Horlacher, 2004a). In contrast to this result, the distribution of the tubers showed a marked differentiation. While the smaller potatoes were mainly to be found in the back and middle parts of the pallet boxes, the big tubers were concentrated in the front part. The medium-sized tubers of the harvested quantity (45 - 55 mm) showed a relatively even distribution across the whole width of the boxes.

There is likewise no change in the distribution of the potatoes and the loose soil in the pallet boxes after removing the box filling chute at the discharge end of the moving floor hopper so that quite probably the distribution of the tubers and the loose soil registered in the boxes will already have been caused in connection with the filling of the hopper by the potato harvester.

The filling process of the different box-filling systems is reflected by the distribution of the tubers in the stationary-filled pallet boxes (Horlacher, 2004b) (Figure 5). The 'tipping box' system starts filling the area near the machine and is then building up a dumping cone at the inside of the pallet box. In connection with the unilateral filling, due to the big tubers' rounder form and their higher own weight they roll down the flank of the potato cone and concentrate

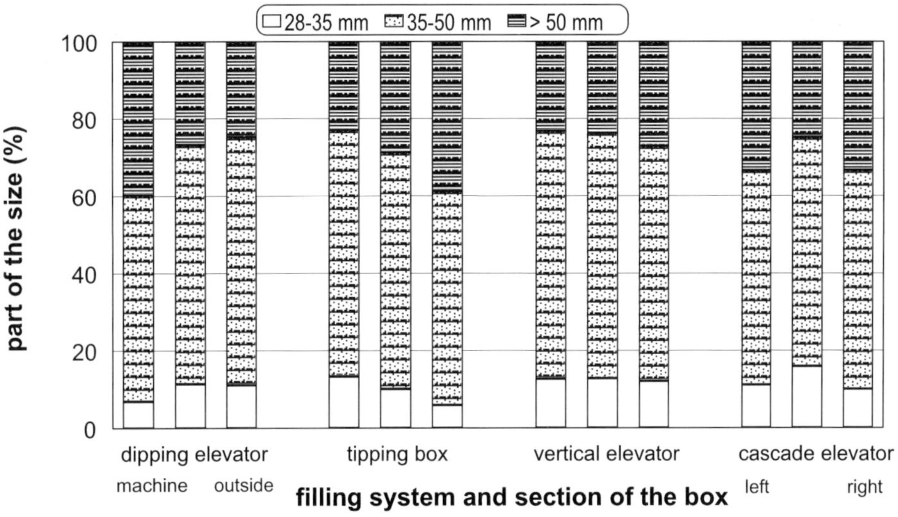

Figure 5. Distribution of potatoes in a box of 1 ton with respect to different box-filling systems.

on the part of the box opposite to the machine. In case of the 'dipping elevator' system, the unilateral filling starts in the most distant area of the pallet box and then the big tubers roll from the dumping cone into the area which is nearer to the machine. Due to the filling in layers by the 'vertical elevator', there is almost no demixing of the basic lot. In contrast to that method, the 'cascade elevator' system shows an uneven distribution of the tubers as the dumping cone is rising in the middle of the box so that the big tubers are once again to be found on the edges of the area.

The tubers of the 28 - 35 mm fraction usually remain where they are dumped in the box and are, therefore, mainly to be found in the area of the respective dumping cone. In the boxes of 1 ton as well as in those of 1.5 tons, the same proportionate shares of distribution can be observed for the long-shaped potatoes and those of the oval type. However, due to the fact that long-shaped tubers are less prone to rolling, the extent of the differences is not so significant.

Labour requirement

Taking into account the conditions of the farm under review (trial 1), the 'bulk/store farm' method required a total of 218 seconds per ton of potatoes for the transfer, transport from and to the field as well as the store-filling (Lüdemann *et al.*, 2002). At the 'box store/farm' method, the required working time rose to 274 seconds per ton of stored potatoes while the total procedure of the 'box store/field' method required 438 seconds per ton. This higher time requirement for the 'box/store field' method is mainly due to the lower amount of potatoes transported on each trailer in pallet boxes.

The required working time for the 'box store/field' method can be reduced by means of the use of boxes with a higher capacity (Horlacher, 2004a). However, the saving of time with the use of the slightly larger boxes of 1.5 tons is insignificant as the filling of such boxes in layers which implies a gentler handling of the tubers requires more time than the continuous filling of the boxes of 1 ton from their upper rim. Only in case of boxes of 3 tons the time saved comes almost up to 50 percent due to the higher amount of potatoes with respect to the operational steps 'Filling in the field' and 'Loading and unloading of the transport vehicles'.

In addition to the width and speed of the filling belt, the performance of the stationary box fillers is conditioned by the actual filling system (Horlacher, 2004b). Because of the one-dimensional movement of their filling belt, the fillers of the 'tipping box' system belong to the top performance segment while machines of the 'cascade elevator' system show a lower performance due to their limited operational capacity. When the system works with a mutual filling of boxes, the actual filling performance is always lower than the performance of the belts as for a certain time no potatoes can be transported during the changing of the working sides.

Economic effectiveness

With respect to the potato harvest, the basis for all three methods was a comparable endowment with machines and personnel in order to obtain a uniform cost unit rate per hour. However, because of the more time-consuming transfer of the potatoes at the 'box store/field' method the digging performance dropped from 30 tons per hour to 28 tons per hour while, simultaneously, the costs for the transport in boxes were slightly higher due to the lower transport capacity of each trailer. On the other hand for the store filling only the costs of the fork lifter must be taken into account while the potatoes supplied in bulk require a feed hopper as well as more conveyors. Based upon the conditions of the trial farm, the operational costs of the 'bulk store/farm' method amounted to € 10.33 per ton while the 'box store/farm' method was slightly more expensive with € 10.73 per ton (Lüdemann *et al.*, 2002). The costs of € 10.54 per ton for the 'box store/field' method ranged between them.

However, upon comparison of the different sizes of boxes within a filling method, cost variances can be noted (Figure 6). The costs for boxes with a capacity of 3 to 5 tons of potatoes are disproportionately high due to their massive type of construction and, therefore, when such box sizes are used the total amount of operational and storage costs is the highest. A more favourable result can be reached with the steel-frame box of 1.5 tons widely used in Germany as the respective costs come up to a rather low level. Nevertheless, despite its substantial wear, a box of 1.5 tons completely made of wood represents an even more favourable alternative, especially in times of high steel prices.

In connection with a cost calculation for the different stationary box fillers it must be taken into consideration that all machines can be used in the store-filling as well as in the grading. Besides, the 'vertical elevator' and 'cascade elevator' systems, respectively, can also be used for the filling of big bags. The price relationship of the 'dipping elevator', 'tipping box', 'vertical elevator' and 'cascade elevator' systems, respectively, comes up to about 1.0 : 1.4 : 1.6 : 0.8.

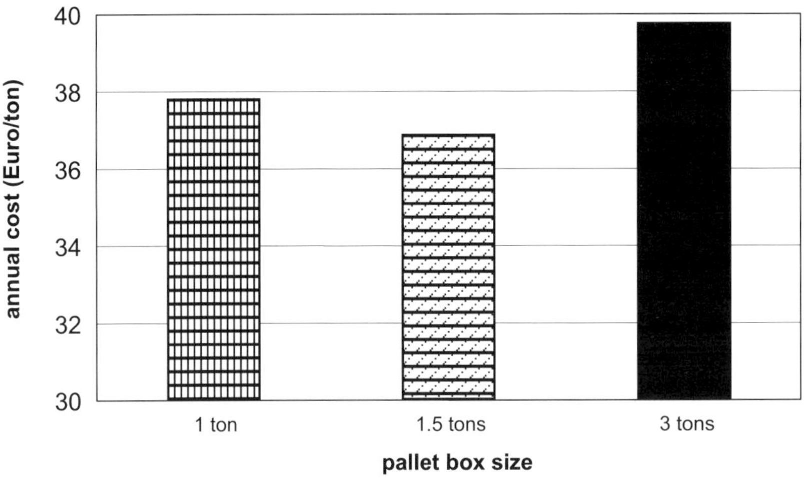

Figure 6. Operational costs of harvest, transport, store filling and storage with respect to different types of boxes.

Discussion

In connection with a rating of the results of the trials 1 and 2 it must be taken into consideration that the studies referred exclusively to the operational steps from the filling of the trailers and pallet boxes, respectively, in the field to the end of the store filling process while, for example, the parameters influencing the respective quality at storage and grading were not taken into account. Considering this background information, the 'bulk store/farm' method represents the preferable solution for the transport and storing of potatoes. The 'bulk store/farm' method showed the lowest rates with respect to mechanical load as well as the tuber damages actually caused thereby. However, a change to bigger box dimensions in case of the 'box store/field' method can have an enduring effect on the reduction of the damage level of the tubers. A box filling chute at the discharge end of the hopper of the harvester provides a gentler filling of the tubers into the pallet boxes in the field without impairing the distribution of the loose soil and the tubers in the boxes.

The 'dipping elevator', 'tipping box' and 'cascade elevator' used as stationary box fillers for the 'box store/farm' method build up dumping cones and, contrary to the 'vertical elevator' system, cause a corresponding demixing of the potatoes in the pallet boxes. While the tubers of the small fraction remain mostly at their respective dumping place in the box, the big potatoes roll down the flanks of the dumping cone. However, a negative influence on the storing due to the uneven filling characteristic is only to be expected if the tubers are covered with soil and thus compact the dumping cone area.

In case of a maximum operational capacity, the tuber damages caused by the box fillers remain at a relatively low level and, therefore, a differentiation of the four systems cannot be supported by statistics. Nevertheless, due to the high sloping of the filling belt special care must be taken for the 'tipping box' system - especially in case of narrow boxes - that the machine is either provided with brake flaps which guarantees the gentle handling of the tubers, an adequate adapted belt surface or the possibility to reduce the speed of the belt.

Considering the operational potato capacity, the 'dipping elevator' and the 'tipping box', respectively, are the most efficient systems. The operational capacity of the two other machine types is restricted by the vertical filling elements, however, due to these features they can also be used for the filling of big bags.

Besides, from the labour economics point of view, the 'bulk store/farm' method is likewise to be preferred to the 'box store/farm' and 'box store/field' methods. However, a change to bigger box dimensions can slightly reduce these differences. And this opinion gains the further in point the more the cultivated areas of a farm are increased and come up to the annual campaign performance of the existing harvest techniques. Despite the widely differing results of the required working time, there is no marked difference between the cost unit rates of the three tested transport and store-filling methods because of the relatively high machine costs of the 'bulk store/farm' and 'box store/farm' methods.

References

Horlacher, T. (2004a). Vergleichsuntersuchung von Kistenfüllsystemen Teil 1 - Befüllung von Großkisten auf dem Feld (Comparison of different store-filling methods part 1 - box filling on the field). Kartoffelbau, 55, 178-183.

Horlacher, T. (2004b). Vergleichsuntersuchung von Kistenfüllsystemen Teil 2 - Befüllung von Großkisten durch stationäre Kistenfüller (Comparison of different store-filling methods part 2 - box filling with stationary box fillers). Kartoffelbau, 55, 426-431.

Lüdemann, K., R. Peters and B. Lehmann (2002). Untersuchung unterschiedlicher Transport- und Einlagerungsverfahren für Speisekartoffeln (Comparison of different transport and store-filling methods for ware potatoes), Kartoffelbau 53, 252-259.

Curriculum vitae – Rolf Peters

Rolf Peters studied agricultural science at the University of Gießen and graduated in 1981 as certified agricultural engineer. After research work at the KTBL-Potato Research Station Dethlingen and the Institute for Agricultural Engineering of the University of Gießen, he obtained the PhD degree of agricultural science in 1984. After a pedagogics training as teacher, in 1986 he took over a research project with respect to single-hill harvest in breeding stations at the KTBL-Potato Research Station. In 1989, he engaged in studies focusing on potato production and mechanization at several universities and research institutions in the USA and Canada. Since 1990 Dr. Peters has been a scientific staff member and since 1992 the head of the KTBL-Potato Research Station. In the past years, he concentrated on the development of sustained production methods, the reduction of mechanical load on tubers in connection with harvest and grading as well as the development and implantation of quality and information management systems with respect to potatoes.

Company profile – The association for Technology and Structures in Agriculture (KTBL)

The association for Technology and Structures in Agriculture (KTBL) is a registered association in Germany. The activities are based on the German Government's agricultural and environmental policies and on the needs of the target groups. The KTBL's main objective is to promote environmentally friendly and socially compatible agriculture in accordance to the needs of consumers, particularly with regard to agricultural engineering techniques. The KTBL-Potato Research Station is located in Dethlingen in the Lüneburg Heath. The goal of the KTBL-Potato Research Station is to make sustainable production possible, and to maintain the best possible potato quality all the way to the consumer. Although process engineering aspects predominate in the work, crop husbandry and ecological, as well as farm and labour management questions are integrated. These objectives result in investigations on process engineering questions on cultivation, storage and grading and the gathering of production engineering data and information on potato technology. The KTBL-Potato Research Station activities are practice oriented. They are in between basic research and advisory service. There is a close co-operation with state research institutes, the chambers or agriculture, universities and the German agricultural society. The results of this work are unbiased, proven and practice-oriented. The products of the KTBL-Potato Research Station are KTBL-data collections, KTBL-papers, -working papers and -working sheets, reports on experiments, specialist publications and lectures and the internet (www.ktbl-kartoffeln.de).

Crop protection

Integrated management of potato tuber moth in field and storage

A. Hanafi
Department of Plant Protection; Institut Agronomique et Veterinaire Hassan II; BP: 12042 Cité Balnéaire Agadir 80.000, Hanafi@iavcha.ac.ma

Abstract

The potato tuber moth (PTM), *Phthorimaea operculella* (Zeller), is a key pest of potatoes, in the field and storage, in North Africa. Nowadays, potatoes are grown year round in some regions of Morocco, Tunisia and Algeria. The availability of potatoes when coupled with favourable weather conditions for the development of PTM even in winter, allow for tremendous build up of populations of PTM. Overlapping of potato growing seasons (winter, fall, spring and summer) creates ideal conditions for this pest to reproduce and migrate beween crops. Migrating adults, coming from surroundings fields are the most important source of PTM invasion into newly planted crops. However, weeds and other crops (tomato or eggplants) are also a potential source of migrating adults. PTM damage to potatoes both in the field and in traditional store is a major constraint for potato production and storage in North Africa and the Middle East.

PTM is active throughout the year in the south of Morocco, but is most active from late April to late August. In winter and early spring, PTM populations remain low and do not cause significant damage to potato crops. However, insect populations develop quickly from April to August and pose a threat to the spring and summer potato crops, the harvest of which continues into July and September, respectively.

It is important to harvest tubers as soon as possible after they have matured. The outcome of delayed harvest in terms of tuber infestation also depends on the number of potato hillings. While using two hillings could prolong the crop cycle, delayed harvest could result in over 20% tuber infestation if only one hilling is practised.

Immediately after harvest, potatoes have to be sorted out, damaged tubers are to be eliminated and the remainder is to be treated with insecticides before transferring them to stores. Both *Bacillus thuringiensis* (BT) and *Granulosis* virus (GV) gave a performance similar to the chemical Deltamethrine and reduced significantly PTM damage in tubers stored in rustic stores for 60 days. An important advantage of BT and GV, which are pathogens of the PTM, is that they pose no health risk to farmers, consumers, or for that matter, other species of animals.

There is a need to integrate pest management in potato fields and stores. Insect damage in stores was found to be more easily controlled if PTM infestation in fields was kept to a minimum. Low infestation at harvest and rapid handling of the potatoes going into storage established good initial storage conditions and decreased the likelihood of post-harvest losses.

Introduction

Potato tuber moth (PTM), *Phthorimaea operculella* (Zeller), is the most damaging insect pest of potatoes in field and storage in regions with a warm climate. In addition to potato, PTM feed on other Solanaceous plants, including tomato, eggplant and pepper (Haines, 1977). PTM damage to potatoes both in the field and in traditional store is a major constraint for potato production and storage in North Africa and the Middle East, where two or three growing seasons per calendar year are common. In fields, larvae bore into leaves, stems and tubers. While the damage caused to plant foliage usually does not reduce yields, infestation in tubers can significantly reduce their market value. Hence, PTM poses a serious threat particularly to the spring crop, before and after harvest. Most damage in the field occurs just prior to harvest, especially when vines are dying naturally (Hanafi, 1999).

Potato tuber moth flight activity and biology

Bioloy and damage potential of PTM are reported by Haines (1977), Von Arx *et al.* (1987) and Hanafi (1999).

PTM is active throughout the year in the Souss Valley in the south of Morocco, but is most active from late April to late August (Figure 1). The PTM females lay eggs directly on leaves, tubers left uncovered in the field, recently harvested, or in storage. Larvae often move from foliage to feed on exposed tubers, or they gain entry to tubers by crawling down through cracks in the soil. When larval development is complete, the larvae spin cocoons on the soil surface or on tuber surface or plant debris.

In the summer, PTM can complete a generation in about 3 weeks. PTM development slows down at cooler temperatures but continues as long as temperatures are above 11 °C.

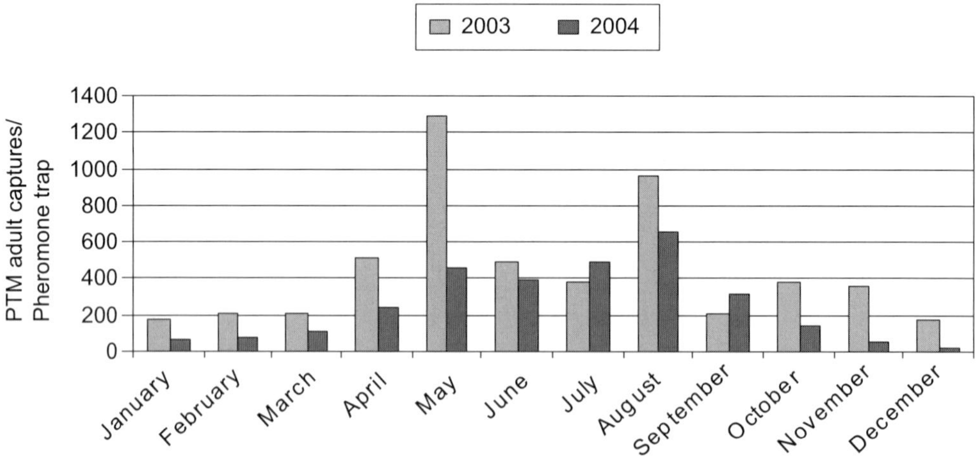

Figure 1. PTM adults captures in pheromone traps at the experimental farm of IAV Hassan II, Agadir in 2003 and in 2004.

In winter and early spring, PTM populations remain low and do not cause significant damage to potato crops. However, insect populations develop quickly from April to August (Figure 1) and pose a threat to the spring and summer potato crops, the harvest of which continues into July and September, respectively.

Cultural practices for the management of PTM

Several cultural practices that are used by farmers to improve the yield and quality of potato can also limit the development of the PTM and minimise damage to tubers (Fuglie *et al.*, 1993, Hanafi, 1999):

1. It is essential to use healthy seed and avoid planting seed tubers infested with PTM.
2. Any practice that reduces the exposure of tubers to egg laying females will limit PTM damage:
   - Potato cultivars that set tubers relatively deep in the soil are less susceptible to infestation than shallow-setting cultivars.
   - Furrow and sprinkler irrigation should be avoided in areas with heavy soil and where the PTM is a problem. Soils tend to crack more under furrow or sprinkler irrigation than under drip irrigation. Drip irrigation helps to keep the soil surface sealed, especially in fine-textured soils.
   - Extra hilling to increase soil coverage of tubers in the field, especially for the spring potato crop. Our experiments conducted in the Souss Valley in the spring of 2004 demonstrated that practising two hillings 30 and 60 days after planting could reduce tuber infestation with PTM by half (Figures 2 and 3).

All of these practices combined or used intelligently are designated to reduce the exposure of tubers to PTM in the field.

3. Harvest measures:
   - It is important to harvest tubers as soon as possible after they have matured. PTM population dynamics studies conducted in Morocco in 2004 (Figures 2 and 3) demonstrate the importance of timely harvest of the spring crop to avoid PTM infestations. The outcome of delayed harvest in terms of tuber infestation also depends on the number of potato hillings. While using two hillings could prolong the crop cycle, delayed harvest could result in over 20% tuber infestation if only one hilling is practised (Figures 2 and 3).
   - It is important to not cover newly dug potatoes with green vines, because as they wilt, PTM larvae leave them and infest the tubers underneath. After harvest, it is essential to discard all tubers showing PTM symptoms, especially if the harvest will be stored, even for short periods.
   - It is also essential that all non harvested or discarded tubers are deeply buried or destroyed.

Management techniques of PTM in stores

Around 60-70% of the spring potato production in Morocco is put in rustic farm stores. About 10% is for seed for the fall crop and the rest of the store is for later market sale.

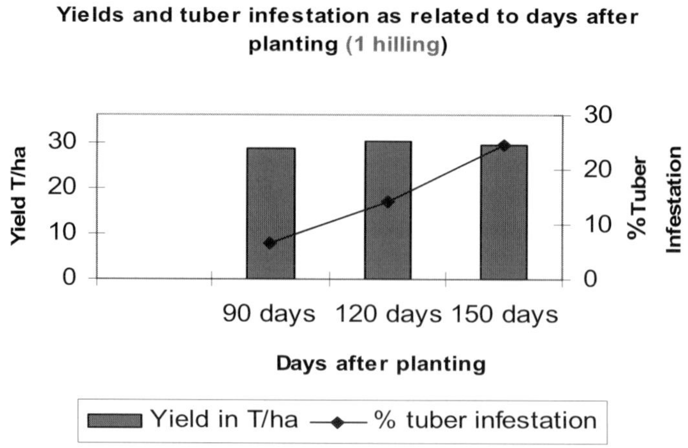

Figure 2. Potato yield (cv. Désirée) and tuber infestation with PTM in a potato field in the Souss Valley, planted on 12 January 2004 with "Désirée" and using one hilling about one month later.

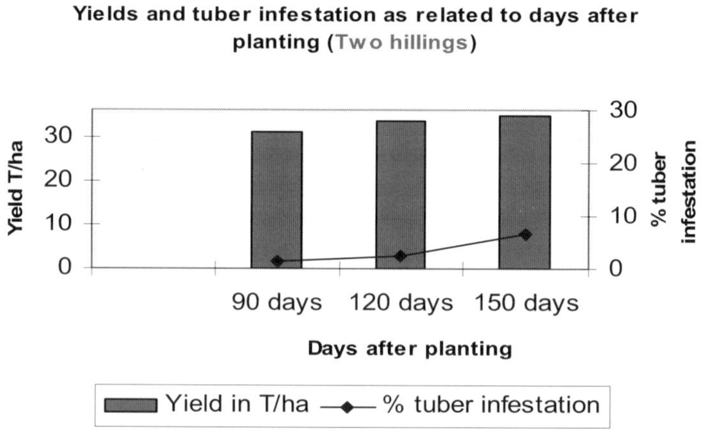

Figure 3. Potato yield (cv. Désirée) and tuber infestation with PTM in a potato field in the Souss Valley, planted on 12 January 2004 with "Désirée" and using two hillings 30 and 60 days after planting.

Throughout the summer PTM poses a substantial threat to the potato store. PTM populations in the surrounding fields remain high and individuals may penetrate the potato stores. PTM may also be introduced from the field at harvest. Even tubers that are seemingly clean at harvest in fact may carry substantial numbers of PTM eggs and first instar larvae which are nearly invisible to the eye.

Several measures could keep PTM damages in traditional stores to a minimum:

1. handling and sorting

Immediately after harvest, potatoes have to be sorted out, damaged tubers are to be eliminated and the remainder is to be treated with insecticides before transferring them to stores.

2. chemical and biological control

The chemicals recommended are pyrethroids if tubers are to be used later for seed. If tubers are to be used for consumption, biological insecticides such as *Bacillus thuringiensis* (BT) and Granulosis virus (GV) are recommended (Ben-Salah and Aalbu, 1992; Hanafi, 1999; Laarif *et al.*, 2004).

In 2004, we have tested the effectiveness of these control methods (Figure 4). Both BT and GV gave performance that was equal to Deltamethrine and reduced significantly PTM damage in tubers stored in rustic stores for 60 days. These results are in agreement with those obtained by Ben-Salah *et al.* (1992) and Laarif *et al.* (2004). An important advantage of BT and GV, which are pathogens of the PTM, is that they pose no health risk to farmers, consumers, or for that matter, other species of animals.

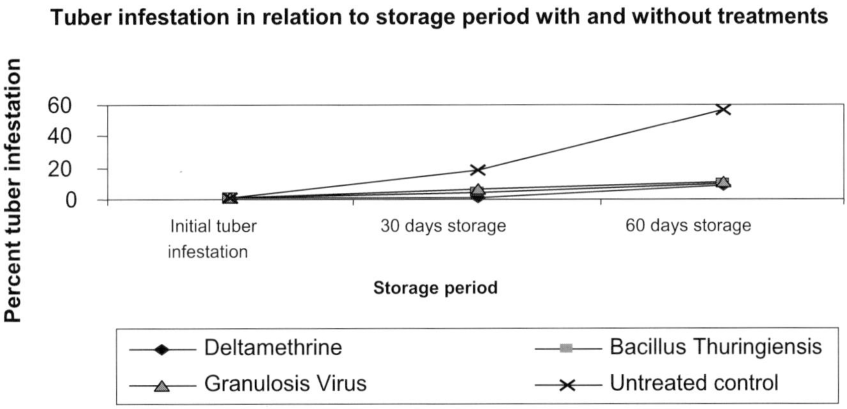

Figure 4. Percent infestation of tubers (cv. Désirée) treated with Deltamethrine (100 g/l), Bacillus thuringiensis var. Kurstaki (Serotype 6 (300 g/l) and Granulosis virus (10 infected larvae/l) as compared to an untreated control, in a traditional store in the Souss Valley, Morocco, 2004.

Farmers' perception of IPM techniques

A thorough understanding of farmers' perceptions of pest problems and their pest management practices is as essential to integrated pest management as is a basic understanding of the insect pest. The full benefit of pest management research is unlikely to be realized unless it leads to improved application of pest management technology by farmers.

While agricultural research has identified several cultural, biological and chemical means of reducing potato tuber moth infestation in potato fields, interviews realized at the level of 17 small farms (3 to 10 ha of potatoes) in the Souss Valley of Morocco have indicated some limitations to the adoption of these IPM packages for PTM management by farmers:

1. The importance of harvest date in avoiding PTM infestation
Farmers clearly recognize the importance of timely harvest in keeping PTM infestation in the field to a minimum. But a primary reason which might preclude farmers from completing their potato harvest before PTM begins to pose a substantial threat is insufficient crop maturity. Since the bulk of the spring crop is not sold immediately but put into rustic farm stores, farmers want the potato skins to be fully mature before harvesting. Moreover, delayed harvest allows the farmers to take advantage of the late season high prices of potatoes on the market. The results of the survey indicated that about 60% of farmers harvest their potatoes about 120 days after planting. The remaining 40% keep their potatoes in the ground for over 120 days after planting and incur significant PTM damage in harvested tubers (Figure 5).

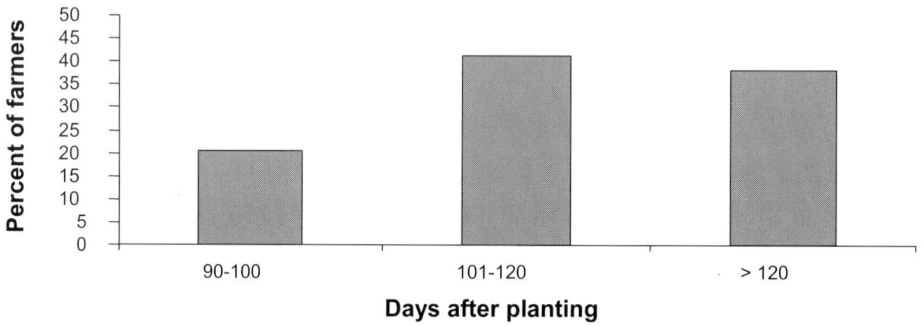

Figure 5. Percent of farmers as related to time of potato harvest (Survey of 17 small farmers in the Souss Valley of Morocco, with an area of potato ranging between 3 and 10 ha of potato, 2004).

2. Hilling as a management tool for PTM
Farmers typically hill up earth around plants once during the potato crop season. The main hilling operation takes place after the potato plants are established, or about 30 days after planting. Our research indicated clearly the benefits of a second hilling to take place about 60 days after planting (Figure 3).

The main objectives of this second hilling are to make sure that the tubers are properly covered and protected from PTM infestation. Our survey indicated that near 25% of farmers practise only one hilling. About 60% of farmers practise two hillings and 15% of farmers practise three hillings (Figure 6). However, as a pest management practice, the effectiveness of hilling is determined not only by when it is carried out or how many times it was done but also by how well it is done.

Conclusion

Ideally, there is a need to integrate pest management in potato fields and stores. Insect damage in stores was found to be more easily controlled if PTM infestation in fields was kept to a

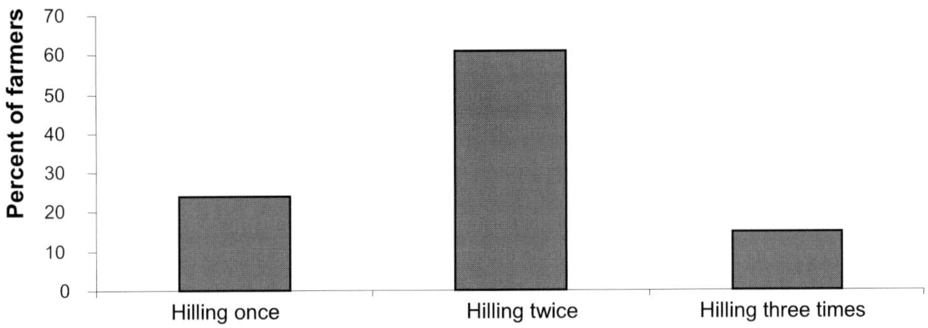

Figure 6. Percent of farmers as related to the number of potato hilling practised (Survey of 17 small farmers in the Souss Valley of Morocco, with an area of potato ranging between 3 and 10 ha of potato, 2004).

minimum. Low infestation at harvest and rapid handling of the potatoes going into storage established good initial storage conditions and decreased the likelihood of post-harvest losses. Biological insecticides in particular were more effective in preventing insect losses in stores in cases where the initial level of infestation was relatively low.

References

Ben-Salah, H. and Aalbu, R. (1992). Field use of Granulosis virus to reduce initial storage infestation of the potato tuber moth, *Phthorimaea operculella* (Zeller), in North Africa. Agriculture, Ecosystems and Environment, 38(3), 119-126.

Fuglie, K., Ben-Salah, H., Essamet, M., Ben-Temime, A. and Rahmouni, A. (1993). The development and adoption of integrated pest management of the potato tuber moth, Phthorimaea operculella (Zeller), in Tunisia. Insect Sci. Appl, 14 (4), 501-509.

Haines, C.P. (1977). The potato tuber moth, *Phthorimaea operculella* (Zeller): a bibliography of recent literature and review of its biology and control on potatoes in the field and in stores. Rep. Trop. Prod. Inst., G112, 15 p.

Hanafi, A. (1999). Integrated Pest Management of Potato Tuber moth in field and storage. Potato Research, 42, 373-380.

Laarif, A., Fattouch, S., Bensalah, H. and Benhamouda, M.H. (2004). Biological control of potato tuber moth in Tunisia using Granulosis virus and methods for the virus detection. P. 206-212. In: Hanafi (Eds) Proceedings of the African Potato Association Congress, Agadir, Morocco, April 5-9, 2004, 414 pp.

von Arx, R., J. Goueder, M. Cheikh and A. Bentamine (1987). Integrated control of potato PTM *Phthorimaea operculella* (Zeller) in Tunisia. Insect Sci. Appl., 4/69, 989-994.

Curriculum vitae – Dr. Abdelhaq Hanafi

Dr. Abdelhaq Hanafi is one of Morocco's leading scientists in integrated pest management (IPM) and biological control. Abdelhaq Hanafi received an M.S. and a Ph.D. in entomology at the University of Minnesota in 1987 and in 1992, respectively. He is currently Professor at Morocco's premier agricultural research institution, the Institut Agronomique et Vétérinaire Hassan II, and is based at the horticultural complex in Agadir within the main horticultural area of Morocco. There he has tested the effectiveness of commercially available biological agents in controlling pests of tomato, peppers, and other crops. This work is credited with the registration of beneficial for use in organic as well as conventional farms. His pioneer work in IPM has contributed significantly to reducing pesticide use in greenhouses in the Souss Valley. He implemented the first IPM network in greenhouse crops in Morocco since 1992 at the level of a few farms. This network is now extended to a few hundred farms in the Souss valley of Morocco. These networks were instrumental in the EUREPGAP certification process of several export farms.

Dr Hanafi has conducted extensive research for the last 20 years on potato pests in Minnesota and in Morocco. His research work focused mainly on the epidemiology of aphid transmitted viruses as well as the management of potato tuber moth in potato field and traditional storage.

Dr Hanafi regularly teaches courses at IAV Hassan II, on IPM, insect vectors of virus diseases, and economic entomology. He has been invited lecturer at the University of Wageningen in the Netherlands and Institute of Agronomy of Bari in Italy (CIHEAM).

He organised four major international conferences on IPM of horticultural crops (Agadir, Morocco, 6-9 May 1997); on Organic Agriculture (Agadir, 7-10 October 2001) and on IPM in greenhouse crops (30 November-5 December 2003) and the African Potato Association Congress (5-10 April 2004). He has been elected convenor of the next International Society of Horticultural Sciences (ISHS International Symposium on Protected Cultivation) to be held in Agadir, Morocco (19-24 February 2006).

Dr Hanafi is involved in several international networks. He is Coordinator of the sub-group Integrated Production and Protection in Greenhouse Crops of the FAO Working Group on Greenhouse Crops in the Mediterranean region. He is the representative of Morocco in the International Society of Horticultural Sciences (ISHS) Council, the representative of Morocco in the IOBC/WPRS working group on IPM in greenhouse crops and the representative of North Africa in the International Whitefly Study Network (IWSN). Dr Hanafi has served his term as President of the African Potato Association (APA) from 2000 to 2004 and is currently a member of the executive committee of APA.

He is author or co-author of over 100 papers and editor or co-editor of 8 books.

Company profile – Institut Agronomique et Vétérinaire Hassan II

The Institut Agronomique et Vétérinaire Hassan II, commonly known as IAV, is Morocco's agricultural university and the country's premier research institution. IAV was founded in 1966 and over the past thirty years has developed an international reputation in agricultural research and teaching. It currently has around 315 faculty members (most with Ph.D.s or the equivalent) and nearly 2,000 students on its two campuses in Rabat and Agadir, including nearly 200 foreign students. IAV places strong emphasis on the integration of research and teaching, and approximately 40 percent of its faculty time is devoted to research. It has ongoing research in all major fields of agriculture, including more than 30 specialised research programmes. Integrated Production and Protection and Sustainable Agriculture are increasingly strong areas of research. Since its founding, IAV has graduated approximately 190 Ph.D. students and 2,830 M.Sc. students.

Purple top disease and beet leafhopper transmitted virescence agent (BLTVA) phytoplasma in potatoes of the Pacific Northwest of the United States

J.E. Munyaneza
USDA-ARS Yakima Agricultural Research Laboratory, 5230 Konnowac Pass Road, Wapato, WA 98951, USA

Abstract

Diseases caused by phytoplasmas are increasingly on the rise in potatoes of the Pacific Northwest of the United States. Recently, a serious epidemic of purple top disease of potato occurred in the Columbia Basin of Washington and Oregon and caused very significant losses to potato fields. There were also indications of reduced tuber quality resulting from infected plants. Potato fields with similar disease symptoms were also reported in some parts of Idaho. In addition, similar disease outbreaks have recently been observed in several vegetable crops grown in the Columbia Basin, including dry beans and radish grown for seed. Using polymerase chain reaction (PCR) technique, DNA analysis of samples of diseased plants collected from potato fields in various locations of the affected area revealed that the beet leafhopper transmitted virescence agent (BLTVA), a clover proliferation group phytoplasma, was the causal agent of the disease. Furthermore, investigation of the insects vectoring this potato purple top disease indicated that the causal phytoplasma was almost exclusively associated with the beet leafhopper, *Circulifer tenellus* Baker, suggesting that this insect is the major vector of the disease in this region. In 2003 and 2004, leafhopper sampling was conducted in the affected area to determine the seasonal occurrence and abundance of the beet leafhopper in this major potato growing region of the United States. The beet leafhopper was very abundant in weeds near potato fields from mid-April to mid-October. This leafhopper moved into potato fields sometime in mid-May and was present in potatoes throughout the remaining of the growing season. This insect was more abundant in potatoes in early summer than in late summer, suggesting that potatoes more likely are infected with the purple top disease during this time of the growing season; however, it is not clear how far into the growing season that potato infection occurs.

Keywords: leafhoppers, *Circulifer tenellus*, phytoplasma, purple top disease, phenology

Introduction

In recent years, diseases caused by phytoplasmas have become increasingly important in vegetable crops of the Pacific Northwest. An epidemic of purple top disease of potato occurred in the Columbia Basin of Washington and Oregon in 2002 and caused significant yield losses to potato fields as well as indications of reduced tuber quality resulting from diseased plants (Jensen, 2002; Hamm *et al.*, 2003; Thomas *et al.*, 2003a,b; Munyaneza, 2004a,b). The disease was also observed in 2003 and 2004, especially in organic potato fields (J.E. Munyaneza, personal observations). Potato fields with similar disease symptoms were also reported in some parts of Idaho (Hamm, 2003). In addition, similar diseases outbreaks have recently been observed in several vegetable crops grown in the Columbia Basin area, including dry beans and radish grown for seed. Symptoms in affected potato plants include a rolling upward of the top leaves with reddish or purplish discoloration, moderate proliferation of buds, shortened internodes, swollen nodes, aerial tubers, and early plant decline. These symptoms resemble very much those of purple top caused by psyllid damage or phytoplasma infection, and in some cases to those caused by potato leafroll virus (PLRV). Early investigation of the cause(s) of the disease indicated that leafhopper transmitted phytoplasmas may have played a significant role in this disease epidemic. Phytoplasmas are microorganisms with characteristics of both bacteria and viruses. The phytoplasma disease complex of potato is poorly understood. Most phytoplasma affected potato plants are broadly termed purple top, and the etiology is attributed to the aster yellows phytoplasma, as it has recently been the case in Mexico (Leyva-López *et al.*, 2002). In response to this disease outbreak, a Washington State and Oregon Potato Commissions funded multi-disciplinary team, mainly made of entomologists and plant pathologists, was formed to investigate various aspects of the problem, including disease causal agent(s) identification, insect(s) vectoring the disease, disease epidemiology, and disease management. This research team includes A.S. Jensen (Washington State Potato Commission, Moses Lake, WA), J.M. Crosslin and P.E. Thomas (USDA-ARS, Prosser, WA), H. Pappu (Washington State University, Pullman, WA), P.B. Hamm (Oregon State University, Hermiston, OR), A. Schreiber (Agriculture Development Group, Inc., Eltopia, WA), and myself. Also, the laboratory of Ing-Ming Lee (USDA-ARS, Beltsville, MD) was involved in the identification of pathogen(s) causing this potato disease.

Disease causal agent identification

During 2003 and 2004, samples of diseased potato plants were collected from potato fields throughout the Columbia Basin and were tested by the USDA-ARS laboratories at Prosser (WA) and Beltsville (MD) for phytoplasmas using the polymerase chain reaction (PCR) technique. Results (Lee *et al.*, 2004a) indicated that all phytoplasmas detected from the diseased plants belong to the clover proliferation group (16SrVI), subgroup A (16SrVI-A). This subgroup currently consists of three phytoplasmas: clover proliferation (CP); potato witches'-broom (PWB); and vinca rosette (VR), a strain of beet leafhopper transmitted virescence agent (BLTVA). The 16S rDNA sequence analysis indicated that the detected phytoplasmas were most closely related to VR with 99.7% gene sequence homology compared to 99.2% with CP and PWB. Also, the results pointed out that the phytoplasmas detected were nearly identical (99.8%) to phytoplasma strains associated with dry bean phyllody disease which recently occurred in the Columbia Basin (Lee *et al.*, 2004b). Furthermore, a similar phytoplasma was identified in infected radish seed from the area (J.M. Crosslin, personal communication). A similar pathogen has previously been reported on potatoes in Utah (Smart *et al.*, 1993) and Korea (Jung *et al.*, 2003). BLTVA was reported on radish seed crops in Idaho (Shaw *et al.*, 1990) and Washington (Schultz and Shaw, 1991). These results verified that the phytoplasma associated with potato purple top disease in the Columbia Basin is different from the potato purple top phytoplasma reported from Mexico (Leyva-López *et al.*, 2002). While the pathogen found in the Columbia Basin is in the clover proliferation group (16SrVI), the one from Mexico is related to the aster yellows group (16SrI). This shows that a similar disease can be caused by two different pathogens. Recently, sequences of three fragments from PCR products obtained from phytoplasma-infected potatoes and leafhoppers were deposited in the GenBank data base as accessions AY692279-AY692281 by J.M. Crosslin. To distinguish it from the potato purple top phytoplasma in Mexico, the accession AY692280 was identified as the "Columbia Basin purple top disease phytoplasma", which is closely related to, or synonymous with, BLTVA.

Disease vector identification

Phytoplasmas are usually transmitted by several groups of insects, including leafhoppers, planthoppers, and psyllids (Lee *et al.*, 1998). Effective management of these phytoplasma related diseases requires knowledge of the biology and ecology of insects vectoring the diseases. In 2003 and 2004, these insects were collected from several sites in the Columbia Basin and identified. Several leafhopper species were found during the sampling and included *Circulifer tenellus*, *Macrosteles* spp., *Ceratagallia* spp., *Dikraneura* spp., *Exitianus exitiosus*, *Ballana* spp., *Colladonus* spp., *Amblysellus* spp., *Paraphlepsius* spp., *Texananus* spp., *Balclutha* spp., *Latalus* spp., *Empoasca* spp., and *Erythroneura* spp. Although various species of psyllids were collected during the sampling, no potato psyllids were found in the samples.

Table 1. Association of phytoplasma with leafhoppers and planthoppers collected in 2003 and 2004[1].

Leafhopper taxon	Collected in 2003	Collected in 2004	Total of both years
Amblysellus	0/4	-[3]	0/4
Balclutha	0/2	-	0/2
Ballana	0/1	0/4	0/5
Ceratagallia	0/7	2/16	2/23
Circulifer tenellus	16/30	31/60	47/90
Colladonus geminatus	0/2	0/2	0/4
Colladonus montanus	0/2	-	0/2
Dikraneura	0/6	-	0/6
Erythroneura	0/2	-	0/2
Exitianus exitiosus	0/8	-	0/8
Latalus	0/4	-	0/4
Macrosteles	0/7	0/11	0/18
Paraphlepsius[2]	0/3	-	0/3
Texananus[2]	0/2	-	0/2
Unknown Cicadellidae	0/1	-	0/1
Unknown Delphacidae	0/2	-	0/2

[1]Insects were captured with sweep nets, identified, and stored in 70% ethanol. Nucleic acid was extracted from groups of five insects. Numbers are the number of groups of five insects positive for phytoplasma over the number tested. PCR primers were rp3/rp4, which amplify a 660 base pair region of the ribosomal protein genes of phytoplasmas in the clover proliferation group (16SrVI). With the exception of the unknown Delphacidae, all of the insects listed belong to the family Cicadellidae.
[2]Due to the low number of available insects and their relatively large size, these insects were tested in groups of three.
[3]None tested.

Collected insects were tested by PCR individually or in groups of 5-10 for the presence of the potato purple top phytoplasma (BLTVA) at the USDA-ARS in Prosser (Crosslin *et al.*, 2005). The phytoplasma was most often detected in *C. tenellus* and less frequently in *Ceratagallia* spp. (Table 1). All other leafhoppers tested negative for the phytoplasma, including *Macrosteles*, the known vector of aster yellows phytoplasma. Because the phytoplasma was almost exclusively associated with the beet leafhopper and this insect was abundant throughout the Columbia Basin, it is likely that this leafhopper is the most important vector of the potato purple top phytoplasma in this region.

Beet leafhopper phenology and population dynamics

To determine the seasonal occurrence and abundance of the beet leafhopper in the affected area, leafhopper monitoring and sampling were conducted from early spring to late fall in 2003 and 2004 at several locations throughout Yakima Valley and the Columbia Basin using a combination of sweep nets and yellow sticky traps. Sweep sampling was conducted in both years and sampling sites were mainly located in the south Columbia Basin and Yakima Valley and included areas near Boardman, Umatilla, and Hermiston in Oregon and Alderdale, Paterson, McNary, Pasco, Wallula, and Moxee in Washington. In addition, a region-wide leafhopper trapping system using yellow sticky traps was also conducted in the Columbia Basin where yellow sticky traps were deployed at 35 locations in Oregon (Umatilla and Morrow counties) in both 2003 and 2004, and at 70 locations in Washington (Adams, Grant, Lincoln, Franklin, Benton, and Walla Walla counties) in 2004. Most of the sampling sites were located in and/or near commercial potato fields that had been significantly affected by the purple top disease in 2002 and along the hills overlooking the Columbia River and Yakima River. Weeds at the sampling sites included grasses, mustards, kochia, filaree, Russian thistle, rabbitbrush, sagebrush, prickly lettuce, hoary cress, and pigweed. Heavy duty sweep nets (BioQuip Products, Inc., Gardena, CA), with a 30 cm net hoop were used for sweep samples. At the sweep locations, at least four 100-sweep samples were taken at each location on each sampling date. Sweep samples were taken weekly, placed in plastic bags, and brought to the laboratory at the USDA-ARS in Wapato, WA, where leafhoppers were sorted, identified, and counted. Yellow sticky traps consisted of 8x13 cm sticky cards (Olsen Products, Inc., Medina, OH) mounted on wooden stakes. The stakes were 30 cm long, with about 8 cm inserted in the ground. The bottom edge of each card was about 5-10 cm above the ground, and the cards were held onto the stake with a large paper-binding clip. All vegetation within 60-90 cm of the trap was kept trimmed to a height of less than 8 cm. Yellow sticky traps were collected and replaced weekly, taken to the Washington State Potato Commission office in Moses Lake, WA, where samples were processed for leafhoppers similarly to the sweep samples. In addition, two potato fields were planted in both 2003 and 2004 at USDA-ARS farms at Moxee and Paterson in Washington to monitor leafhopper populations and purple top incidence. These fields were left untreated with insecticides. Leafhopper sampling in these fields was also conducted using sweep nets and samples were processed as previously described.

Results are summarized in Table 2 and Figures 1-4. As mentioned earlier, a number of leafhopper species were found in collected samples (Table 2). Leafhopper species composition was similar at most of the sampled locations but the abundance of leafhopper species varied depending on types of vegetation at the sampling locations and growing season during which the samples

Table 2. Abundance of the leafhopper species commonly found in weeds near potato fields in the Columbia Basin of Washington and Oregon. Leafhoppers were collected on yellow sticky traps and in sweep net samples in 2003 and 2004.

Leafhopper species	Average number of leafhoppers/10 sticky traps	Average number of leafhoppers/100 sweeps	
	2004 season	2003 season	2004 season
Circulifer tenellus	1921.9	22.0	36.9
Macrosteles spp.	470.3	8.0	17.6
Exitianus exitiosus	409.3	15.0	18.2
Ballana venditaria	361.2	17.1	10.8
Empoasca spp.	293.8	5.7	3.4
Ceratagallia spp.	206.2	38.0	29.4
Dikraneura spp.	144.2	63.8	21.3
Latalus spp.	115.5	8.3	7.7
Amblysellus spp.	112.3	8.2	13.1
Paraphlepsius spp.	100.7	4.3	6.2
Colladonus montanus	24.1	1.2	0.7
Ballana sp.	17.1	0.8	0.3
Texananus spp.	15.4	2.1	1.8
Balclutha neglecta	12.3	0.7	1.5
Colladonus geminatus	3.8	2.3	2.7
Balclutha impicta	3.5	0.3	0.6
Erythroneura spp.	–	0.8	0.4

were collected. Most of the leafhopper species found in weeds and crops in the vicinity of potatoes were also present within potato fields. Although leafhoppers were observed in weeds near potato fields very early in spring, most leafhopper species seemed to invade potatoes in early summer as weeds matured and died. The beet leafhopper was by far the most abundant species on yellow sticky traps (Table 2), and was also one of the most abundant species in the area during the sweep sampling (Table 2). *Dikraneura* species which were by far the most abundant in 2003 sweep samples are mainly grass feeders and have not been implicated in phytoplasma transmission. The beet leafhopper was very abundant in weeds near potato fields from mid-April to mid-October (Figs. 1-3) and had at least 3 generations per year. The beet leafhopper moved into potato fields sometime in mid-May and was present in potatoes throughout the remaining of the growing season (Figure 4). This leafhopper was more abundant in potatoes in early summer than in late summer, suggesting that potatoes most likely are infected with the purple top disease during this time of the growing season; however, it is not clear how far into the growing season that potato infection occurs.

Conclusion

Information from the present study indicated that the potato purple top disease in the Columbia Basin of Washington and Oregon is caused by the BLTVA phytoplasma and not aster yellows phytoplasma. The beet leafhopper is likely the major vector of the potato purple top

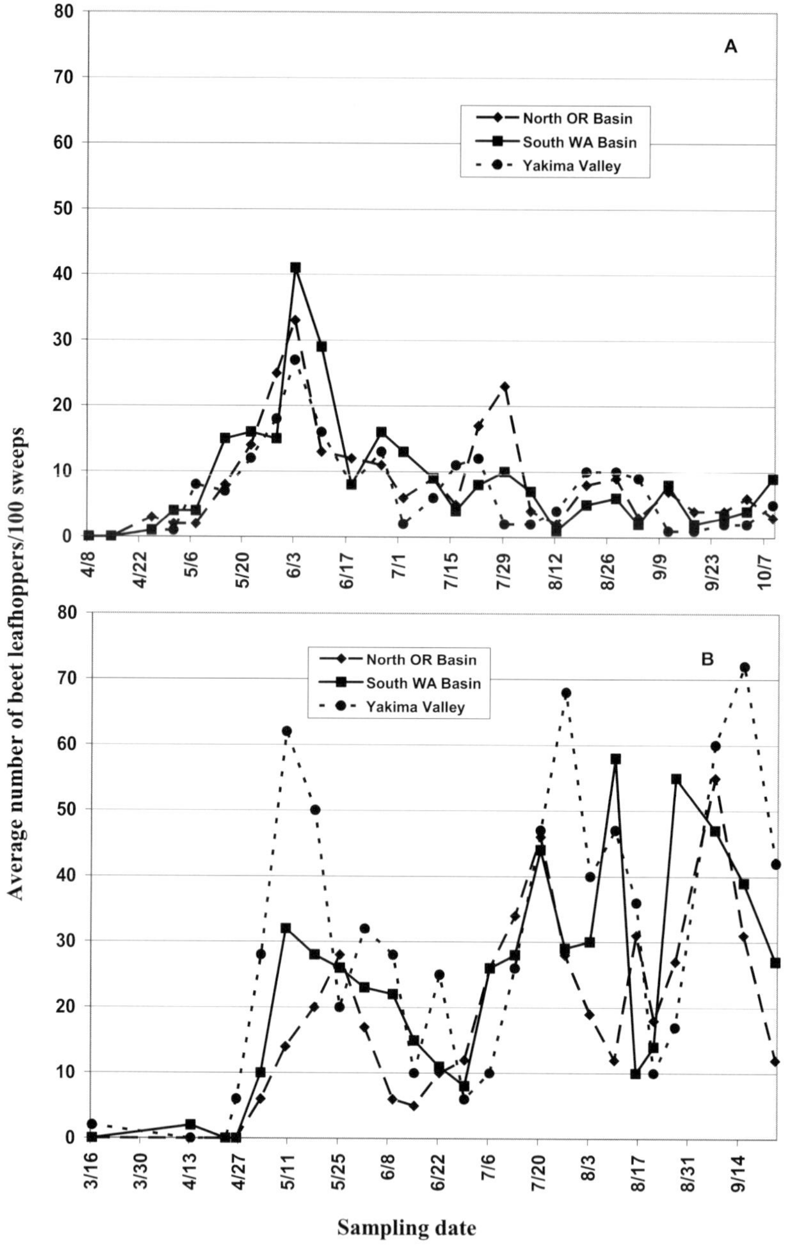

Figure 1. Average number of beet leafhoppers per 100 sweeps on each sampling date in 2003 (A) and 2004 (B). Sampling was conducted weekly in weeds near potatoes.

phytoplasma in this important potato growing region. It has become evident that the management of this emerging potato disease will not be realized until effective approaches to control the beet leafhopper in the Pacific Northwest potatoes are developed. Weeds

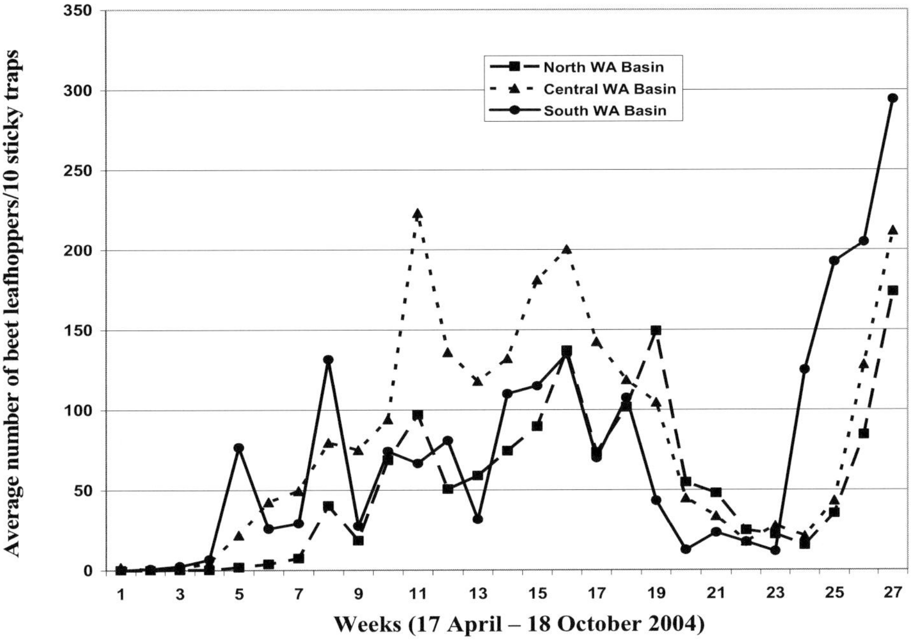

Figure 2. Average number of beet leafhoppers per 10 yellow sticky traps collected weekly at various locations in the Columbia Basin of Washington. The trapping was conducted in 2004 only and traps were placed near potato fields.

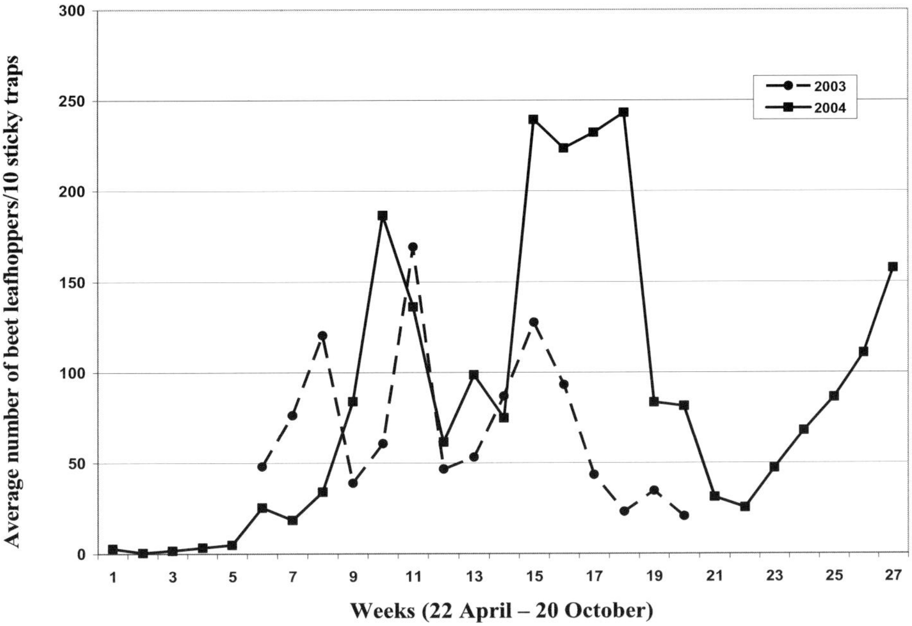

Figure 3. Average number of beet leafhoppers per 10 yellow sticky traps collected weekly in both 2003 and 2004 at various locations in the Columbia Basin of Oregon. The traps were located near potato fields.

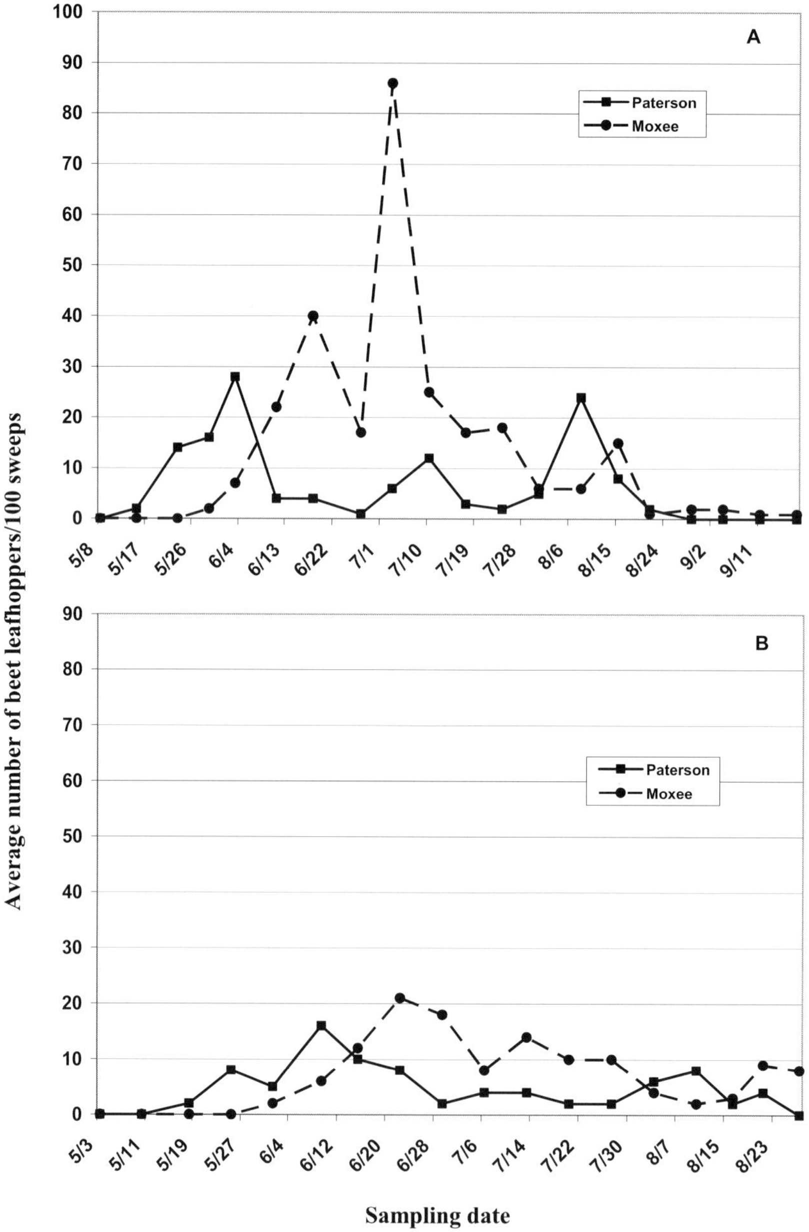

Figure 4. Average number of beet leafhoppers per 100 sweeps on each sampling date within potato fields at Paterson and Moxee in 2003 (A) and 2004 (B) growing seasons.

immediately surrounding potato fields play an important role in the dispersal of the beet leafhopper and epidemiology of the potato purple top disease. Leafhoppers seem to invade the Columbia Basin potato fields around mid-May to mid-June. More research on the epidemiology and management of the potato purple top disease in the Columbia Basin is underway. Future research objectives include investigating the importance of weeds as hosts to both beet leafhopper and BLTVA phytoplasma, determining the sources of infective leafhoppers, investigating the susceptibility of different potato cultivars and plant growth stages to BLTVA, determining the effects of BLTVA on potato tubers, and establishing action thresholds for the disease.

Acknowledgements

We are grateful to Jeff Upton and Launa Hamlin for invaluable technical assistance. We also wish to thank Stuart H. McKamey, Systematic Entomology Laboratory, USDA, Beltsville, MD, for his assistance in the identification of collected leafhoppers. Financial support for this work was partially provided by the Washington State Potato Commission and the Oregon Potato Commission.

References

Crosslin, J.M., J.E. Munyaneza, A. Jensen and P.B. Hamm (2005). Association of the beet leafhopper (Hemiptera: Cicadellidae) with a clover proliferation group phytoplasma in the Columbia Basin of Washington and Oregon. Journal of Economic Entomology, 98, 279-283.

Hamm, P.B. (2003). Potato virus Y and purple top, what do these problems look like, what effect do they have on tuber quality and yield, and how to manage them. In: 'Proceedings of the University of Idaho Winter Commodity Schools - 2003', pp. 105-109. Pocatello, ID.

Hamm, P.B., J.M. Crosslin, G. Pelter and A. Jensen (2003). Potato purple top or psyllid yellows: what was the problem in 2002, and how might it be controlled? Potato Progress, Vol. 3, No.1, pp. 1-3.

Jensen, A. (2002). Potato psyllids and psyllid yellows. Potato Progress, Vol. 2, No. 10, pp. 1-2.

Jung, H.-Y, Y.I. Hamh, J.-T. Lee, T. Hibi and S. Namba (2003). Characterization of a phytoplasma associated with witches' broom disease of potatoes in Korea. Journal of General Plant Pathology, 69, 87-89.

Lee, I.-M., K.D. Bottner, J.E. Munyaneza, G.A. Secor and N.C. Gudmestad (2004a). Clover proliferation group (16SrVI) Subgroup A (16SrVI-A) phytoplasma is a probable causal agent of potato purple top disease in Washington and Oregon. Plant Disease, 88, 429.

Lee, I.-M, K.D. Bottner, P.N. Miklas and M.A. Pastor-Corrales (2004b). Clover proliferation group (16SrVI) subgroup A (16SrVI-A) phytoplasma is a probable causal agent of dry bean phyllody disease in Washington. Plant Disease, 88, 429.

Lee, I.-M., D.E. Gundersen-Rindal and A. Bertaccini (1998). Phytoplasma: ecology and genomic diversity. Phytopathology, 88, 1359-1366.

Leyva-López, N.E., J.C. Ochoa-Sánchez, D.S. Leal-Klevezas and J.P. Martínez-Soriano (2002). Multiple phytoplasmas associated with potato diseases in Mexico. Canadian Journal of Microbiology, 48, 1062-1068.

Munyaneza, J.E. (2004a). Leafhopper-transmitted diseases: emerging threat to Pacific Northwest potatoes. In: 'Proceedings of the University of Idaho Winter Commodity Schools - 2004', pp. 141-150. Pocatello, ID.

Munyaneza, J.E. (2004b). Leafhopper population dynamics in the south Columbia Basin. In: 'Proceedings of the 43rd Annual Washington State Potato Conference and Trade Show', pp. 51-58. Moses Lake, WA.

Schultz, T.R. and M.E. Shaw (1991). Occurrence of the beet leafhopper-transmitted virescence agent in red and daikon radish seed plants in Washington State. Plant Disease, 75, 751.

Shaw, M.E., D.A. Golino and B.C. Kirkpatrick (1990). Infection of radish in Idaho by beet leafhopper transmitted virescence agent. Plant Disease, 74, 252.

Smart, C.D., S.V. Thomson, K. Flint and B.C. Kirkpatrick (1993). The beet leafhopper transmitted virescence agent is associated with diseased potatoes in Utah. Phytopathology, 83, 1399 (Abstr.).

Thomas, P., G. Reed, K. Richards, B. Kirkpatrick and J. Crosslin (2003a). Evidence that the beet leafhopper-transmitted virescence agent caused the 2002 epidemic of potato yellows disease in the Columbia Basin. Potato Progress, Vol. 3, No. 3, pp. 1-4.

Thomas, P., J. Crosslin, K. Pike, A. Schreiber, A. Jensen, J. Munyaneza, P. Hamm, M. Nielsen and J. Upton (2003b). Source and Dissemination of the BLTVA in Potatoes. Potato Progress, Vol. 3, No. 7, pp. 3-4.

Curriculum vitae – Joseph E. Munyaneza

Joseph E. Munyaneza received a B. S. degree in Biology and a Master's degree in Zoology from the National University of Rwanda in 1984 and 1986, respectively. He later moved to the United States where he received a Master's degree and a PhD in Agricultural Entomology (minor in Plant Pathology) from Southern Illinois University (1993) and Iowa State University (1996), respectively. After 2 years as a Postdoctoral Research Associate in the Department of Entomology at Iowa State University in Ames, Iowa, he joined the Department of Entomology at the University of Minnesota in St Paul, Minnesota, to work as a potato Research Entomologist. In 2002, Dr. Munyaneza was appointed Research Entomologist with the United States Department of Agriculture (USDA)/Agricultural Research Service (ARS) at Yakima Agricultural Research Laboratory near Wapato, Washington. Dr. Munyaneza primarily works on potatoes. His current research interests include management of insect pests vectoring potato diseases, potato viruses and phytoplasmas in particular. Dr. Munyaneza is also an Adjunct Professor in the Department of Entomology at Washington State University in Pullman, Washington. Furthermore, Dr. Munyaneza currently serves as Vice-Chair of the Plant Protection Section of the Potato Association of America (2005-2006).

Company profile – USDA, ARS, Yakima Agricultural Research Laboratory Profile

The Yakima Agricultural Research Laboratory is a fruit and vegetable insect research unit of the United States Department of Agriculture (USDA)/Agricultural Research Service (ARS). It is located near Wapato, Washington. Its mission is to seek new and more effective means to control insect pests of deciduous tree fruit and vegetable crops. Research emphases include development of biological control and bio-intensive methods for insect management and the expansion of fundamental knowledge of the biology of the major pests. The unit research programme seeks novel approaches and techniques for integrated pest management systems that will reduce pesticide use. These approaches include the application of biotechnology and other molecular methods, pathogens, parasites, predators, and semiochemicals of pest insects. The vegetable research programme at Yakima Agricultural Research Laboratory presently targets insect pests of potato in the Pacific Northwest, including the green peach aphid, wireworms, potato tuber moth, leafhoppers, potato flea beetle, Colorado potato beetle, and several species of moths.

Survival and disease suppression of potato brown rot in organically and conventionally managed soils

N.A.S. Messiha[1,2], J.D. Janse[2], A. van Diepeningen[1], F.G. Fawzy[3], A.J. Termorshuizen[1] and A.H.C. van Bruggen[1]
[1]Wageningen University Dept. of Biological farming systems BFS, Marijkeweg 22, 6709 PG Wageningen, the Netherlands
[2]Plant Protection Service, Dept. of Bacteriology, Geertjesweg 15, 6706 EA Wageningen, the Netherlands
[3]Potato Brown Rot Project (Phase II), Dokki, Cairo, Egypt

Abstract

Survival of *Ralstonia solanacearum* race 3 biovar 2, the causative agent of potato brown rot disease was studied in differently managed soils, namely four pairs of soils from conventional and organic farms: two pairs of sandy desert and clay soils from Egypt and two pairs of sandy and clay soils from Europe (The Netherlands). Survival of the pathogen was lowest in the Egyptian desert soils, 60 days for conventional and 21 days for organic soils. The desert soils are known for their low nutrient and organic matter content, which may explain this short survival period. Survival of the pathogen was the longest in the Dutch clay soils, 160 days for the organic and conventional soils followed by the Dutch sandy soils, 120 days for the conventional and 140 for the organic. The survival of the pathogen was 120 days for all Egyptian clay soils.

The severity of brown rot disease was studied for the same soils. Disease severity was highest in Dutch sandy soil with 88% infection for the conventional soil and 91% infection for the organic soil. The Total Organic Carbon (TOC) content was relatively high in Dutch sandy soils compared to Egyptian sandy soils. Thus, disease severity was likely related to substrate availability. The disease was found to be suppressed in the Dutch clay soils with 22% of infection for the conventional soil and 35% for the organic soil. The calcium content was highest in those soils, and may have made the plants more resistant. Disease severity was intermediate in the Egyptian soils with 56 to 25% infection in conventional and organic desert soils, respectively, and 69 to 50% in conventional and organic clay soils, respectively.

Keywords: *Ralstonia solanacearum*, survival, disease severity, organic, conventional

Introduction

Potato brown rot (bacterial wilt) disease, caused by *Ralstonia solanacearum* (*Rsol*) is primarily an economic disease in tropical regions of the world, where race 1, 2, 4 and 5 of the pathogen are prevalent. However, since the fourties of last century, the disease has also been reported in temperate regions, where race 3, biovar 2 appeared to be prevalent (Janse, 1996), Potato brown rot is an economically damaging disease in Egypt, where consumption potatoes are grown for the European market. *Rsol* is a quarantine organism for Europe. Survival of the

pathogen has been studied by different authors (Janse *et al.*, 1998), but a distinction between survival in organically and conventionally managed soils has not been made. Other soil-borne plant pathogens are frequently suppressed in organically compared to conventionally farmed land (Van Bruggen and Termorshuizen, 2003).
The aim of this study was to compare survival of *Rsol* and potato brown rot suppression in conventionally and organically managed soils from European and Egyptian origin and ultimately, to investigate the mechanism of pathogen suppression. The earlier the pathogen disappears from contaminated soils (due to suppression, antagonism and starvation) the earlier these soils can be re-introduced in the so-called Pest-Free-Area (PFA-) system that is in place in Egypt. Only from PFA's export of consumption potatoes to the EU is allowed.

Materials and methods

Survival of *R. solanacearum* in different soil types with various management strategies

Survival of *Rsol* in Egyptian and Dutch sandy and clay soils from paired organic and conventional farms was studied *in vitro* at 15 °C and 28 °C. This experiment was conducted three times in June 2003, July 2003 and March 2004 for five months each time at the Bacteriology department in the Plant Protection Service (PD) in Wageningen, The Netherlands. The pathogen was inoculated into the soil as a suspension of 10^7 $cfu.g^{-1}$ soil. The pathogen was monitored daily from the first day of inoculation by plating on modified Selective Medium South Africa (SMSA), (Anonymous, 1998). There were three replications, and the experiment was conducted three times. Temporal declines in cfu's of the pathogen were fitted separately to an exponential decay model with an asymptote:

$$Ct = a + (m-a) * e^{(-b*t)} \qquad (1)$$

Where Ct = number of bacteria (cfu g^{-1} of dry soil for plate counting), a = asymptote (m = initial count of the pathogen, b = decrease rate ($days^{-1}$), and t = time (days). The estimated parameter values for the four pairs of organically and conventionally managed soils were subjected to multivariate analysis of variance (MANOVA, using SPSS v 12) to detect differences between soil types and management strategies.

Disease suppression of potato brown rot in different soils and management regimes

This experiment was conducted in the quarantine greenhouse of the Plant Protection Service (PD) in Wageningen, The Netherlands for two times in August 2003 and March 2004. Tubers of brown rot susceptible potato variety Nicola were planted in the same soils as used in the *in vitro* experiments mentioned above 1 (two tuber eyepieces per pot with 1.25 kg soil each). Disease development was monitored daily from the start of first wilting symptoms. Wilt was scored from 0 to 5, 0=no visible symptoms; 1=wilt in one leaf; 2=wilt in 2 or 3 leaves; 3=most of leaves are wilted; 4=all leaves are wilted and 5=plant is dead, (He *et al.*, 1983). Wilt severity was calculated for each pot per treatment following the equation:

$$\text{Wilt severity} = \frac{\text{no. of plants} * \text{wilt score}}{5 * \text{total no. of plants}} * 100 \qquad (2)$$

Latent infection in the lower stem area of each plant was checked at the end of the experiment (35 days). Latency was tested by plating the extracted weighted and surface sterilized plant tissues from the crown area (in phosphate buffer saline PBS 0.01 M) on SMSA plates. Plating was made with three serial dilutions starting from 100 µl per plate and the recovered *Rsol* bacteria were counted after 5 to 7 days. Population densities of the pathogen were also determined in rhizosphere soil of potato plants and in bulk soil at the end of the experiment. Soil suspension was made by adding 10 soil to 90 ml of sterile phosphate buffer (0.05 M). In case of rhizospere soil, 1 g was added to 99 ml of sterile buffer. After shaking for 2 hours at 15 °C, 100 µl were spread on the surface of the agar plates with consideration of one-tenth dilution in another set of plates. Incubation was made at 30 °C for 5-7 days. Colonies developed at the surface of agar plates were tested using Immunofluorescence Antibody Staining, IFAS, (Janse, 1988). There were six pots per treatment, and the experiment was carried out twice. Non parametric analyses were conducted using SPSS (v 12) to compare wilt severity and area under a disease progress curve (AUDPC) between different soil sources (country), types and management. The same analysis was conducted to compare cfu of the pathogen in soil, rhizosphere and endophytic plant in different soils.

Results

Survival of *R. solanacearum* in different soil types with various management strategies

There was a significant difference in pathogen survival rate between soils from different countries (P < 0.001). The pathogen survived much longer in European (Dutch) than in Egyptian soils (Figure 1). Within the Egyptian soils, the pathogen survived longer in clay than in sandy

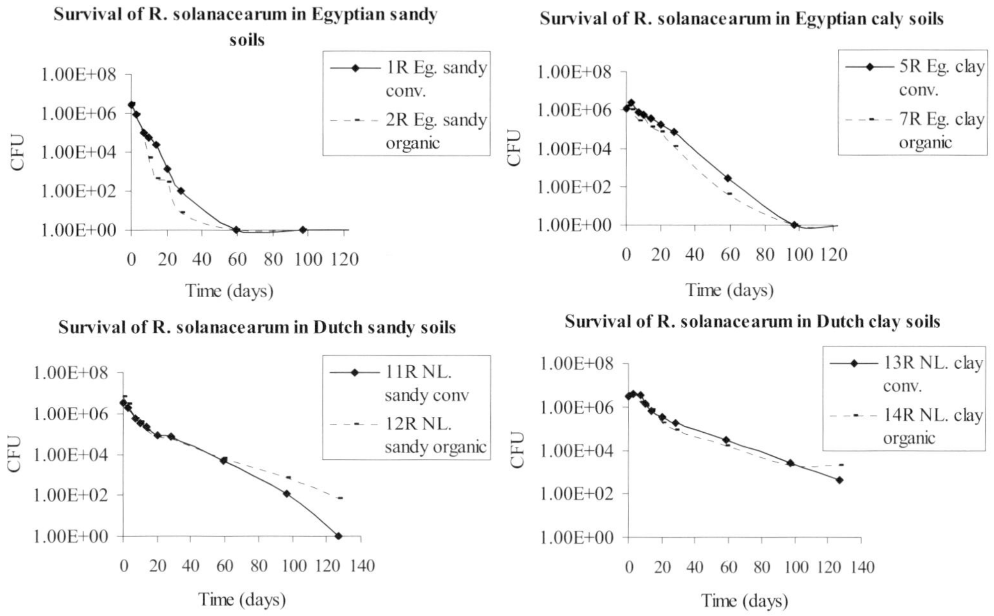

Figure 1. Decline in density (cfu g^{-1} dry soil) of Ralstonia solanacearum *added to different soil types and management regimes.*

soils (P<0.001). There was a significant interaction between soil type and management type (P=0.05) for sandy soils, in the sense that the pathogen was suppressed in the organically farmed compared to the conventionally farmed Egyptian sandy soil (P=0.05), but there was no difference between organic and conventional clay soils used in this study. For Dutch soils, no significant differences were found between different soil types or management regimes.

Disease suppression of potato brown rot in different soils and management regimes
Wilt severity in potato cv. 'Nicola' was highest in Dutch sandy soils (from 43.8 to 55.5) compared to all other soils (Dutch and Egyptian), with 88.2 to 91.1% of the stems infected (Table 1). Wilt severity and percentage of infection were lowest in Dutch clay soils, ranging from 1.2 to 20.7 and from 22.1 to 35.1%, respectively. Organic management had no effect on disease development in Dutch soil, but reduced wilt severity from 23.6 to 8.7 and infection from 56.3 to 24.9% in Egyptian sandy soil, and from 27.5 to 12.3 and from 68.8 to 50%, respectively, in Egyptian clay soil.

Table 1. Wilt severity and percentage of potato stems infected by R. solanacearum 35 days after inoculation of various soils with the pathogen.

Soil type treatment	Wilt severity	% infection
Egyptian sandy conventional	23.6	56.3
Egyptian sandy organic	8.7	25
Egyptian clay conventional	27.5	68.8
Egyptian clay organic	12.3	50
Dutch sandy conventional	43.8	88.2
Dutch sandy organic	55.5	91.1
Dutch clay conventional	1.2	22.1
Dutch clay organic	20.7	35.1

The Mann-Whitney test (a non-parametric test in which sets of data are ranked and the ranks are compared) showed a significant difference in AUDPC between soils from Egypt and from the Netherlands (P=0.02) with higher AUDPCs (and ranks) in the Dutch soils. A chi-square test showed a significant difference in AUDPC between Egyptian sandy, Egyptian clay, Dutch sandy and Dutch clay soils (P<0.001) with the highest AUDPC for Dutch sandy soils. The chi-square test also showed a significant difference between differently managed soils (P<0.001) with the highest AUDPC for Dutch sandy organic soil (table 2). Statistical analysis for cfu's from bulk soil, rhizosphere and surface-sterilized plant tissues showed similar results.

Discussion

No effects of management type on survival of *R. solanacearum* and brown rot severity or stem infection were found for the different soils used, except for sandy soil from the new production areas in the Egyptian desert. In the latter soil type the disease was suppressed in the organic managed field soil as compared to the conventional managed field soil. Conversion of those poor desert soils to organic management may have increased microbial diversity and activity,

Table 2. Non-parametric analysis for the AUDPC from different managed soil types from different countries and soil types

Soil type treatment	N (no. of plants)	Mean Rank
Egyptian sandy conventional	18	31.83
Egyptian sandy organic	18	26.67
Egyptian clay conventional	18	35.75
Egyptian clay organic	18	29.52
Dutch sandy conventional	18	51.71
Dutch sandy organic	18	57.15
Dutch clay conventional	18	24.06
Dutch clay organic	18	35.31
Total	144	

leading to a more stable soil ecosystem that is better able to suppress invading pathogens (Van Bruggen and Termorshuizen, 2003). Low nutrient availability and high competition for nutrients may explain the failure of the pathogen to establish itself in the soil in the absence of plants, and suppress the pathogen in the rhizosphere. Survival of *Rsol* and disease severity were highest in Dutch soils compared to all other soils. Chemical analysis revealed that the Total Organic Carbon (TOC) was highest in those soils, suggesting high nutrient availability and better survival chances for *Rsol.*

The disease was suppressed in the Dutch conventional and organic clay soils and in the Egyptian organic soils in one of the two experiments. The calcium content was relatively high in those soils (about 25 g/kg soil). Calcium is known for enhancing host defence mechanisms through repressing pehA endopolygalacturonase gene expression which was found one of the most virulence determinants of *Erwinia carotovora* (Flego *et al.*, 1997). Also protein kinases which are calcium dependent play an important role in signalling during pathogenicity and hence activating the plant defence mechanism (Romeis, 2001).

Conclusions

In our experiments, the main factor found to influence the survival of the pathogen (in absence of plants) was the availability of the DOC, which was found to be higher in the Dutch soils compared to the Egyptian soils. On the other hand, calcium content was the main suppressive factor of the disease, which was highest in Dutch clay soils. Organic management was a disease suppressive factor in case of organic managed Egyptian desert soil compared to the conventional one as a result of higher microbial diversity and hence more competition for nutrients in the organic soil compared to the conventional soil.

Acknowledgements

This project is funded by the EU through the EU-Egypt Potato Brown Rot Project Phase II (SEM03/220/51A / EGY 1B/1999/0192). We appreciate the advisory and technical help from the team at the Department of Bacteriology of the Plant Protection Service (PD), Wageningen.

We are also grateful to the AIO group and the technicians at the Department Biological Farming Systems of Wageningen University for technical assistance and advisory help via the discussion group meetings. We are thankful to drs. Eelco Franz of the Biological Farming Systems Group for assistance with the statistical analyses.

References

Anonymous (1998). Interim testing scheme for the diagnosis, detection and identification of *Ralstonia solanacearum* (Smith) Yabuuchi *et al.* in potatoes. Annex II to the Council Directive 98/57/EC of 20 July 1998 on the control of *Ralstonia solanacearum* (Smith) Yabuuchi *et al.* Publication 97/647/EC, Official Journal European Communities No. L 235, 8-39.

Flego D., M. Pirhonen, H. Saarilahti, T.K. Palva and E.T. Palva (1997). Control of virulence gene expression by plant calcium in the phytopathogen Erwinia carotovora. Mol. Microbiol., 25, 831-8.

He, L.Y., L. Sequeira and A. Kelman (1983). Characteristics of *Pseudomonas solanacearum* from China. Plant Dis., 67, 1357-1361.

Janse, J.D. (1988). A detection method for *Pseudomonas solanacearum* in symptomless potato tubers and some data on its sensitivity and specificity. Bull. OEPP/EPPO Bull., 18, 343-351.

Janse J.D. (1996). Potato brown rot in western Europe - history, present occurrence and some remarks on possible origin, epidemiology and control strategies. *Bull. OEPP/EPPO Bull.*, 26, 679-695.

Janse, J.D., F.A.X. Arulappan, J. Schans, M. Wenneker and W. Westerhuis (1998). Experiences with bacterial brown rot *Ralstonia solanacearum* biovar 2, race 3 in The Netherlands. Pages 146-152 in: Bacterial wilt disease. Molecular and ecological aspects. Prior, P., Allen, C. and Elphinstone, J. eds. Springer-Verlag, Heidelberg, Germany.

Van Bruggen, A.H.C. and A.J. Termorshuizen (2003). Integrated approaches to root disease management in organic farming systems. Aust. Plant Pathol., 32, 141-156.

Romeis T. (2001). Protein kinases in the plant defence response. Cuur Opinion in Plant Biol, 4, 407-414.

Curriculum vitae – Nevein A.S. Messiha

Nevein A. S. Messiha studied botany and got her B.Sc. from Cairo University in 1992. In 1994, she started her M.Sc. at the Department of Microbiology in Zagazig University in Egypt. Since 1996 she was employed as an assistant researcher at the Potato Brown Rot Project (PBRP) Institute in the Agricultural Research Center, Cairo. She received an intensive training course in detection and identification of plant pathogenic bacteria at the Plant Protection Service (PD) in NL and at Central Science Lab (CSL) in UK in 1996. In 2001, she obtained her M.Sc. with the research subject "Biological control of potato brown rot disease". In 2003 she started a PhD study in the framework of the EU-Egypt Potato Brown Rot Project Phase II (SEM03/220/51A / EGY 1B/1999/0192) at the "Biological Farming Systems" Group of the Plant Science Department at Wageningen University, NL with the topic "Disease suppression and biological control of potato brown rot in organically and conventionally managed soils".

Company profile – Potato Brown Rot Project Institute

Since the late seventies of the past century, the Potato Brown Rot Project Institute in Cairo is involved in testing imported and exported potatoes for the occurrence of the quarantine disease potato brown rot, caused by *Ralstonia solanacearum*. The institute participates since

1996 in the EU-Egypt Potato Brown Rot Project Phase I (1997-2000) and Phase II (2002-2006). The main purpose of this project is to eliminate the pathogen from the production column and to sustain pest free areas (PFA's) for safe exportation of potatoes to the European markets. During these last nine years, a testing laboratory was established, extensive surveys of the disease were made through testing potato, soil, water and different host weeds for the occurrence of the pathogen. Maps were drawn for the occurrence of the pathogen and for the PFA's. In addition, epidemiological research on survival and biological control of the pathogen was executed, extension, scholarships and training courses were provided in European universities and institutes for the staff of the project.

Survival of *Ralstonia solanacearum* biovar 2 in canal water in Egypt

D.T. Tomlinson[1], J.G. Elphinstone[1], M.S. Hanafy[2], T.M. Shoala[2], H. Abd El-Fatah[2], S.H. Agag[2], M. Kamal[2], M.M. Abd El-Aliem[2], H. Abd El-Ghany[2], S.A. El-Haddad[2], F.G. Fawzi[2] and J.D. Janse[3]
[1]Central Science Laboratory, Sand Hutton, York, YO41 1LZ, UK
[2]Potato Brown Rot Project (Phase II), Dokki, Cairo, Egypt
[3]Plant Protection Service, Geertjesweg 15, 6700 HC, Wageningen, The Netherlands

Abstract

Surveys over three seasons of irrigation, drainage and artesian well water throughout the major potato growing areas of Egypt indicated that *Ralstonia solanacearum* biovar 2 was limited to the canals of the traditional potato growing areas in the Nile delta region with positive findings more commonly associated with the smaller waterways. The pathogen was not detected in irrigation or drainage water associated with potato cultivation in newly reclaimed desert areas, designated as Pest Free Areas. Naturally occurring pathogen populations in the canals of the delta were generally low and variable throughout the year with presence usually linked to potato cultivation in the immediate area. Few or no visible symptoms were observed in potato crops irrigated with water periodically contaminated with low populations of *R. solanacearum* and tests on harvested tubers failed to detect the pathogen although it could be detected in the soil in which the potatoes were grown. Results of laboratory studies showed that temperature, biological activity and inoculum pressure were the main factors affecting survival of the pathogen. In experiments at temperatures of 4, 15, 28 and 35 °C, survival was longest at 15 °C and shortest at 35 °C. Survival at 4 and 28 °C was intermediate between these extremes and there was little effect of temperature fluctuations on an approximately 24 h cycle. Survival in autoclaved or filter sterilised canal water was always longer than in untreated water irrespective of other factors and in this study survival time at 28 °C was generally closely linked to inoculum pressure. Aeration, solarisation and pH variation between 5 and 9 had little effect on survival. The maximum survival time in non-sterilised Egyptian waters was estimated to be below 250 days at optimum temperature for survival and high inoculum pressure.

Keywords: *Ralstonia solanacearum*, survival, canal water, Egypt

Introduction

R. solanacearum biovar 2 is a major constraint to Egypt's potato export trade with the European Union. The disease is endemic in parts of the traditional potato growing areas in the Nile Delta so pest-free areas have been designated in non-traditional areas for multiplication of imported seed potatoes and production of the ware crop for export to the EU. A research project has been initiated through EU Technical Assistance programme EGY1B/1999/0192 to study the epidemiology of potato brown rot under Egyptian conditions. Since the pathogen has become established in a number of European waterways and can spread to potato crops during irrigation, experiments and surveys have been undertaken to assess the risks of spread of the pathogen in Egyptian irrigation canals.

Methods

Populations of *R. solanacearum* were determined by dilution plating on selective SMSA medium.

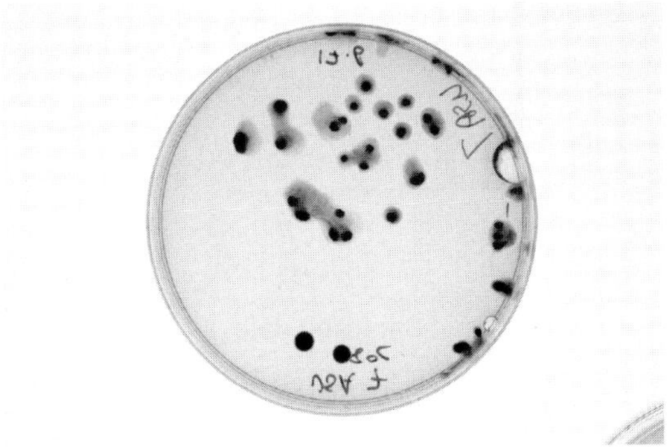

Figure 1. Dilution plating on selective SMSA medium.

Results

Table 1. Test results from water survey samples collected over three seasons.

Season	No. water samples tested[1]	% containing *R. solanacearum*
Pest Free Areas[2]		
02/03	283	0
03/04	296	0
04/05	207	0
Traditional Nile Delta potato growing areas[3]		
02/03	162	20.0
03/04	114	23.3
04/05	169	4.7

[1] Six sub-samples tested per sample. Samples collected from each sampling point on at least 4 different dates during the year.
[2] Samples tested from irrigation and drainage canals or artesian wells at 64 locations in 21 pest free areas in 3 Governorates (Ismailia, Nubaria and Sharkia).
[3] Samples tested from irrigation and drainage canals at 27 locations in 4 Governorates (El-Daqahliya, El-Gharbiya, Minufiya and Nubaria).

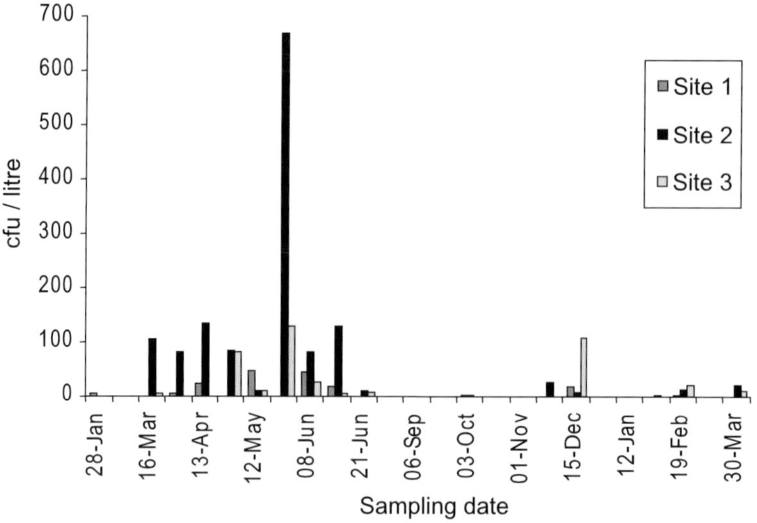

Figure 2. R. solanacearum *population variation (cfu per litre) at three sites on the El-Naggar Canal, (Monofiya Governorate) Jan. 04 - Mar. 05.*

Effect of different factors on survival of R. solanacearum *in inoculated canal water samples under laboratory conditions*

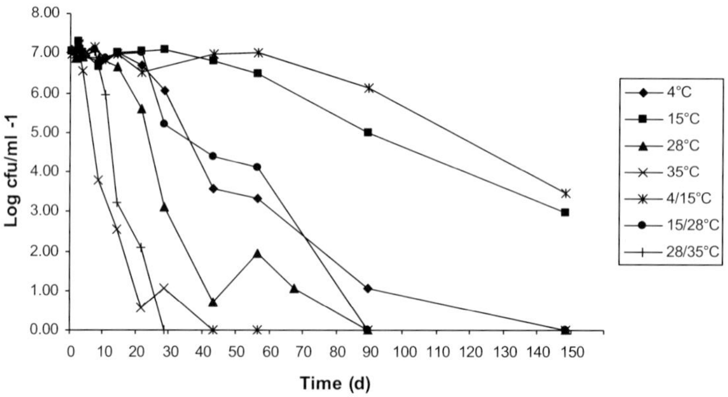

Figure 3a. Temperature effect on survival of R. solanacearum *in inoculated canal water samples under laboratory conditions.*

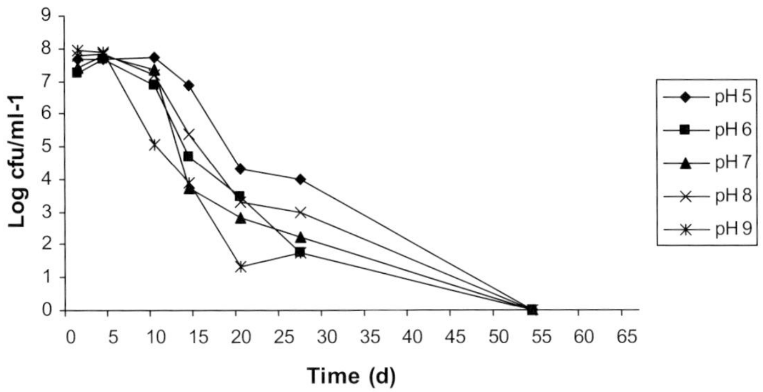

Figure 3b. pH effect on survival of R. solanacearum *in inoculated canal water samples under laboratory conditions.*

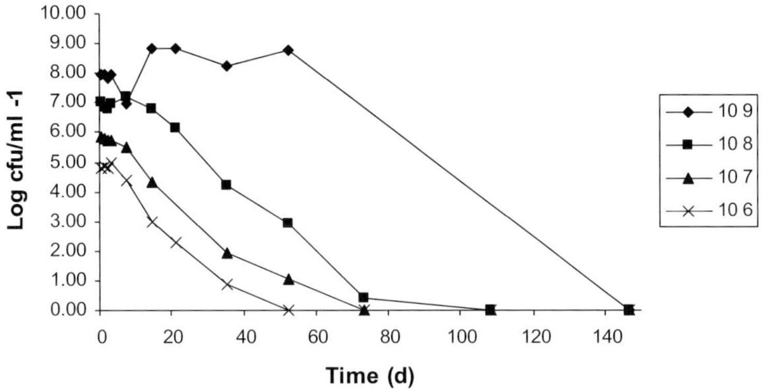

Figure 3c. Inoculum load effect on survival of R. solanacearum *in inoculated canal water samples under laboratory conditions.*

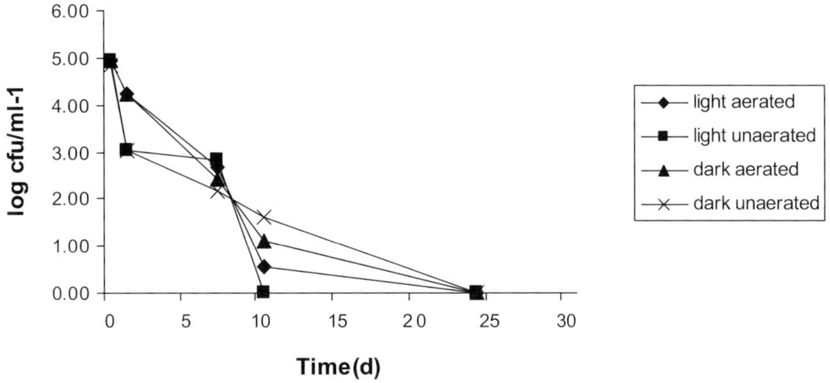

Figure 3d. Aeration and solarisation effect on survival of R. solanacearum *in inoculated canal water samples under laboratory conditions.*

Conclusions

Surveys over three seasons of irrigation, drainage and artesian well water throughout the major potato growing areas of Egypt indicated that *Ralstonia solanacearum* biovar 2 was limited to some canals of the traditional potato growing areas in the Nile delta region with positive findings more commonly associated with smaller waterways.

The pathogen was not detected in irrigation or drainage water associated with potato cultivation in designated Pest Free Areas over 3 seasons, even in those associated with interceptions of brown rot in EU member states.

The maximum survival time in non-sterilised Egyptian waters was estimated to be below 250 days, even at optimum temperature for survival and high inoculum pressure.

Naturally occurring pathogen populations in the canals of the delta were generally low and variable throughout the year with presence usually linked to potato cultivation in the immediate area.

Spread of *R. solanacearum* in irrigation water does not play a significant role in the epidemiology of brown rot during potato production in the pest free areas in Egypt.

Acknowledgements

Research was conducted through EU Technical Assistance programme EGY1B/1999/0192 at the Potato Brown Rot Project (Phase II), Dokki, Giza, Egypt. EU partners involved are Plant Protection Service (PD), Wageningen, The Netherlands (Co-ordinator), Central Science Laboratory (CSL), York, UK, Departement Gewasbescherming (DGB), Merelbeke, Belgium and Laboratoire de Protection des Végétaux (LNPV), Angers, France.

Survival of the potato brown rot bacterium (*Ralstonia solanacearum* biovar 2) in Egyptian soils

D.T. Tomlinson[1], J.G. Elphinstone[1], H. Abd El-Fatah[2], S.H.A. Agag[2], M. Kamal[2], M.Y. Soliman[2], M.M. Abd El-Aliem[2], H. Abd El-Ghany[2], S.A. El-Haddad[2], Faiza G. Fawzi[2] and J.D. Janse[3]
[1]Central Science Laboratory, Sand Hutton, York, YO41 1LZ, UK
[2]Potato Brown Rot Project (Phase II), Dokki, Cairo, Egypt
[3]Plant Protection Service, Geertjesweg 15, 6700 HC Wageningen, The Netherlands

Abstract

Survival of *R. solanacearum* biovar 2 in different Egyptian soils has been studied under both laboratory and field conditions. Survival in inoculated soil samples in the laboratory was found to vary to some extent with temperature, soil type and differences in moisture content between 50 and 100% of water holding capacity. The effects of temperature and moisture content were variable in different experiments but survival was generally longer at 15 °C than at 28 °C and at 100% rather than 50% of water holding capacity. In laboratory experiments, maximum survival times in both clay and sandy soil samples from the Nile Delta region were around 180 days at 15 °C and 100% water potential whilst at 50% water potential and 15 °C survival was around 120 days in sandy soil and around 100 days in clay soil. At 28 °C, survival was shorter in both soil types and both water potentials. The maximum survival time recorded over all soil types and conditions during *in vitro* studies was around 200 days. Similar survival times were also observed in inoculated sandy and clay soils held in vertical columns under field conditions at previously brown rot affected sites in the Nile Delta region. Recoverable populations of the bacterium on selective SMSA medium declined from 10^8 cfu per ml to zero in around 70-90 days, regardless of the depth of the soil column from 10 - 50 cm below the soil surface. Furthermore, at the end of the experiment, no detectable infections occurred in tomato seedlings up to 10 weeks after transplanting into the same soil samples under quarantine glasshouse conditions at 25 °C. Extensive surveys conducted in both Nile Delta and reclaimed desert areas were conducted in order to assess the distribution of *R. solanacearum* in Egyptian potato growing areas. Results collected over 2 years (2003-2005) showed that the pathogen was rarely isolated from soil samples except occasionally in the Nile Delta area when collected during or immediately after a potato crop affected by bacterial wilt. Testing of samples of representative weed spp. and volunteer potatoes collected from potato fields also showed that the pathogen was only detected during or soon after infected potato crops. No evidence was found for the existence of naturally infected perennial weed hosts of the bacterium which could act as long term reservoirs of the pathogen, nor was there any evidence that the pathogen was surviving outside of the host crop in potato-growing regions designated as pest free areas. It was therefore concluded that potato brown rot could be reliably controlled through use of adequate rotation periods and pathogen-free seed.

Keywords: *Ralstonia solanacearum*, potato brown rot, biovar 2, survival, soil

Introduction

R. solanacearum biovar 2 is a major constraint to Egypt's potato export trade with the European Union. The disease is endemic in parts of the traditional potato growing areas in the Nile Delta so pest-free areas have been designated in non-traditional areas for multiplication of imported seed potatoes and production of the ware crop for export to the EU. A research project has been initiated through EU Technical Assistance programme EGY1B/1999/0192 to study the epidemiology of potato brown rot under Egyptian conditions. Since the pathogen has often been regarded as soil borne, experiments and surveys were undertaken to assess the likelihood of survival in Egyptian field soils from one potato crop to the next.

Methods

Populations of *R. solanacearum* were determined by dilution plating on selective SMSA medium. Isolates were identified by immunofluorescence microscopy and specific PCR assays.

Results

Table 1. Test results from soil survey samples collected over three seasons.

Season	No. soil samples tested[1]	% containing *R. solanacearum*[2]
Pest Free Areas		
02/03	127	0
03/04	204	0
04/05	364	0
Traditional Nile Delta potato growing areas		
02/03	117	6.8
03/04	64	13.3
04/05	71	0

[1]Two sub-samples (10 g each) tested per sample of 600 g collected from a total of 180 sites per ha at a depth of 20 cm. Samples tested from 1 ha areas randomly chosen within each field or pivot.
[2]All positive findings associated with brown rot affected crops.

Table 2. Test results from weed survey samples collected over three seasons.

Season	No. weed samples tested[1]	% containing *R. solanacearum*[2]
Pest Free Areas[3]		
02/03	349	0.29
03/04	1500	0.53
04/05	793	0
Traditional Nile Delta potato growing areas[4]		
02/03	1042	9.1
03/04	191	4.2
04/05	130	0

[1]Root and collar (stem base) sections from composite samples (up to 25 plants per weed spp. per 10 fd area) were homogenised and tested.
[2]All *R. solanacearum* positive findings were associated with brown rot affected potato crops.
[3]Weeds testing positive in PFA areas were all from fields associated with brown rot interceptions in EU member states. Positive samples were identified either as potato volunteers (02/03 and 03/04) or (in 03/04 season only) as *Amaranthus cruenta, A. viridis, Convolvulus rafanistrum* and *Cyperus difformis*.
[4] Weeds testing positive in Nile Delta areas included: *Amaranthus sylvestris, Capsella bursa-pastoris, Cenchrus ciliaris, Chenopodium album, C. murale, Cichorium pamilium, Convolvulus arvensis, Cyperus difformis, C. rotundus, Datura stramonium, Euphorbia pelus, Lycopersicon esculentum* (volunteer tomato), *Malva parviflora, Portulaca oleracea, Raphanus raphanistrum, Rumex dentatus, Senecio vulgaris, Sida alba, Solanum melongena* (volunteer eggplant), *Solanum tuberosum* (volunteer potato), *Sonchus oleraceus* and *Urtica urens*.

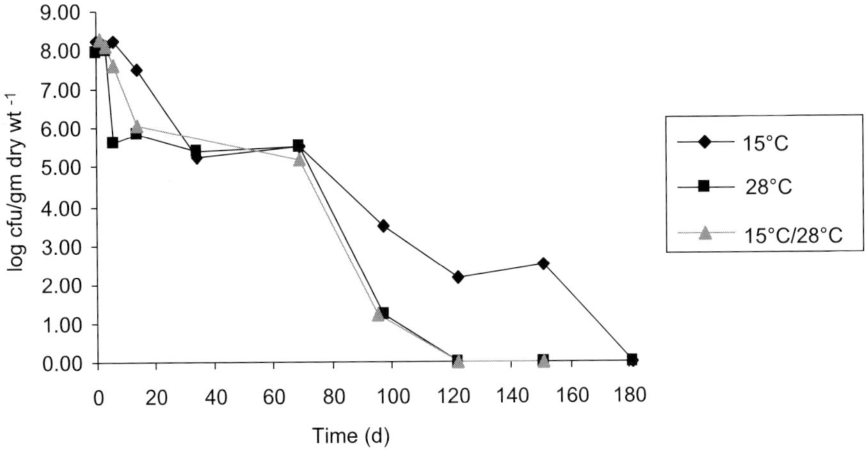

a) Sandy soil, 100% water potential

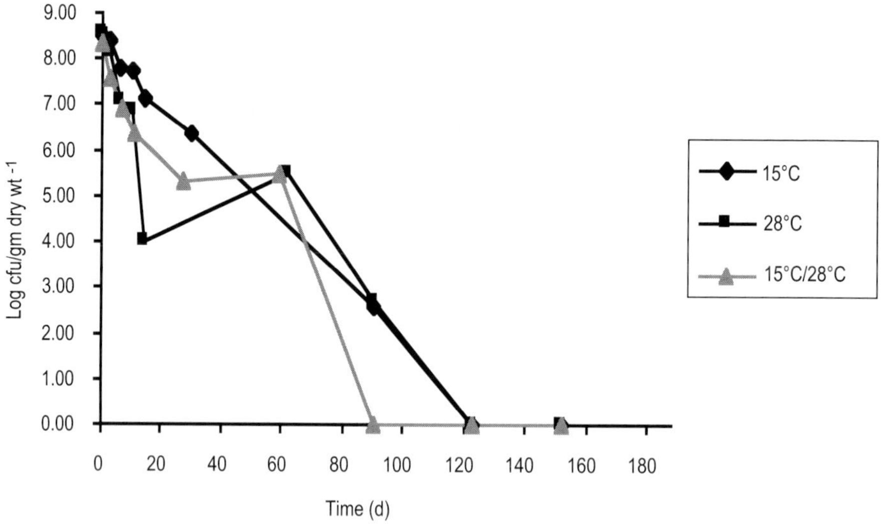

b) Sandy soil, 50% water potential

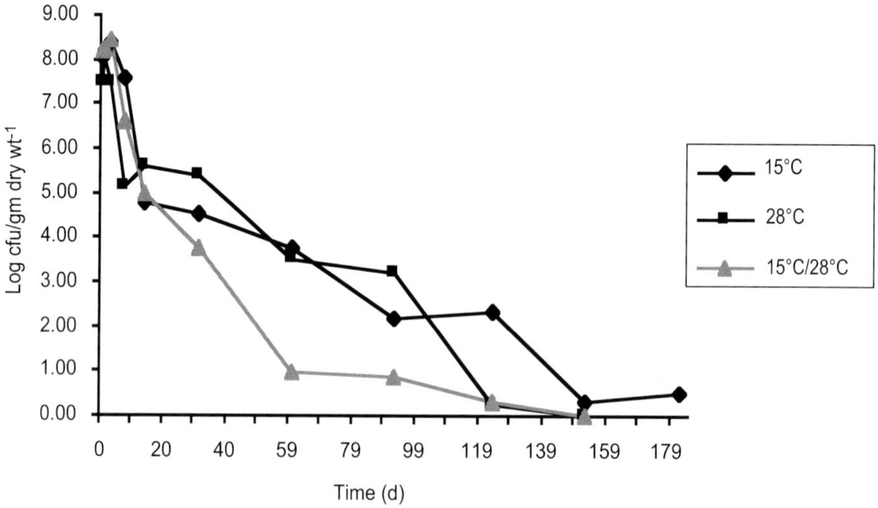

c) Clay soil, 100% water potential

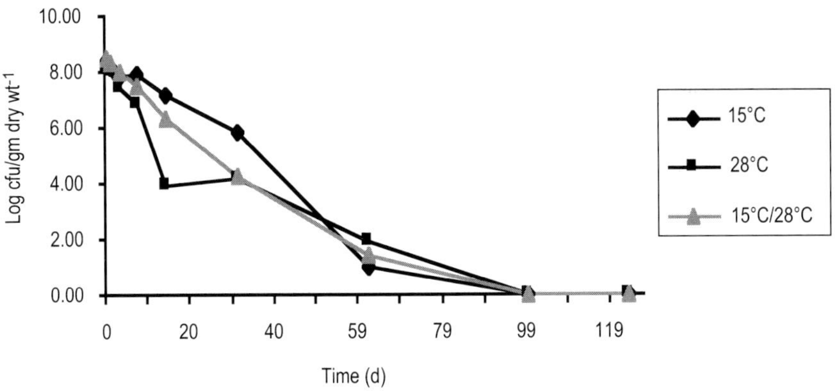

d) Clay soil, 50% water potential

Figure 1. Effect of temperature, soil type and water potential on survival of R. solanacearum *in soil samples under laboratory conditions.*

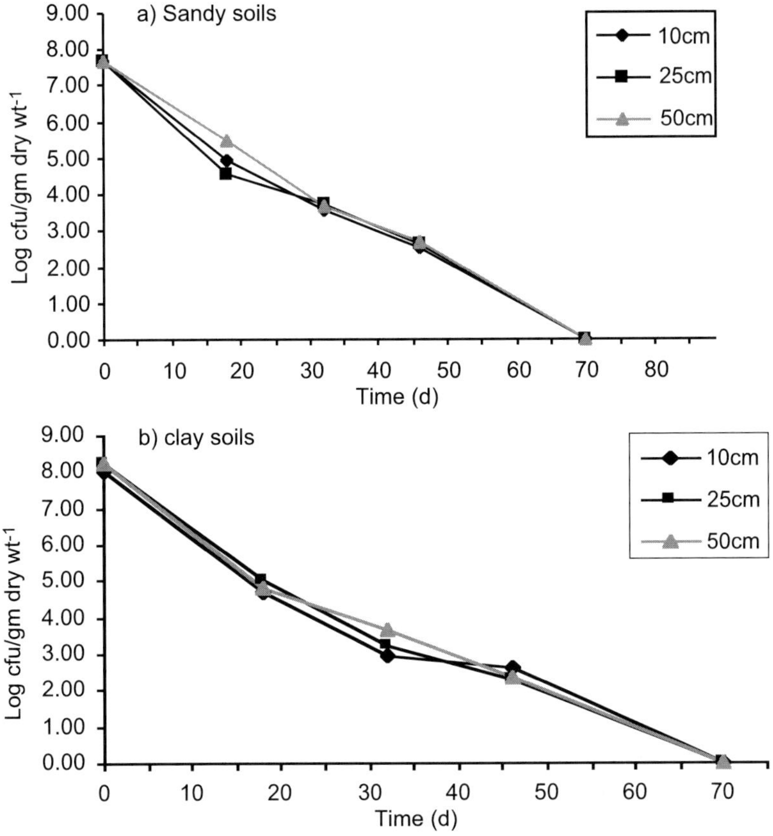

Figure 2. Survival of R. solanacearum *at different depths in columns of soil in field plots in Menufiya Governerate.*

Conclusions

No evidence was found for the existence of naturally infected perennial weed hosts of the bacterium, which could act as long term reservoirs of the pathogen.

No evidence was found that the pathogen was surviving outside of the host crop in potato-growing regions designated as pest free areas.

The bacterium survived in laboratory experiments for a maximum of 200 days whilst in field experiments survival was less than 100 days.

Potato brown rot should therefore be reliably controlled through use of adequate rotation periods and pathogen-free seed.

Acknowledgements

Research was conducted through EU Technical Assistance programme EGY1B/1999/0192 at the Potato Brown Rot Project (Phase II), Dokki, Giza, Egypt. EU partners involved are Plant Protection Service (PD), Wageningen, The Netherlands (Co-ordinator), Central Science Laboratory (CSL), York, UK, Departement Gewasbescherming (DGB), Merelbeke, Belgium and Laboratoire de Protection des Végétaux (LNPV), Angers, France.

The influence of *Solanum sisymbriifolium* on potato cysts nematode population reduction

Elzbieta Malinowska[1], Jozef Tyburski[2], Bogumil Rychcik[2] and Jadwiga Szymczak-Nowak[1]
[1]Institute of Plant Breeding and Acclimatization, Bydgoszcz, Poland
[2]University of Warmia and Mazury, Olsztyn, Poland

Introduction

Potato cyst nematodes (PCN) are a serious pest of potato. There are only a few methods to control PCN, either chemical or non-chemical ones. In Poland, growing resistant potato cultivars is the most important method of control. This method has advantages: it is cheap and safe for the environment. Unfortunately, it is not very reliable because there is a danger that the resistance will be broken (Zawislak *et al.*, 1999). Dr K. Scholte of Wageningen University began research on alternative methods of PCN control in the late 1990s. He examined 90 Solanaceae species to test their resistance against PCN and to monitor their abilities to induce hatching of larvae (Scholte, 2000b). He investigated in detail two varieties of *Solanum nigrum* and the species *Solanum sisymbriifolium*. The latter was most promising. From an agronomic point of view the following *Solanum sisymbriifolium* features are most significant: it has a high tolerance for hot weather conditions, it is resistant to night frosts, and it grows very well on acid soils (Scholte, 2000a, c). This species is resistant to all pathotypes of *Globodera rostochiensis* (golden PCN) and *Globodera pallida* (white PCN). The additional advantage of *Solanum sisymbriifolium* is its resistance to the root-knot nematodes *Meloidogyne hapla* and *M. chitwoodi* (Fassulotis, 1973; Ali *et al.*, 1992; Scholte and Vos, 2000). Moreover *Solanum sisymbriifolium* is resistant to *Verticillium dahliae* (Alconero *et al.*, 1988).

The aim of the study was to evaluate the hatching potential of *Solanum sisymbriifolium* under the climate conditions in Poland.

Methods

In 2003 a field experiment to evaluate the potential of *Solanum sisymbriifolium* to induce hatching of PCN was carried out in Bydgoszcz, Poland. The experiment was established on a light, slightly acid soil of ca. 40 points on a 100 point scale, with high nutrient content. The experiment was a randomized block design with microplots of 1 m^2 each and 4 replications. The plants were fertilized with mineral NPK in doses of respectively 90 kg N, 50 kg P and 30 kg K per ha. The following planting densities were included: 1, 5, 10, 15, 20, 25, 30, 100, and 200 plants per m^2. The experimental field was naturally infested with golden PCN. Soil samples to examine the level of PCN infestation were taken twice: in spring before planting of *Solanum sisymbriifolium* (May 20) and in autumn after its harvest (September 28). Each soil sample was air-dried and cysts were separated from the soil. The numbers of unhatched juveniles remaining in the cysts were estimated by crashing the cysts and making a dilution count.

Results and discussion

The results of microplots experiment showed good effectiveness of *Solanum sisymbriifolium* as a trap crop in controlling PCN. PCN population reduction was not as high as in the case of chemical method or in cropping of PCN resistant potato cultivars. In our study the average PCN population reduction due to one year *Solanum sisymbriifolium* cropping was 72% (Table 1). According to Scholte and Vos (2000) *Solanum sisymbriifolium* PCN control effectiveness depends on the initial PCN infestation level and is higher in moderately infested soil than in heavily infested soil (respectively 77 and 52%). We found that *Solanum sisymbriifolium* PCN control effectiveness depended also on plant density of the crop. The best results were obtained when 25-30 *Solanum sisymbriifolium* plants were grown per m^2. The lower and higher planting densities resulted in a slight decrease of effectiveness of *Solanum sisymbriifolium* as a trap crop.

Table 1. Planting density of Solanum sisymbriifolium and effectiveness of PCN control.

Solanum sisymbriifolium density, no./m^2	Number of "alive" cysts, in 100 g of air-dried soil		Number of juveniles in 100 g of air-dried soil		PCN population reduction,%
	sowing	harvest	sowing	harvest	
1	230	110	14760	5758	61.0
5	218	120	12320	3890	68.4
10	282	120	16214	4974	69.3
15	248	112	17110	5220	69.5
20	252	120	17036	4700	72.4
25	362	114	21662	4688	78.4
30	290	94	16155	3566	77.9
100	256	142	19000	4812	74.7
200	282	136	17170	4378	74.5
Average	269	119	16821	4665	72.3

Conclusions

Solanum sisymbriifolium was found an effective trap crop in PCN control in northern Poland. An average PCN juveniles population reduction of 72% was achieved. It showed the best effectiveness when grown in a planting density of 25-30 plants per m^2.

References

Alconero, R., R.W. Robinson, B. Dicklow and J. Shail (1988). Verticilium wilt resistance in eggplant, related Solanum species, and interspecific hybrids. Horticultural Science, 23 (2), 388-390.

All, M., N. Matsuzoe, H. Okubo and K. Fujeda (1992). Resistance of non-tuberous Solanum to root-knot nematode. Journal of Japan Society of Horticultural Sciences, 60 (4), 921-926.

Fassuliotis, G. (1973). Suseptibility of eggplant, Solanum melogena, to root-knot nematode, Meloidogyne incognita. Plant Disease Reporter, 57 (7), 606-608.

Scholte, K. (2000a). Effect of potato used as a trap crop on potato nematodes and other soil pathogens on the growth of a subsequent main potato crop. Annals of Applied Biology, 136, 229-238.
Scholte, K. (2000b). Screening of non-tuber bearing Solanaceae for resistance to and induction of juvenile hatch of potato cyst nematodes and their potential for trap cropping. Annals of Applied Biology 136: 239-246.
Scholte, K. (2000c). Growth and development of plants with potential use as trap crops for potato cyst nematodes and their effects on the number of juveniles in cysts. Annals of Applied Biology, 137, 031-042.
Scholte, K. and J. Vos (2000). Effect of potential trap crops and planting date on soil infestation with potato cyst nematodes and root-knot nematodes. Annals of Applied Biology, 137, 153-164.
Zawislak, K., E. Adamiak and B. Rychcik (1999). [Possibilities of limiting of negative biotype factors in monocultural plants cropping]. Relation from the realization of the research project KBN 5 P06B 004 10/95. Manuscript 105.

Late blight

Late blight: the perspective from the pathogen

Francine Govers
Laboratory of Phytopathology, Wageningen University, Binnenhaven 5, NL-6709 PD Wageningen, The Netherlands

Abstract

Late blight is a destructive disease found in nearly all areas of the world where potatoes are grown. Typical symptoms are the water-soaked lesions on the leaves that rapidly enlarge resulting in dark, blighted areas and, eventually, in total destruction of the plant. Without chemical control and when cool and wet weather conditions prevail, late blight kills all plants in a field within 10 to 14 days. The causal agent is *Phytophthora infestans,* a spore producing, filamentous eukaryote classified as an oomycete. Oomycetes resemble fungi but have an independent evolutionary history. In the 1980s and 1990s a worldwide displacement of the pathogen population caused a re-emergence of late blight.

In recent years various tools and resources for molecular and genetic analyses of *P. infestans* have been developed. Gene silencing has facilitated studies of genes involved in growth, development, and pathogenicity, and high throughput sequencing has provided a large EST database that represents the majority of the genes present in this organism. The challenge is to exploit these tools and resources to elucidate cellular and metabolic processes that are vital and unique for *Phytophthora infestans,* and to uncover new potential targets for disease control.

Keywords: *Phytophthora infestans*, oomycetes, late blight, pathogenicity, genomics

Introduction

Late blight is the number one disease in potato. Costs of losses and crop protection are estimated at US$3.25 billion per annum worldwide. The first reports of late blight epidemics date from the 1840s when the disease devastated the foliage and tubers of potato in nearly all of Europe and the northeastern United States and set the stage for the disaster recorded in history as the Irish potato famine. Ireland, at that time dependent on potato as the chief staple, lost nearly half of its population; over a million Irish starved to death and many more emigrated. This epidemic occurred prior to the development of the germ theory and as a result it was not until 1867 that Anton de Bary was the first to demonstrate conclusively that a water mold, which he named *Phytophthora infestans*, was the causal agent of the disease; hence the official name *Phytophthora infestans* (Mont.) de Bary (Large, 1940).

More than 100 years later, it became clear that *Phytophthora* is not a true fungus but instead belongs to a unique eukaryotic lineage, called oomycetes, that is evolutionarily unrelated to fungi. Features such as distinct cell wall compositions and metabolic pathways (Erwin and Ribeiro, 1996; Latijnhouwers *et al.*, 2003) partly explain the differences in sensitivity between fungi and oomycetes to certain classes of fungicides and justify the introduction of the term 'oomicides' as a generic name for biocides effective against oomycetes (Govers, 2001).

Phytophthora (Greek for 'plant destroyer') is undoubtedly the best-studied oomycete genus since it encompasses species causing severe diseases on economically important crops such as soybean, cocoa and potato and on valuable tree species such as Californian oak, Australian jarrah trees and European alder trees. Over 65 *Phytophthora* species have been identified and regularly new species are discovered. The host range of *P. infestans* is limited to the Solanaceae with potato (*Solanum tuberosum*) and tomato (*Lycopersicon esculentum*) as the most prominent hosts (Erwin and Ribeiro, 1996). However, many *Phytophthora* species have a broad host range and thousands of plant species can be infected by these damaging pathogens.

The disease cycle

At the start of the growing season spores may be produced from diseased potato sprouts arising from infected tubers. Well-known sources from where inoculum can easily develop and spread are cull piles or dumps, volunteer plants, and seed tubers. Also the sexual spores, called oospores, can act as an inoculum source but this is limited to areas where the two mating types that are required for completion of the sexual life cycle are present. Oospores are dormant structures that can survive in the soil for many years in the absence of the host. When oospore germination is triggered, the germ tube develops into a sporangium that can easily enter the infection cycle (Figure 1).

The disease spreads optimally under cool, moist conditions. The asexual spores called sporangia are dispersed by wind or in water drops. Infection generally occurs in the foliage and occasionally on stems. When sporangia land on a host they either germinate directly or undergo

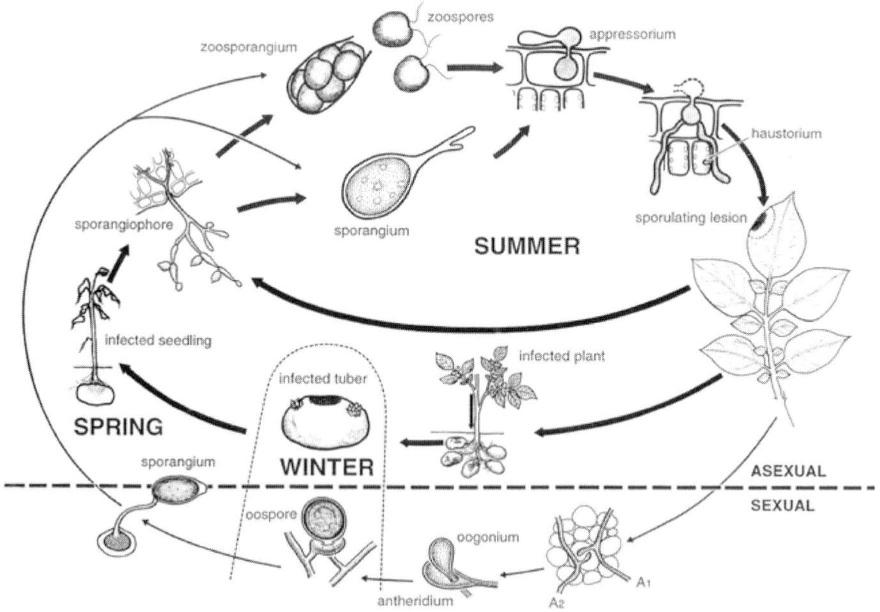

Figure 1. The late blight disease cycle.

cytoplasmic cleavage to form 7 to 8 swimming zoospores. Zoospores are attracted to the host where they halt, retract their flagella, and secrete materials to form a cell wall. This process is referred to as encystment. A sporangium or cyst germinates and the germ tube tip differentiates into an appressorium, a spherical or elliptical structure from which a penetration peg emerges to breach the plant cuticle and cell wall. After having passed the epidermis, the hyphae grow mainly intercellularly and form haustorium-like feeding structures that protrude into mesophyll cells. This biotrophic phase is an essential step in the disease cycle. In later stages, the pathogen adopts a necrotrophic lifestyle: it kills plant cells and feeds on dead tissue. Once in this phase, sporangia are formed on branched hyphae called sporangiophores that emerge from the stomata and cover the lesion surface. Leaf symptoms consist of water-soaked, dark lesions that expand rapidly, resulting in total destruction of the plant in a few days. Individual lesions can produce 100,000 to 300,000 sporangia per day. In favourable weather the period from infection to sporangia formation may be as short as four days and, therefore, many generations of asexual spores can be formed and dispersed in one growing season, leading to late blight epidemics.

Tubers may be infected whenever sporangia and tubers come into contact. Infections can occur on both developing and mature tubers. Sporangia are washed from lesions on stems and foliage to the soil and then through the soil to the tubers. They can survive for days or weeks in soil and therefore tubers can still become infected for a period of time after lesions on the foliage are no longer producing sporangia. At first infected tubers show dark blotches. The invasion proceeds into the outer layers of the tuber tissue and the affected areas become firm and dry and somewhat sunken (dry rot). The rot continues to develop after the tubers are harvested. Infected tubers are especially susceptible to secondary infections by fungi and bacteria, resulting in soft rot. During storage soft rot can spread to previously healthy tubers.

Sexual reproduction only occurs when a leaf or a stem is invaded by two *P. infestans* strains, each with a different mating type. *P. infestans* is an outcrossing heterothallic species with two known mating types, A1 and A2. When mycelium of an opposite mating type grows in the vicinity, specific hormones are produced that trigger formation of oogonia (♀) and antheridia (♂) on the opposite mating type. Subsequently, the nuclei in the gametangia undergo meiosis and via a fertilization tube a male haploid nucleus is deposited in the oogonium, which then develops into a thick walled oospore. The nuclei fuse during maturation of the oospore. When blighted potato plants decompose, numerous oospores are released and these are able to overwinter in the soil.

Population displacement and re-emergence of late blight

Central Mexico is considered to be the origin of *P. infestans* and the region where *Solanum* and *P. infestans* co-evolved. Prior to 1980 the A1 mating type was distributed all over the world whereas the A2 mating type was confined to Mexico. Mexican populations of the pathogen consist of many different genotypes and with a high variation in race structure. In contrast, one single genotype, designated US-1, populated the rest of the world and all isolates seemed to belong to this clonal lineage. This situation changed in the 1980s when A2 mating type isolates were discovered in Europe. This was due to new migrations presumably initiated in 1976 by a large shipment of potatoes from Mexico to Europe. In the 1980s these new and more diverse genotypes gradually migrated eastwards to Eurasia and further to the Far East and Japan resulting in a significant increase in the genetic variation and in coexistence of the two mating types within populations. Sexual reproduction and hence new recombinant types may account for the observed increase in genetic variation and, indeed, occasionally oospores are detected in the field. Also in the US and Canada new genotypes that migrated from Mexico replaced the US-1 genotype. Only a few clonal lineages, however, settled successfully and the genetic variation is much less diverse than in Europe. Overall, the new genotypes of *P. infestans* seem more aggressive than old US-1 isolates and as a result, the epidemics have become more severe (Smart and Fry, 2001).

Late blight control

Integrated management practices are necessary to control late blight. Sanitary measures such as eliminating primary sources of inoculum, and cultural practices including early planting, can lower the chance of infections early in the growing season. Also continuous scouting for disease foci in potato fields is critical. In the industrialized countries regular spaying with agrochemicals is common practice and this keeps losses limited to about 15% (Garelik, 2002). The basic strategy is to prevent the establishment of any infection within the crop. For the timing of applications, decision support systems (DSS) are used, giving recommendations based on expected crop development, the epidemiology of *P. infestans*, climatic conditions, infection pressure, and the costs and efficacy of crop protection measures. A substantial contribution to late blight control would be the availability of resistant cultivars that also meet production and processing requirements. In the first half of the twentieth century, late blight resistance derived from *Solanum demissum* and conferring resistance based on the gene-for-gene interaction, were bred into *S. tuberosum*. Unfortunately, each of the eleven genes succumbed to attack by *P. infestans* within a few years, demonstrating the capacity of this pathogen to quickly circumvent recognition by its host. Currently, a wide variety of wild *Solanum* species are being explored for novel resistance genes. Deploying resistance genes using genetic modification may bring us closer to the resistant potato cultivar with the desired characteristics (Allefs *et al.*, 2005).

A molecular toolbox for *Phytophthora*

In recent years knowledge on the biology and pathology of oomycetes has gradually increased but it is still just a tip of the iceberg (Judelson and Blanco, 2005; Latijnhouwers *et al.*, 2003). Pinpointing effective and specific targets for control in *P. infestans* requires a major input in unravelling the genetic and biochemical networks that underlie the synthesis of pathogenicity factors and other effectors. The ability to study these networks depends on the technology available to analyse regulation and function of genes encoding these factors. For a long time, molecular genetic studies on oomycetes were hampered by the lack of molecular tools for in-depth studies. The last decade several research groups have focussed on creating a molecular toolbox for *P. infestans* and, as a result, *P. infestans* has become a model oomycete species that can now be exploited to study aspects of pathogenesis of oomycete plant pathogens (Kamoun, 2003). A major accomplishment was the development of a DNA transformation system in the early 1990s (Judelson *et al.*, 1991). A new milestone was reached when we demonstrated that targeted gene inactivation in *P. infestans* is feasible (van West *et al.*, 1999). We introducted extra copies of the elicitin gene *inf1* into *P. infestans* thereby inducing homology dependent gene silencing. This resulted in stable INF1 deficient mutants that appeared to have an extended host range. The mutants remained pathogenic on potato and tomato, but were also able to induce disease lesions on *Nicotiana benthamiana* hence demonstrating that INF1 functions as a species specific avirulence factor (Kamoun *et al.*, 1998b). Thus far, the function of a handful of *P. infestans* genes has been deciphered by analysing the phenotypes of mutants in which the target gene has been silenced (Latijnhouwers and Govers, 2003; Latijnhouwers *et al.*, 2004; Fong and Judelson 2003; Blanco and Judelson, 2005). More recently a transient RNAi silencing method for *P. infestans* was described which could be useful for a high throughput screening for genes involved in, for example, sporulation, germination and colony morphology (Whisson *et al.*, 2005).

Other useful tools for studying expression of *P. infestans* genes *in vivo* are reporter genes such as GUS and GFP. A gene promoter-GUS fusion introduced into *P. infestans* demonstrated that *ipiO*, an *in planta* induced gene of *P. infestans*, is highly expressed in the hyphal tips at the edge of the expanding lesion where the pathogen is invading healthy plant cells (van West *et al.*, 1998). This suggests that IPI-O is localised at the interface between the invading hyphae and the plant cells, and could play a role in pathogenicity. The characterisation of a plant plasma membrane protein that binds IPI-O with high affinity supports this hypothesis (Senchou *et al.*, 2004). *Phytophthora* transformants carrying reporter genes regulated by a constitutive promoter can be used to monitor host invasion and to quantify resistance levels. Both, GUS and GFP reporter genes have been used for that purpose (Kamoun *et al.*, 1998a; Si-Ammour *et al.*, 2003; van West *et al.*, 1999).

An alternative approach for functional analysis of *Phytophthora* effector genes is *in planta* expression. The gene of interest is cloned in *Agrobacterium tumefaciens* T-DNA vectors or in binary PVX vectors and expressed in plant cells. Based on differential defence responses in different potato clones or cultivars, or in different plant species the activity of the effector can be monitored. This approach was used, for example, to identify *Avr3a,* a cultivar specific avirulence gene that is recognized by the resistance gene *R3a* (Amstrong *et al.*, 2005). Genetically, the *Solanum demissum* derived *R3a* gene and *Avr3a* interact according to the gene-for-gene model.

Phytophthora genomics

For unravelling the genetic and biochemical networks that underlie the synthesis of pathogenicity factors and effectors the genes involved have to be available. The number of *P. infestans* genes that has been cloned by traditional methods is limited. Strategies exploited are differential screening to select *P. infestans* genes expressed *in planta* (*ipi* genes) (Pieterse *et al.*, 1993), heterologous probes or PCR approaches in the case of conserved sequences, reverse genetics (from protein to gene), and positional cloning. The latter is also called map-based cloning and is traditionally used for genes which, based on segregation of the phenotype in a mapping populations can be positioned on a molecular genetic linkage map. We used this strategy to clone avirulence genes, the genes responsible for the production of race-specific effectors, and that interact in a gene-for-gene fashion with major resistance genes. Loss of production of these effectors correlates with the appearance of virulent races. We constructed molecular genetic linkage maps of *P. infestans* based on AFLP markers and high-density linkage maps of two chromosomal regions that contain avirulence genes (van der Lee *et al.*, 2001, 2004). AFLP markers were then used to select clones from a *P. infestans* BAC library and to physically map BAC contigs containing avirulence genes (Whisson *et al.*, 2001). In parallel, we performed cDNA-AFLP analysis on the mapping population resulting in a number of Transcript Derived Fragments (TDFs) that represent genes differentially expressed in strains with different AVR phenotypes. By marker landing on the BACs and genome walking using the cloned TDFs we could identify two candidate avirulence genes. The functional analysis is in progress.

To further increase the arsenal of candidate pathogenicity and avirulence genes high throughput cDNA sequencing of *P. infestans* was performed (Kamoun *et al.*, 1999; Randall *et al.*, 2005). Currently a unique database of over 75,000 Expressed Sequence Tag's (ESTs) is available (www.pfgd.org and deposited in GenBank). The ESTs are derived from 20 different cDNA libaries and represent over 18,000 unique sequences expressed in different developmental stages and growth conditions. In the coming year the *P. infestans* genome, which is relatively large compared to genomes of other oomycetes (i.e. 240 Mb), will be sequenced at the Broad Institute (http://www.broad.mit.edu/) (Zody *et al.*, 2005). In 2004 the genome sequences of two *Phytophthora* species were released (http://www.jgi.doe.gov/). *P. sojae,* a soybean root and stem rot pathogen, and *P. ramorum,* the causal agent of Sudden Oak Death, have smaller genome sizes (95 and 65 Mb, respectively) but nevertheless the two assembled draft genome sequences are very useful for comparative analysis and instrumental for genomic analysis of *P. infestans.*

Pathogenicity factors as targets for control

One of our approaches to shed light on the pathology of *P. infestans* is by studying the role of conserved signalling pathways in development and virulence. One such pathway is the heterotrimeric G-protein pathway via which external signals perceived by membrane bound G-protein-coupled receptors, are transferred to the nucleus thus resulting in up- or down-regulation of a wide range of genes. This enables cells to respond to changes in the environment. G-protein complexes, composed of an α, β, and γ subunit, are key players in cellular signalling in virtually all eukaryotic taxa. *P infestans* possesses one gene encoding a G-protein α subunit (*Pigpa1*) and one encoding a G-protein β subunit (*Pigpb1*) (Laxalt *et al.*, 2002) and the highest expression of both genes is found in sporangia, the air-borne asexual spores that are important for dissemination and infection. To study the involvement of heterotrimeric G-proteins in the disease cycle of *P. infestans* we generated PiGPA1- and PiGPB1-knock-down mutants by homology-dependent gene-silencing. PiGPB1-knock-down mutants formed an abnormally dense mat of aerial mycelium and produced very little or no sporangia (Latijnhouwers and Govers, 2003). PiGPA1-knock-down mutants grew normally and produced sporangia but the zoosporogenesis and zoospore release were much less efficient. Moreover, the zoospores had lost the ability to aggregate and the virulence of the PiGPA1-knock-down mutants on potato leaves was strongly reduced (Latijnhouwers *et al.*, 2004). Because of their obvious hypovirulent phenotype we exploited the PiGPA1-knock-down mutants to identify downstream targets of the G-protein signalling pathway. We used expression profiling to identify transcripts that are differentially expressed in the mutants compared to the wild-type strain.

cDNA-AFLP analysis (Dong *et al.*, 2004) and microarray analysis (unpublished) resulted in the identification of a large number of transcripts that may be derived from genes that are positively or negatively regulated by the G-protein signaling pathway. Many of these have no homology to known genes and a subset is currently analysed further. Assigning a function to those genes may reveal novel pathogenicity determinants that are specific for *Phytophthora* or oomycetes in general, and may lead to novel, potential targets for control.

Concluding remarks

By molecular-genetic dissection of *P. infestans* and the subsequent unravelling of the genetic and biochemical networks that underlie the synthesis of pathogenicity factors and effectors we will get to know the secrets of this notorious pathogen. We can now exploit structural genomics to quickly isolate and identify genes, and gene silencing to determine their function. Within a few years we hope to know what the vital links are in the lifecycle of *P. infestans* and this knowledge will be very useful, maybe even essential, in designing molecular breeding strategies and in developing novel control agents for late blight.

Every strategy to develop a durable solution for the late blight problem, either by crop protection or via resistant cultivars, should consider both the role of the pathogen and the host. In recent years large genomics resources have been developed for potato and *P. infestans* and the challenge for the near future is to explore and exploit these resources to unravel infection strategies of the pathogen and resistance mechanisms that can defeat the pathogen. This knowledge would help establish a more rational design of oomicides and more strategic breeding programmes. For the time being, however, growers will have to cope with this devastating pathogen.

References

Allefs, J.J.H.M., M.W.M. Muskens and E.A.G. van der Vossen (2005). Breeding for foliage late blight resistance in the genomics era. In: A.J. Haverkort and P.C. Struik (eds.) Potato in progress: science meets practice, Wageningen Academic Publishers, The Netherlands, 368 p.

Armstrong, M.R., S.C. Whisson, L. Pritchard, J.I.B. Bos, E. Venter, A.O. Avrova, A.P. Rehmany, U. Bohme, K. Brooks, I. Cherevach, N. Hamlin, B. White, A. Frasers, A. Lord, M.A. Quail, C. Churcher, N. Hall, M. Berriman, S. Huang, S. Kamoun, J.L. Beynon and P.R.J. Birch (2005). An ancestral oomycete locus contains late blight avirulence gene Avr3a, encoding a protein that is recognized in the host cytoplasm. *Proceedings of the National Academy of Sciences of the United States of America,* 102, 7766-7771.

Blanco, F.A. and H.S. Judelson (2005). A bZIP transcription factor from *Phytophthora* interacts with a protein kinase and is required for zoospore motility and plant infection. *Molecular Microbiology,* 56, 638-648.

Dong, W.B., M. Latijnhouwers, R.H.Y. Jiang, H.J.G. Meijer and F. Govers (2004). Downstream targets of the *Phytophthora infestans* G alpha subunit PiGPA1 revealed by cDNA-AFLP. *Molecular Plant Pathology,* 5, 483-494.

Erwin, D.C. and O.K. Ribeiro (1996). *Phytophthora* diseases worldwide. APS Press, St. Paul, Minnesota, 562 pp.

Fong, A.M.V.A. and H.S. Judelson (2003). Cell cycle regulator Cdc14 is expressed during sporulation but not hyphal growth in the fungus-like oomycete *Phytophthora infestans. Molecular Microbiology,* 50, 487-494.

Garelik, G. (2002). Agriculture - Taking the bite out of potato blight. *Science,* 298, 1702-1704.

Govers, F. (2001). Misclassification of pest as 'fungus' puts vital research on wrong track. *Nature,* 411 633-633.

Judelson, H.S. and F.A. Blanco (2005). The spores of *Phytophthora*: Weapons of the plant destroyer. *Nature Reviews Microbiology,* 3, 47-58.

Judelson, H.S., B.M. Tyler and R.W. Michelmore (1991). Transformation of the oomycete pathogen, *Phytophthora infestans. Molecular Plant-Microbe Interactions,* 4, 602-607.

Kamoun, S. (2003). Molecular genetics of pathogenic Oomycetes. *Eukaryotic Cell,* 2, 191-199.

Kamoun, S., P. van West and F. Govers (1998a). Quantification of late blight resistance of potato using transgenic *Phytophthora infestans* expressing beta-glucuronidase. *European Journal of Plant Pathology,* 104, 521-525.

Kamoun, S., P. van West, V.G.A.A. Vleeshouwers, K.E. de Groot and F. Govers (1998b). Resistance of Nicotiana benthamiana to *Phytophthora infestans* is mediated by the recognition of the elicitor protein INF1. *Plant Cell,* 10, 1413-1425.

Kamoun, S., P. Hraber, B. Sobral, D. Nuss and F. Govers (1999). Initial assessment of gene diversity for the oomycete pathogen *Phytophthora infestans* based on expressed sequences. *Fungal Genetics and Biology,* 28, 94-106.

Large, E.C. (1940). The advance of the fungi. First published in 1940 by Jonathan Cape Ltd., London; in 2003 re-issued by APS Press, St. Paul, Minnesota, 510 pp.

Latijnhouwers, M. and F. Govers (2003). A *Phytophthora infestans* G-protein beta subunit is involved in sporangium formation. *Eukaryotic Cell,* 2, 971-977.

Latijnhouwers, M., P.J.G.M. de Wit and F. Govers (2003). Oomycetes and fungi: similar weaponry to attack plants. *Trends in Microbiology,* 11, 462-469.

Latijnhouwers, M., W. Ligterink, V.G.A.A. Vleeshouwers, P. van West and F. Govers (2004). A G alpha subunit controls zoospore motility and virulence in the potato late blight pathogen *Phytophthora infestans. Molecular Microbiology,* 51, 925-936.

Laxalt, A.M., M. Latijnhouwers, M. van Hulten and F. Govers (2002). Differential expression of G protein alpha and beta subunit genes during development of *Phytophthora infestans. Fungal Genetics and Biology,* 36, 137-146.

Pieterse, C.M.J., M.B.R. Riach, T. Bleker, G.C.M. van den Berg-Velthuis and F. Govers (1993). Isolation of putative pathogenicity genes of the potato late blight fungus *Phytophthora infestans* by differential hybridization of a genomic library. *Physiological and Molecular Plant Pathology,* 43, 69-79.

Randall, T.A., R.A. Dwyer, E. Huitema, K. Beyer, C. Cvitanich, H. Kelkar, A.M.V.A. Fong, K. Gates, S. Roberts, E. Yatzkan, T. Gaffney, M. Law, A. Testa, T. Torto-Alalibo, M. Zhang, L. Zheng, E. Mueller, J. Windass, A. Binder, P.R.J. Birch, U. Gisi, F. Govers, N.A. Gow, F. Mauch, P. van West, M.E. Waugh, J. Yu, T. Boller, S. Kamoun, S.T. Lam and H.S. Judelson (2005). Large-scale gene discovery in the oomycete *Phytophthora infestans* reveals likely components of phytopathogenicity shared with true fungi. *Molecular Plant-Microbe Interactions,* 18, 229-243.

Senchou, V., R. Weide, A. Carrasco, H. Bouyssou, R. Pont-Lezica, F. Govers and H. Canut (2004). High affinity recognition of a *Phytophthora* protein by Arabidopsis via an RGD motif. *Cellular and Molecular Life Sciences* 61: 502-509.

Si-Ammour, A., B. Mauch-Mani and F. Mauch (2003). Quantification of induced resistance against *Phytophthora* species expressing GFP as a vital marker: beta-aminobutyric acid but not BTH protects potato and Arabidopsis from infection. *Molecular Plant Pathology,* 4, 237-248.

Smart, C.D. and W.E. Fry (2001). Invasions by the late blight pathogen: renewed sex and enhanced fitness. *Biological Invasions,* 3, 235-243.

van der Lee, T., A. Robold, A. Testa, J.W. van 't Klooster and F. Govers (2001). Mapping of avirulence genes in *Phytophthora infestans* with amplified fragment length polymorphism markers selected by bulked segregant analysis. *Genetics,* 157, 949-956.

van der Lee, T., A. Testa, A. Robold, J. van 't Klooster and F. Govers (2004). High-density genetic linkage maps of *Phytophthora infestans* reveal trisomic progeny and chromosomal rearrangements. *Genetics,* 167, 1643-1661.

van West, P., S. Kamoun, J.W. van 't Klooster and F. Govers (1999). Internuclear gene silencing in *Phytophthora infestans. Molecular Cell,* 3, 339-348.

van West, P., A.J. de Jong, H.S. Judelson, A.M.C. Emons and F. Govers (1998). The ipiO gene of *Phytophthora infestans* is highly expressed in invading hyphae during infection. *Fungal Genetics and Biology*, 23, 126-138.

Van West, P., B, Reid, T.A. Campbell, R.W. Sandrock, W.E. S. Kamoun, S. and N.A.R. Gow (1999). Green fluorescent protein (GFP) as a reporter gene for the plant pathogenic oomycete *Phytophthora palmivora*. *FEMS Microbiology Letters*, 178, 71-80.

Whisson, S.C., A.O. Avrova, P. Van West and J.T. Jones (2005). A method for double-stranded RNA-mediated transient gene silencing in *Phytophthora infestans*. *Molecular Plant Pathology*, 6, 153-163.

Whisson, S.C., T. van der Lee, G.J. Bryan, R. Waugh, F. Govers and P.R.J. Birch (2001). Physical mapping across an avirulence locus of *Phytophthora infestans* using a highly representative, large-insert bacterial artificial chromosome library. *Molecular Genetics and Genomics*, 266, 289-295.

Zody M.C., K. O'Neill, B. Handsaker, E. Karlsson, F. Govers, P. van de Vondervoort, R. Weide, S. Whisson, P. Birch, L. Ma, B. Birren, J. Ristaino, W. Fry, H. Judelson, S. Kamoun and C. Nusbaum Sequencing the *Phytophthora infestans* Genome: Preliminary Studies. Abstract book 'Phytophthora Molecular Genetics Network Workshop'. Asilomar March 13-15 (2005). Accessible at http://pmgn.vbi.vt.edu/

Curriculum vitae – Francine Govers

Francine Govers is Associate Professor in Phytopathology at the Plant Sciences Group of Wageningen University, the Netherlands, and staff member at the Graduate School Experimental Plant Sciences. She received an M.Sc. degree in Plant Pathology from the Wageningen Agricultural University in 1982 and she then joined the research group of Dr. Ton Bisseling at the same university to study nodulin gene expression in developing pea root nodules. In 1987 she received a Ph.D. degree in Plant Molecular Biology. In 1990 she moved to the Laboratory of Phytopathology where she is in charge of a research group that focusses on the biology and pathology of the potato late blight pathogen *Phytophthora infestans*. The main interest of the group is (i) on effectors from *Phytophthora* that act as virulence and avirulence factors in the interaction with host and non-host plants, and (ii) on signal transduction pathways underlying pathogenicity. Results from these areas of research are directed towards designing rational control strategies for late blight and other diseases caused by oomycete pathogens. She teaches introductory plant pathology and advanced plant-microbe interaction courses to undergraduates and more specialised, thematic courses to graduate students. Since 2001 Dr. Govers is senior editor of the journal Molecular Plant-Microbe Interactions. As member of the Steering Committee of the Global Initatitive on Late Blight (GILB) (since 2001) she was co-organizer of the GILB conference in Hamburg in 2002. She participates in various national and international networks dealing with late blight and *Phytophthora*, in particular genomics of *Phytophthora* and other oomycetes, and she is member of the steering committee of the NSF *Phytophthora* Molecular Genetics Research Network.

Company profile – Wageningen University

Wageningen University was founded in 1918. In recent decades it has evolved into one of the world's leading education and research centres in the plant, animal, environmental, agrotechnological, food and social sciences.

The way food is grown has undergone rapid change in the last few decades. All over the world, agriculture is becoming more and more market driven, while pressures on the environment are increasing.
For many years, Wageningen scientists have been aware of the ultimate importance of maintaining a habitable planet earth, as well as ensuring reliable supplies of safe, high quality food, while maintaining the biodiversity of natural habitats and conserving natural resources. Wageningen University has therefore formulated its mission accordingly: "Wageningen University wishes to develop and disseminate the scientific knowledge needed to sustainably supply society's demand for sufficient, healthy food and a good environment for humans, animals and plants".

Recognising the importance of disseminating knowledge and expertise, we offer students from around the world a wide range of educational programmes. Wageningen University offers students from abroad a special international Master of Science Programme as well as a post-graduate programme.

Breeding for foliage late blight resistance in the genomics era

J.J.H.M. Allefs[1], M.W.M. Muskens[1] and E.A.G. van der Vossen[2]
[1]Agrico Research, P.O. Box 40, 8300 AA Emmeloord, the Netherlands
[2]Plant Research International, P.O. Box 16, 6700 AA Wageningen, the Netherlands

Abstract

Based on an analysis of 8333 potato related papers published between 1995 and 2004 in scientific journals, late blight can be considered to be the most important disease in potato production. The cost of fungicides required for an optimal spraying regime on a susceptible variety is estimated to be 1.7 billion euro per year for the top 38 potato producing countries in the world. Under conditions of organic production in the Netherlands, with no application of protectants at all, the average yield loss for the most important varieties was 45 percent in the period from 2002 to 2004. This illustrates the failure of a century of breeders' attempts to significantly add to late blight disease management.

We propose a new paradigm to replace the classic division of resistance types found in potato germplasm. It is argued that data in the scientific literature comprising blight scores or AUDPC values for potato clones obtained in field trials, can be explained by a basic maturity related level of resistance on the one hand and the mode of resistance (r)-genes on the other. The R-genes function in a compatible or incompatible mode, depending on the virulence spectrum of the *Phytophthora infestans* strain(s) present.

Data are summarised that show that *P. infestans* isolates carry unnecessary virulence factors to the R-genes present in the differential set containing R1 to R11 and that the factors for R5, R6, R8 and R9 are underrepresented. Furthermore, the virulence spectra in *P. infestans* isolates show spatio-temporal variation similar to what has been reported for isolate specific genotypic markers. It is suggested that this spatio-temporal variation opens a possibility to develop a decision support system that employs "strong" R-genes in combination with real time monitoring for matching virulences in *Phytophthora*.

Keywords: *Phytophthora infestans,* R-genes, maturity, *Solanum* spp., avirulence

Introduction

It may be surmised that scientific interest in crop specific topics is somehow related to the importance of these topics as felt in agricultural practice. A database of over 8000 records extracted from Current Contents for Agriculture, Biology & Environmental Sciences (Thomsom ISI, Philadelphia, USA), covering a 10-year period from 1995 to 2004, all containing the keywords potato or *Solanum tuberosum* in any of the searchable fields, was screened for appearance of terms used for the major potato pests and diseases in the paper titles (Table 1). Roughly 1800 potato disease related records were retrieved. Among these, fungal diseases

Table 1. Number of scientific papers retrieved from a potato literature database containing papers from 1995 to 2004 with appearance of disease names or causal organisms in their title.

Fungi	Fungal diseases	No. of records	Percentage
Phytophthora infestans	Late or tuber blight	302	17
Verticillium spp.	Early dying disease	47	3
Rhizoctonia solani	Black scurf	35	2
Fusarium spp.	Dry rot	34	2
Helminthosporium solani	Silver scurf	31	2
Alternaria solani	Early blight	25	1
Colletotrichum coccodes	Black dot	23	1
Spongospora subterranea	Powdery scab	17	1
Phytophthora erytroseptica	Pink rot	10	1
Phoma foveata	Gangrene	7	<1
Total fungi		531	29
Insects		407	22
Viruses		350	19
Bacteria		280	15
Nematodes		247	14
Total		1815	100
Total number of records analysed 1995-2004		8333	

formed the largest group, covering 29 percent of the records. Late or tuber blight was by far the most important fungal disease covering 17 percent of all disease related papers in the period analysed. Based on this analysis, it is fair to consider late and tuber blight, caused by *Phytophthora infestans*, as the most important potato disease.

Figures on the economic impact of late blight in potato production today are available only for specific regions or countries but these are rather hard to compare (see the contribution "How bad is blight? - the social impact and economic importance of late blight" on http://gilb.cip.cgiar.org/index.php for an overview). An often cited figure for annual financial damage in developing countries was produced by CIP in 1996. It summed up to 2.75 billion dollar and was based on an estimate of 15 percent production loss caused by late blight (Anonymous, 1997). Based on the data from Hijmans *et al.* (2000), who predicted the optimum number of sprays on a susceptible crop, we have estimated late blight related fungicide costs in the 38 major potato producing countries in the world (Table 2). The total cost of fungicides required in this group of countries summed up to 1.7 billion euro. Additional costs of the disease comprise application costs and revenue losses due to yield and quality decrease, storage loss and price adjustment (Guenthner *et al.*, 2001).

Table 2. Estimation of fungicides needed and the costs for optimal control of potato late blight in 38 major potato producing countries.

Continent	No. of countries	Optimal no. sprays[a]	Potato area[b] (x 1000 ha)	Costs range[c] in euro/ha/yr	Total costs in euro/yr (x 10^6)
Africa	9	3.6 - 10.2	591	87 - 247	115
Asia	7	1.5 - 10.3	5210	36 - 249	867
Europe	11	6.8 - 16.5[d]	1096	165 - 400	359
Middle East	2	0.4 - 3.1	370	10 - 75	17
North America	1	8.2[d]	561	199	111
South + Centr. America	8	3.2 - 15.6	1016	78 - 378	246
Total costs					1,716

[a] Optimal numbers of sprays are based on the Simcast prediction model according to Hijmans *et al.* (2000).

[b] Average annual potato area for 1996-1998, as estimated by the Food and Agriculture Organisation of the United Nations (www.apps.fao.org). Aggregate values only include countries with more than 30.000 ha of potato.

[c] Value is calculated using the average costs (€24.22 per spray per ha) of crop protection products in the Netherlands as summarized by Schepers and Wustman (2003).

[d] The optimal number of sprays in the USA and the UK was based on a mean value calculated for three different areas in North-America (Maine, Red River Valley, Washington) and in the UK (Northern Ireland, England and Wales, Scotland).

Failure of breeding for resistance as a part of an effective disease management

Since its start, around the turn to the 20th century, breeding for resistance to late blight in potato has hardly been successful in an agronomic sense. Early achievements obtained with 11 major R-genes (referred to as R1 to R11) from *S. demissum* did not hold their promise for large scale potato production in Europe since soon after, or even before introduction of newly bred varieties, races of *P. infestans* emerged that were able to cause heavy infections on varieties which carried such R-genes. This apparently discouraged potato breeders to further explore R-gene sources from wild potato species. Already in the 1950s the idea was launched to seek a solution in breeding for partial resistance, aiming at pyramiding of minor genes that were thought to ultimately yield more durable types of resistance (see Bradshaw *et al.*, 1995). However, the development and wide scale adoption by growers of fungicides, effective both in terms of disease control and costs, minimalised attempts by commercial breeders to enhance late blight resistance in the next two decennia. The threat of late blight seemed to be diminished decisively by the crop protectants industry until pressure groups in the early 1970s started to express concerns about their ecological side effects and their risks for consumer health. This ongoing debate stimulated breeders to reconsider their breeding goals and subsequently, efforts towards improving late blight resistance were gradually strengthened. The scientific community continued advocating the breeding path towards increasing partial resistance by using race-nonspecific sources of resistance. Even the prevention of adventitious use of known and unknown major genes was advocated. However, under long day conditions, breeders using this strategy have achieved little progress, the major draw-back being the strong linkage between foliage resistance and late foliage maturity. This inevitable relation causes the contradicting phenomenon that varieties with better scores for foliage resistance in variety description lists require more fungicides in ware potato production than early maturing, more susceptible varieties. The longer growing period to crop maturity and hence the need for longer preventive spraying to protect the crop, outweighs the reduction in fungicide use that potentially can be obtained due to the higher level of foliage resistance of these late maturing varieties.

The failure of breeders to substantially add to the management of the disease is most clearly seen under circumstances where application of fungicide is completely absent. Such conditions exist in Dutch and Scandinavian organic farming systems where even the application of copper based products is prohibited. A severe reduction of yield is recorded in organic ware potato production as compared to conventional production of the same varieties (Table 3).

Table 3. Mean yield[a], averaged across cropping seasons from 2002 to 2004, for the three potato varieties in Dutch ware potato production, mean percent yield reduction of organic relative to conventional production and rating for foliar late blight resistance[a] and maturity[a].

Variety	Organic (mt/ha)	Conventional (mt/ha)	Yield reduction (%)	Late blight resistance[b]	Maturity[b]
Agria	25.4	50.8	50	6	5.5
Ditta	26.5	46.3	43	6	6
Santé	27.4	45.8	40	6	6
Mean	26.4	47.7	45		

[a] Source: Agrico Cooperative, Emmeloord, Netherlands. [b] Rating on a scale from 3 to 8 where 8 is most resistant, earliest.

Although some effect of suboptimal crop nutrition cannot be ruled out, this reduction is mainly due to the need for haulm destruction at a premature stage of crop development at some time point during the growing season when the infection level by late blight in farmers' fields rises above a threshold that is no longer considered acceptable. It should be noted that the varieties listed in Table 3 are currently the most important in Dutch organic potato production and reflect not only their agronomic suitability but also the commercial quality required in terms of storability, processing and consumers' preference. Given this set of prerequisites, potato breeders apparently do not have more to offer to the Dutch organic growers, and consequently to conventional growers, in terms of late blight resistance than the acceptance of a 45% yield loss in unsprayed crops.

A new paradigm for types of late blight resistance under long day conditions

Since the early 1960s, terminology evolved to describe two types of resistance that were thought to exist in potato germplasm (reviewed in Umaerus *et al.*, 1983). Table 4 lists the

Table 4. Terms used for describing types of resistance to foliar late blight in potato.

Resistance type 1	Resistance type 2
race specific	race non-specific
vertical	horizontal
immunity	field resistance
qualitative	quantitative
complete	partial
monogenic	polygenic or multigenic
R-gene or major gene	minor gene
specific (compatible versus incompatible)	general (rate reducing)
This paper	
R-gene based	maturity related

terms used to describe the contrast between race specific resistance on the one hand and race non-specific resistance on the other. Race specific resistance is believed to be genetically based on major R-genes that show classic monogenic or qualitative Mendelian segregation in crossing experiments when plant material is tested with an avirulent race of *Phytophthora*. Race non-specific resistance is generally considered to be inherited in a polygenic fashion and is considered more durable. In fact, under field conditions, race non-specific resistance behaves as a grade of susceptibility and is often expressed as an area under the disease progress curve (AUDPC) value that is only informative in comparison with a set of genotypes for which the relative AUDPC values are known. Furthermore it is thought to be composed of several distinguishable components such as latency period, infection efficiency, lesion growth rate and sporulation capacity. Under long day conditions, race non-specific resistance is strongly correlated with foliar maturity, i.e., late maturing genotypes generally show lower AUDPC values than earlier genotypes. The separation between the two types of resistance as listed in Table 4 is widely adopted in contemporary scientific literature although exceptions to nearly all describing terms are known.

Since the cloning of the first R-gene in higher plants, some 10 years ago, knowledge on the role and function of R-genes to late blight has also rapidly increased. Additional R-genes, different from the ones known from *S. demissum*, have been discovered (Figure 1). Also, several studies, aiming at mapping quantitative trait loci (QTLs) for partial resistance to both foliar and tuber blight have been published (reviewed in Gebhardt and Valkonen, 2001). Here it is argued that, in view of these numerous recent studies, breeders should adopt a new paradigm that replaces the long standing hypothesis that R-gene mediated race specific resistance is of a fundamentally different nature than race non-specific resistance in terms of genetics, histology and durability. We hypothesise that under long day conditions, any level of resistance to foliar late blight in potato genotypes that is higher than expected on the basis of maturity class is R-gene based and that the expression of such R-gene based resistance is either complete or partial, depending on the virulence spectrum of the *P. infestans* strain(s) present. Briefly, this hypothesis is based on the following observations:

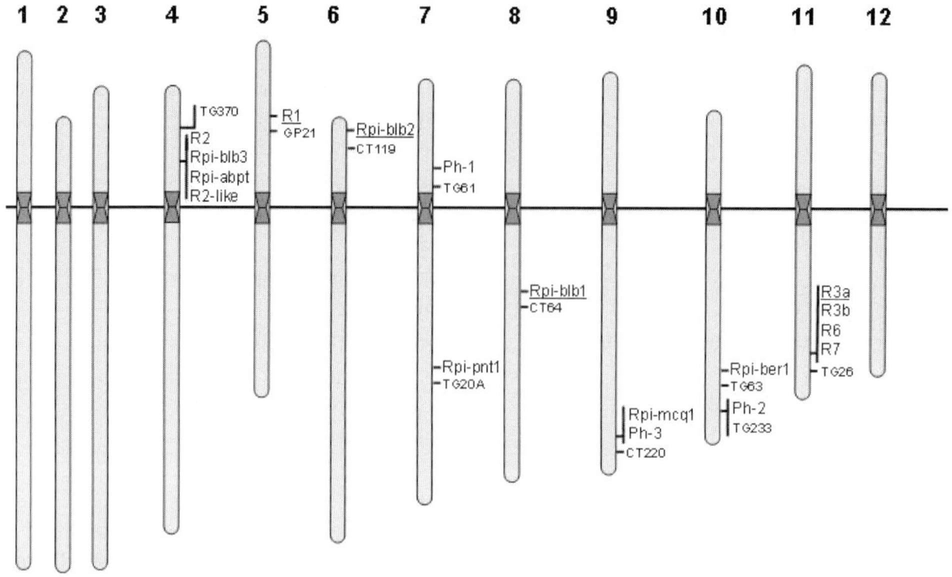

Figure 1. Idiogram of the potato and tomato genome with mapped R-genes in potato and tomato (Ph) and linked anchor markers (see Sol Genomics Network, www.sgn.cornell.edu). The idiogram was adopted from Chang (2005). Centromeres are aligned to a line. Inversions between both genomes were ignored. Cloned R-genes are underlined. Three letter codes indicating potato species are according to Hawkes (1990). References: R2, Rpi-blb3, Rpi-abpt, R2-like (Park et al.*, 2005); R1 (Ballvora* et al.*, 2002), Rpi-blb2 (van der Vossen* et al.*, 2005); Ph-1 (Zhang* et al.*, 2002); Rpi-pnt1, originally described as Rpi1 (Kuhl* et al.*, 2001); Rpi-blb1 (van der Vossen* et al.*, 2003). Rpi-blb1 is functionally allelic to the RB gene (Song* et al.*, 2003); Rpi-mcq1, originally described as Rpi-moc1 (Smilde* et al.*, 2005); Ph-3 (Chunwongse* et al.*, 2002); Rpi-ber1 (Ewing* et al.*, 2000); Ph-2 (Moreau* et al.*, 1998); R3a, R3b (Huang* et al.*, 2005); R6, R7 (El-Kharbotly* et al.*, 1996).*

- When groups of offspring clones with or without one of the R-genes R1, R10 or R11 were tested with a virulent isolate in field trials, the group harbouring the R-gene had slightly but consistently higher (i.e. more resistant) blight scores than the control groups without the R-gene (Stewart *et al.*, 2003). This phenomenon is also described as the residual effect of an R-gene.
- QTLs for maturity-corrected resistance to foliar late blight in populations derived from the R-gene carrying varieties Escort and Nikita mapped to regions of the genome where known R-gene clusters reside (Bormann *et al.*, 2004).
- Differentials for R2 and R11 gave blight scores of 7.0 and 4.7, respectively, in a field trial where inoculation was performed with an avirulent isolate (Stewart and Bradshaw, 2001). In other words, the level of resistance governed by R11 is different from that of R2 and could be easily mistaken for some level of partial resistance as meant in Table 4.
- When a new R-gene from *S. berthaultii* was mapped in a first backcross population (Figure 1), the clones of the sub-population missing this R-gene appeared to show a more or less normally distributed histogram for AUDPC values. Mapping of these data yielded markers at positions that previously had been associated with QTLs for foliage maturity (Ewing *et al.*, 2000).

These observations easily form sources of variation that can give the typical data sets obtained when screening potato germplasm for late blight resistance in field trials which, when analysed, reveal genotype times year, site and isolate interactions. It is our opinion that the new paradigm suits to explain most if not all data from literature in those cases where AUDPC or AUDPC-like resistance values are accompanied with multi-year maturity scores for the clones tested.

For breeders, *P. infestans* is a fast moving target

Swiezynsky *et al.* (2000) analysed published data on the virulence spectra found in *P . infestans* isolates in different countries in the world and concluded that all 11 virulence factors were found, both in so called old and new populations of *P. infestans* but not with equal frequency. Compatible reactions with differentials carrying R5, R6 and R9 were less frequently reported than with the other differentials. Table 5 lists recent, additional studies on the virulence spectra

Table 5. Mean number of virulences detected in isolates of P. infestans on potato differentials carrying R1 to R11. Studies are summarized that were not reviewed by Swiezynski et al. *(2000).*

Country, collection period	Nr[a]	Host[b]	Mean nr of virulence factors	ND[c]	Rare virulences[d]	Ref.[e]
Finland, 90-96	269	P	5.3	R9	2,5,6,8	1
Norway, 93-96	105	P	5.8	R9	(2,8),5,6	1
France+	134	P	7.6		1,2,3,4,5	2
Switzerland, 96-97	42	T	3.4		2,6,8 and 5,9 not found	2
Poland, 85-91	95	P	5.5 to 6.7	R6	5,8 and 9 not found	3
Russia, 97-98	191	P+T	5.5 to 10[f]	R9	5,6,8[f]	4
Israel, 93-00	244	P	6.8 and 6.3[g]		2,5,9 or 6,11[g]	5
Nepal, 99-00	251	P	2.3	R8	5,6,9 not found	6
US (WA), 98-99	49	P,T.S	8.2 - 9.3 (US8)		8,9	7
		P	5.4 - 6.3 (US11)		(3,10,11),8,9	
Canada, 97-99	59	T	3.0 - 6.1[h]	R8	5,9	8
	28	T	3.1 - 5.6[h]	R8	5,9	
Uruguay, 98-99	25		6.7 - 7.8[i]		5,6,9 not found	9

[a] Number of isolates analysed.

[b] Plant species from which isolates were collected. P: potato, T: tomato, S: different Solanaceae including *Solanum sarachoides, S. crispum, S. dulcamara, Petunia hybrida*

[c] Differentials not included in the set.

[d] Virulence factors between brackets were found to be slightly underrepresented but less rare than the other factors.

[e] 1; Hermansen *et al.*, 2000, 2; Knapova and Gisi, 2002, 3; Sujkowski *et al.*, 1996, 4; Elansky *et al.*, 2001, 5; Cohen, 2002, 6; Ghimire *et al.*, 2001, 7; Derie and Inglis, 2001, 8; Daayf and Platt, 2004, 9; Deahl *et al.*, 2003.

[f] Depending on the region in Russia that was sampled.

[g] Depending from sampling period being 1993 to 1998 and 1999 to 2000 respectively.

[h] Depending on the glucose phosphate isomerase (Gpi) genotype.

[i] In 1998 and 1999 respectively.

of *P. infestans*, again from diverse countries in the world. The mean number of virulence factors per isolate varied from 2.3 to 10 and were often found to be in the range from 5 to 7. Again, differentials carrying R5 and R9 and to a lesser extent R6 and R8 were less frequently found compatible with the isolates collected. In larger countries, such as Canada and Russia, race structures vary per region sampled (Elansky *et al.*, 2001; Daayf and Platt, 2004). Within countries, the mean number of virulence factors per isolate may vary also per collection period, per clonal lineage, per glucose phosphate isomerase (Gpi) genotype or host type (Table 5).

In Europe, mainly R1, R3 and R10 and to some extent R2 have been used in potato breeding. The current presence of the R-genes in the potato acreage however is unknown. In Russia, R1 to R4 are employed in potato cultivars (Elansky *et al.*, 2001). In Nepal, the use of foreign cultivars is very limited and the old, supposed R-gene free cultivars dominate (Ghimire *et al.*, 2001). In the US and Canada, no known R-genes are present in the major potato cultivars with the exception of R1 (Kennebec). The picture which emerges from all the studies reviewed by Swiezynsky *et al.* (2000) and in Table 5 is that *P. infestans* carries unnecessary virulence factors. This phenomenon is remarkably ignored by the scientific community and it does not fit in a theoretical hypothesis that maintenance of multiple virulence factors imposes a fitness cost to the pathogen and should be the outcome of continuous selection by the host population carrying corresponding R-genes in either an "arms-race" or a "trench warfare" model (Stahl *et al.*, 1999). In fact, *P. infestans* is a very dynamic pathogen in respect to its capacity to change its virulence spectrum. This was convincingly shown by Abu-El Samen *et al.* (2003) in asexually obtained single zoospore isolates. In an extreme case, none of 28 single zoospore isolates asexually derived from a parental isolate showed the same virulence spectrum as the parent and both loss and gain of virulence factors was observed. For the breeder, such findings make worries regarding "new populations", introduction of A2 strains, and hence, increase of genetic variability due to sexual recombination among A1 and A2 isolates, as well as formation of oospores, irrelevant. Other important conclusions can be drawn as well:

- The terminology by which the observation is described that an R-gene carrying potato variety can show full compatibility, being that such an R-gene is "broken" or "defeated" or "succumbed" or "overcome" is suggestive. It suggests that the varieties or hosts that also contain the R-gene have imposed a selection pressure on the pathogen as a result of which variants in the pathogen population that underwent a mutation from avirulent to virulent were able to prevail over the wild type. Whereas it would be more correct to describe the observation as an R-gene carrying variety that, in a specific part of a growing season, in a specific region of potato cultivation, has just encountered a *P. infestans* strain with a matching virulence factor.
- It should be realised that the so called Black differential set that has been used by researchers world wide for over 50 years to determine the virulence spectra (R1 to R11) of *P. infestans* isolates is likely to have an arbitrary composition. Any potato genotype carrying an additional R-gene (Figure 1) could be included as a differential and it is likely that corresponding virulence factors are detected in collections of isolates irrespective of whether such a gene has been applied in varieties on an agronomic scale or not. Similarly, it is unlikely that true null isolates exist that carry no virulence factors at all. Also, the mean number of virulence factors found in *P. infestans* (Table 5) will probably increase if the set of differentials would be broadened.

- One could argue that if selection in favour of matching virulence factors by the host is less strict than anticipated, sometimes, R-gene carrying varieties have to show their promising, complete R-gene based resistance in farmers' fields as if they possess a fresh, unexploited ("unbroken") R-gene. Indeed, this is seen occasionally in organically grown crops when R-gene carrying varieties such as Fresco, Escort, Appell and Stirling are grown. The reverse is sometimes seen when promising breeder's clones, possessing newly introgressed R-genes for late blight resistance, are further tested in other growing regions than where the breeding activities took place. Occasionally, testing on mini plot scale, e.g. with 10 plants in total, thus yields fully compatible reactions in unsprayed trials (unpublished data).

Studies on the genotypic constitution of *P. infestans* populations in Europe have revealed large, so-called, spatio-temporal variation. Rapid changes in the dominating clonal lineages between regions, even at a fine scale, and between seasons within regions are reported (Cooke and Lees, 2004; Day *et al.*, 2004). Given the situation regarding virulence spectra that are also found to be variable over space and time, we speculate on the possibility that in any growing season and in any potato growing region, further referred to as a *year-regions*, a defined, but still dynamic condition for late blight management exists that determines whether or not the R-genes of a host will function in agronomic sense or not.

Breeding for late blight resistance in the genomics era

We consider maturity related resistance inadequate for substantial contribution to integrated late blight management strategies under long day conditions and think that exploration of R-genes ultimately could prove to be the better option. Fortunately, the reservoir of known and unknown R-genes present in the existing modern and primitive varieties as well as in wild potato species seems to be large and will be further disclosed shortly. Once R-genes are identified, preferentially by mapping on the *Solanum* genome or, even better, by cloning, which allows identification of functional paralogs and/or alleles within an R-gene cluster, e.g. the R3 complex locus on the long arm of chromosome 11 (Figure 1), research can start to detect differences in the frequency of year-regions in which a given R-gene shows full incompatibility (or compatibility). The observation that virulence factors compatible to R5, R9 and maybe also R6 and R8 are somehow underrepresented in the *P. infestans* populations worldwide (Swiezynsky *et al.*, 2000; Table 5), could be an example of R-genes that show higher frequencies of incompatibility in a series of year-regions and hence, such genes have a higher potential for application by plant breeders than for example R1 to R4. Now that a first avirulence gene *Avr3a* from *P. infestans* has been cloned (Armstrong *et al.*, 2005), and that the *P. infestans* genome will be sequenced, the way is being paved to clone more avirulence genes matching well characterised R-genes. This will greatly speed up, and make far more accurate, the possibilities to understand the mechanisms in the pathogen population causing the observed spatio-temporal variation in virulence spectra. Consequently, the window of opportunities for development of management support systems that are based on real time monitoring of virulence factors in year-regions can be elucidated. If, on the long term, sufficient R-gene carrying potato varieties are available, a warning system could possibly be developed that allows farmers to start protective spraying of only those varieties at certain time points in year-regions, for which a matching virulence is detected.

Conclusions

Genomics, being the development of techniques and studies of organisms at the DNA level, already enabled important progress in understanding the nature of the interaction between potato and *P. infestans*. It is anticipated that breeding for resistance in potato and tomato, aiming at substantially contributing to disease management, requires, by one way or another, the deployment of known and yet unknown R-genes. Further genomics research will give insight in the frequency at which (combinations of) different R-genes provide full resistance to the pathogen. This on the long term potentially opens a way for breeders to significantly add to disease management. In light of this, and given the small set of R-genes actually used in potato varieties and the vast array of R-genes yet to be discovered in the wild potato species, we conclude that breeding for late blight resistance in potato has only just begun.

References

Abu-El Samen, F.M., G.A. Secor and N.C. Gudmestad (2003). Variability in virulence among asexual progenies of *Phytophthora infestans*. Phytopathology, 93 (3), 293-304.

Anonymous (1997). CIP in 1996. The International Potato Center Annual Report. International Potato Center, Lima, Peru. 59 pp.

Armstrong, M.R., S.C. Whisson, L. Pritchard, J.I. Bos, E. Venter, A.O. Avrova, A.P. Rehmany, U. Bohme, K. Brooks, I. Cherevach, N. Hamlin, B. White, A. Fraser, A. Lord, M.A. Quail, C. Churcher, N. Hall, M. Berriman, S. Huang, S. Kamoun, J.L. Beynon and P.R. Birch (2005). An ancestral oomycete locus contains late blight avirulence gene Avr3a, encoding a protein that is recognized in the host cytoplasm. Proceedings of the National Academy of Sciences USA.,102 (21), 7766-7771.

Ballvora, A., M.R. Ercolano, J. Weiss, K. Meksem, C.A. Bormann, P. Oberhagemann, F. Salamini and C. Gebhardt (2002). The R1 gene for potato resistance to late blight (*Phytophthora infestans*) belongs to the leucine zipper/NBS/LRR class of plant resistance genes. The Plant Journal, 30 (3), 361-371.

Bormann, C.A., A.M. Rickert, R.A.C. Ruiz, J. Paal, J. Lubeck, J. Strahwald, K. Buhr and C. Gebhardt (2004). Tagging quantitative trait loci for maturity-corrected late blight resistance in tetraploid potato with PCR-based candidate gene markers. Molecular Plant-Microbe Interactions, 17 (10), 1126-1138.

Bradshaw, J.E., R.L. Wastie, H.E. Stewart and G.R. Mackay (1995). Breeding for resistance to late blight in Scotland. In: L.J. Dowley, E. Bannon, L.R. Cooke, T. Keane and E. O'Sullivan (eds.) *Phytophthora infestans* 150, European Association for Potato Research, Boole Press Ltd, Dublin, 382 pp.

Chang, S. (2005). Cytogenetic and molecular studies on tomato chromosomes using diploid tomato and tomato monosomic additions in tetaploid potato. Thesis Wageningen Agricultural University, Wageningen, 128 pp.

Chunwongse, J., C. Chunwongse, L. Black and P. Hanson (2002). Molecular mapping of the Ph-3 gene for late blight resistance in tomato. The Journal of Horticultural Science & Biotechnology, 77(3), 281-286.

Cohen, Y. (2002). Populations of *Phytophthora infestans* in Israel underwent three major genetic changes during 1983 to 2000. Phytopathology, 92 (3), 300-307.

Cooke, D.E.L. and A.K. Lees (2004). Markers, old and new, for examining *Phytophthora infestans* diversity. Plant Pathology, 53, 692-704.

Day, J.P., R.A.M. Wattier, D.S. Shaw and R.C. Shattock (2004). Phenotypic and genotypic diversity in *Phytophthora infestans* on potato in Great Britain, 1995-98. Plant Pathology, 53, 303-315,

Daayf, F. and H.W. Platt (2004). Changes in race structure of Gpi 100:111:122 and Gpi 100:100:111 populations of *Phytophthora infestans* in Canada during 1997-1999. Canadian Journal of Plant Pathology, 26 (4), 548-554.

Deahl, K.L., M.C. Pagani, F.L. Vilaro, F.M. Perez, B. Moravec and L.R. Cooke (2003). Characteristics of *Phytophthora infestans* isolates from Uruguay. European Journal of Plant Pathology, 109 (3), 277-281.

Derie, M.L. and D.A. Inglis (2001). Persistence of complex virulences in populations of *Phytophthora infestans* in western Washington. Phytopathology, 91 (6), 606-612.

Elansky, S., A. Smirnov, Y. Dyakov, A. Dolgova, A. Filippov, B. Kozlovsky, I. Kozlovskaya, P. Russo, C. Smart and W. Fry (2001). Genotypic analysis of Russian isolates of *Phytophthora infestans* from the Moscow region, Siberia and Far East. Journal of Phytopathology, 149 (10), 605-611.

El-Kharbotly, A., C. Palomino-Sanchez, F. Salamini, E. Jacobsen and C. Gebhardt (1996). R6 and R7 alleles of potato conferring race-specific resistance to *Phytophthora infestans* (Mont) de Bary identified genetic loci clustering with the R3 locus on chromosome XI. Theoretical and Applied Genetics, 92 (7), 880-884.

Ewing, E.E., I. Simko, C.D. Smart, M.W. Bonierbale, E.S.G. Mizubuti, G.D. May and W.E. Fry (2000). Genetic mapping from field tests of qualitative and quantitative resistance to *Phytophthora infestans* in a population derived from Solanum tuberosum and Solanum berthaultii. Molecular Breeding, 6 (1), 25-36.

Gebhardt, C. and Valkonen, J.P.T (2001). Organization of genes controlling disease resistance in the potato genome. Annual Review of Phytopathology, 39, 79-102.

Ghimire, S.R., K.D. Hyde, I.J. Hodgkiss and E.C.Y. Liew (2001). Race diversity and virulence complexity of *Phytophthora infestans* in Nepal. Potato Research, 44, 253-263.

Hawkes, J.G. (1990). The potato. Belhaven Press, London, 259 pp.

Hermansen, A., A. Hannukkala, R.H. Naerstad and M.B. Brurberg (2000). Variation in populations of *Phytophthora infestans* in Finland and Norway: mating type, metalaxyl resistance and virulence phenotype. Plant Pathology, 49 (1), 11-22.

Hijmans, R.J., G.A. Forbes and T.S. Walker (2000). Estimating the global severity of potato late blight with GIS- linked disease forecast models. Plant Pathology, 49 (6), 697-705.

Huang, S.W., E.A.G. van der Vossen, H.H. Kuang, V.G.A.A. Vleeshouwers, N.W. Zhang, T.J.A. Borm, H.J. van Eck, B. Baker, E. Jacobsen and R.G.F. Visser (2005). Comparative genomics enabled the isolation of the R3a late blight resistance gene in potato. The Plant Journal, 42 (2), 251-261.

Knapova, G. and U. Gisi (2002). Phenotypic and genotypic structure of *Phytophthora infestans* populations on potato and tomato in France and Switzerland. Plant Pathology, 51 (5), 641-653.

Moreau, P., P. Thoquet, J. Olivier, H. Laterrot and N.H. Grimsley (1998). Genetic mapping of Ph-2, a single locus controlling partial resistance to *Phytophthora infestans* in tomato. Molecular Plant-Microbe Interactions, 11 (4), 259-269.

Park, T., J. Gros, A. Sikkema, V.G.A.A. Vleeshouwers, M. Muskens, S. Allefs, E. Jacobsen, R.G.F. Visser and E. van der Vossen (2005). The late blight resistance locus Rpi-blb3 from *Solanum bulbocastanum* belongs to a major late blight R gene cluster on chromosome 4 of potato. Molecular Plant-Microbe Interaction, in press.

Schepers, H. and R. Wustman, 2003. *Phytophthora* 2003: middelen en aanpak (*Phytophthora 2003: Fungicide compounds and application*). Informa, June 6th 2003.

Smilde, W.D., G. Brigneti, L. Jagger, S. Perkins and J. D. G. Jones (2005). *Solanum mochiquense* chromosome IX carries a novel late blight resistance gene Rpi-moc1. Theoretical and Applied Genetics, 110 (2), 252-258.

Song, J., J.M. Bradeen, S.K. Naess, J.A. Raasch, S.M. Wielgus, G.T. Haberlach, J. Liu, H. Kuang, S. Austin-Phillips, C.R. Buell, J.P. Helgeson and J. Jiang (2003). Gene RB cloned from *Solanum bulbocastanum* confers broad spectrum resistance to potato late blight. Proceedings of the National Academy of Sciences USA, 100 (16), 9128-9133.

Stahl, E.A., G. Dwyer, R. Maurico, M. Kreitman and J. Bergelson (1999). Dynamics of disease resistance polymorphism at the Rpm1 locus of *Arabidopsis*. Nature, 400, 667-671.

Stewart, H.E. and J.E. Bradshaw (2001). Assessment of the field resistance of potato genotypes with major gene resistance to late blight (*Phytophthora infestans* (Mont.) de Bary) using inoculum comprised of two complementary races of the fungus. Potato Research, 44, 41-52.

Stewart, H.E., J.E. Bradshaw and B. Pande (2003). The effect of the presence of R-genes for resistance to late blight (*Phytophthora infestans*) of potato (*Solanum tuberosum*) on the underlying level of field resistance. Plant Pathology, 52, 193-198.

Sujkowski, L.S., S.B. Goodwin and W.E. Fry (1996). Changes in specific virulence in Polish populations of *Phytophthora infestans*: 1985-1991. European Journal of Plant Pathology, 102 (6), 555-561.

Swiezynski, K.M., L. Domanski, H. Zarzycka and E. Zimnoch-Guzowska (2000). The reaction of potato differentials to *Phytophthora infestans* isolates collected in nature. Plant Breeding, 119, 119-126.

Umaerus, V., M. Umaerus, L. Erjefält and B.A. Nilsson (1983). Control of *Phytophthora* by host resistance: problems and progress. In: D.C. Erwin, S. Bartnicki-Garcia and P.H. Tsao (eds.) *Phytophthora*, its biology, taxonomy, ecology and pathology. American Phytopathological Society, St. Paul, Minnesota, pp. 315-326.

van der Vossen, E., A. Sikkema, B.T.L. Hekkert, J. Gros, P. Stevens, M. Muskens, D. Wouters, A. Pereira, W. Stiekema and S. Allefs (2003). An ancient R gene from the wild potato species *Solanum bulbocastanum* confers broad-spectrum resistance to *Phytophthora infestans* in cultivated potato and tomato. The Plant Journal, 36, 867-882.

van der Vossen, E., J. Gros, A. Sikkema, M. Muskens, D. Wouters, P. Wolters, A. Pereira and S. Allefs (2005). The *Rpi-blb2* gene from *Solanum bulbocastanum* is a *Mi-1* ortholog conferring broad spectrum late blight resistance in potato and tomato. The Plant Journal, in press.

Zhang, L.P., A. Khan, D. NinoLiu and M.R. Foolad (2002). A molecular linkage map of tomato displaying chromosomal locations of resistance gene analogs based on a *Lycopersicon esculentum* x *Lycopersicon hirsutum* cross. Genome, 45, 133-146.

Curriculum vitae – Sjefke J.H.M. Allefs

Sjefke J.H.M. Allefs studied Plant Breeding at the Wageningen Agricultural University and graduated in 1988. In 1989 a civil service project was carried out at the former Foundation for Plant Breeding (SVP) in Wageningen on molecular marker techniques in wheat and potato. From 1990 to 1994 a research project was carried out for Agrico Research at the former DLO-Centre for Plant Breeding and Reproduction Research (CPRO-DLO) in Wageningen. The project aimed at developing screening methods suitable for breeding potato varieties with enhanced levels of resistance to *Erwinia* spp. Also, sources of resistance were sought in wild potato species or by making use of genes coding for broad spectrum antibacterial peptides. A thesis was written for which he received the degree of PhD in 1995. In 1994, he was appointed as research scientist and breeder at the potato breeding station of Agrico Research in Bant, Noordoostpolder, the Netherlands. From 1999, he serves as director of Agrico Research.

Company profile – Agrico Research

Agrico Research is a full daughter company of the cooperative Agrico U.A in Emmeloord, the Netherlands. Its main activity is the breeding of new potato varieties that are also subject of own agronomic research which is carried out to suit the demand of the member growers of the cooperative for adequate information. The breeding work is carried out on experimental fields and in greenhouses or laboratories. In the greenhouses, about 1000 crosses are made each year and 200.000 individual potato plants are raised from the botanic seed obtained. This first clonal generation is the starting point of variety selection on the experimental fields in a series of nine successive years. This selection is carried out in collaboration with a group of about 35 small scale private breeders and a few professional breeding companies.

In the labs, several trait characters of the selected clones are being determined such as resistance to potato cyst nematodes, wart disease, and late blight. In the quality lab, suitability of the selected clones as table potato or for processing to French fries and precooked potato products (CêlaVîta) is determined. In a biotech lab molecular and in vitro techniques are carried out. Also, Agrico Research has a farm with 96 hectares of arable land with a clay soil type.

The most important goals of the breeding programme for new potato varieties are improvement of disease resistances, quality for fresh consumption and suitability for processing to French fries and precooked potato products.
Apart from the breeding programme for commercially grown varieties, almost 40 percent of Agrico Research' own breeding efforts is directed to the development of superior breeding parents. Often, traits from primitive cultivated potatoes or even wild potato species are exploited.

More information can be found on www.agrico.nl

Control of *Phytophthora infestans* in potato

H.T.A.M. Schepers
Applied Plant Research, P.O. Box 430, 8200 AK Lelystad, The Netherlands

Abstract

Within the theme "Phytophthora Toolbox" of the Netherlands Initiative of Late Blight (NILB) results from all projects will be translated into practical solutions resulting in an integrated control strategy for *P. infestans* with a minimal input of fungicides. The campaign in The Netherlands to reduce the role of primary and secondary inoculum sources is described. The possibilities to reduce fungicide input in resistant cultivars and the strategy to convince growers of the robustness of this approach are presented. By combining the characteristics of the fungicides with infection pressure and crop growth their use can be optimised. Even more important than carrying out the trials is transferring this knowledge to farmers. The role of decision support systems in the integrated control of late blight will be presented as well as possibilities to improve them.

Keywords: late blight, fungicides, integrated control, decision support system, Phytophthora

Introduction

Late blight caused by *Phytophthora infestans* is generally considered as the most devastating disease of potato. In the Netherlands Initiative of Late Blight (NILB) important themes are addressed to find short term and long term leads to control this disease (Boonekamp, 2003). Within the theme "Phytophthora Toolbox", results from the other five themes of NILB will be translated into practical solutions resulting in an integrated control strategy for *P. infestans* with a minimal input of fungicides. The strategy will have to be adapted to the specific Dutch conditions regarding crop intensity, varieties, pathogen, population, pathogen epidemiology, available fungicides, effectiveness of fungicides during the growth cycle, and weather conditions. Integrated practical knowledge will be generated by experimental applied research, on-farm research using different farming systems and by evaluation of farmers' results after applying the new knowledge.

In this paper some examples and NILB-projects are presented of elements that can be part of such an integrated control strategy of potato late blight. The campaign in The Netherlands to reduce the role of primary and secondary inoculum sources is described as well as the possibilities to reduce fungicide input in resistant cultivars. The targeted use of fungicides by combining the characteristics of the fungicides with infection pressure and crop growth is also presented. The role of DSSs in the integrated control of late blight will be presented as well as possibilities to improve them.

Primary and secondary sources of inoculum

The first step in integrated control is reducing the primary sources of inoculum. In The Netherlands it has been shown that in most years blight epidemics start from infected plants on dumps (Zwankhuizen, 2000). Therefore, farmers were intensively informed about a nationwide regulation to cover dumps before April 15. This campaign organised by the Masterplan Phytophthora, launched by the Agricultural and Horticultural Organisation LTO-Nederland in 1999, resulted in a significant reduction in the number of uncovered dumps (Schepers *et al.*, 2000). The better "control" of dumps led to an increased importance of (latently) infected seed tubers as a source of primary inoculum (Turkensteen *et al.*, 2002). The use of (healthy) certified seed is therefore very important to delay the onset of the epidemic as long as possible. Early crops covered with perforated polythene, volunteer plants and oospores can also act as (primary) inoculum sources. To further reduce disease pressure the regulation has been extended to control of volunteers and excessive late blight foci. Volunteers have to be controlled after July 1 when more than 2 plants are present per m^2 on a part of the field of 300 m^2. In one of the NILB-projects an automated system for detection and control of potato volunteers is developed in sugar beet. A field is considered to contain an excessive amount of blight when more than 1000 infected leaflets on 20 m^2 or 2000 infected leaflets on 100 m^2 are observed. The regulation forces growers to take measures to control this disease either by spraying eradicant fungicides or by desiccation of the crop with propane burners (organic growers). In one of the NILB-projects the efficiency of propane burners to kill late blight spores and mycelium was investigated.

Cultivar resistance

Both partial resistance (lower susceptibility) and fungicides can slow down the development of late blight. Several reports show that partial resistance in the foliage may be used to complement fungicide applications to allow savings of fungicide by reduced applications rates or extended intervals between applications (Nærstad, 2002). Nærstad showed that exploiting high foliage resistance to reduce fungicide input was risky when field resistance to tuber blight was low. When field resistance to tuber blight was high, a medium-high resistance in the foliage could be exploited to reduce the fungicide input. In a number of European countries trials are carried out in which the possibilities of a reduced input in resistant varieties are investigated. In Western Europe resistant cultivars are not grown on a large scale because commercially important characteristics such as quality, yield and earliness are usually not combined in the same cultivar with late blight resistance. In the grower's perspective, the savings in fungicide input that can be achieved in resistant cultivars are not in balance with the higher (perceived) risk for late blight. In countries where fungicides are not available or very expensive, the use of resistant cultivars is one of the most important ways to prevent too much damage from late blight. In future certification schemes, requested by governments or supermarkets, with strict rules for input of fungicides might provide additional motives for growers to reduce fungicide input. As a part of NILB, field trials are carried out to estimate the infection efficiency and the minimum dose rate of fluazinam required to achieve protection, of the 30 most important cultivars in the Netherlands (Kessel *et al.*, 2004; Spits *et al.*, 2004). The recommendations to reduce fungicide input in resistant cultivars will have to be validated and demonstrated in a range of practical situations (with low and high disease pressure) to convince growers of their robustness.

Fungicide characteristics

Fungicides play a crucial role in the integrated control of late blight. In order to use fungicides in the most optimal way it is important to know the effectiveness and action mode of the active ingredients used in fungicides to control late blight. What is their effectiveness on leaf blight, stem blight and tuber blight and do they protect the new growing point? Are the fungicides protectant, curative or eradicant? What is their rainfastness and mobility? During the yearly workshops on integrated control of potato late blight, the fungicide characteristics of the most important fungicide active ingredients used for control of late blight in Europe, are discussed and ratings are given. The ratings are based on the consensus of experience of scientists in countries present during the workshop (Tables 1 and 2). The characteristics of the fungicides can be used to optimise their efficacy by combining their strong points with specific situations in the growing season concerning infection pressure and plant growth. In a number of NILB-projects the efficacy of fungicides in specific situations is investigated. In one of the projects the efficacy of fungicides on stem lesions (originating from infected seed) is tested. Another projects looks into the ability of the fungicides to prevent tuber blight and the optimal timing of these fungicides. Different strategies, in which efficacy, costs and environmental side-effects are recorded, are tested on five regional farms situated in important potato growing regions in the Netherlands. During the growing season everything happening in these trials, including types of products used, timing and occurrence of late blight, can be followed on the internet (http://www.kennisakker.nl). The results of these trials are used in discussions with farmers and comparisons are made between the trial results and their spraying schedules regarding efficacy, costs and environmental side-effects. Each year the control strategy is discussed with all stakeholders in potato and published in farmers journals and in a brochure that is sent to all potato growers and advisors.

Decision Support Systems

Decision Support Systems (DSSs) integrate and organise all available information on the life cycle of *Phytophthora infestans*, the weather (historical & forecast), plant growth, fungicides, cultivar resistance and disease pressure, required for decisions to control late blight. Computer-based DSS that require weather information and regular late blight scouting inputs have been developed and validated in a number of European countries (Hansen *et al.*, 2002). Six different DSSs were tested in validation trials in 2001: Simphyt (D), Plant-Plus (NL), NegFry (DK), ProPhy (NL), Guntz-Divoux/Milsol (F) and PhytoPre+2000 (CH). The use of DSSs reduced fungicide input by 8-62% compared to routine treatments. The level of disease at the end of the season was the same or lower using a DSS compared to a routine treatment in 26 of 29 validations. DSS can deliver general or very site-specific information to the users by extension officers, telephone, fax, e-mail, SMS, PC and websites on the internet. Web-blight, an international collaboration on information and DSS for potato late blight, provides online warning and prognosis systems for late blight in the Nordic and Baltic countries and Poland (http://www.web-blight.net). In the UK, the Blightwatch site can be used in conjunction with the British Potato Council's 'Fight against Blight' site (http://www.potato.org.uk,

Table 1. The effectiveness of fungicide products for the control of P. infestans *in Europe.*
These ratings are the opinion of the Fungicides Sub-Group (both independent scientists and delegates from the crop protection industry) at the Jersey late blight workshop (2004) and are based on field experiments and experience of their performance when used in commercial conditions (Bradshaw, 2004).

Product [1]	Effectiveness				Mode of action			Rainfastness	Mobility in the plant
	Leaf blight	New growing point	Stem blight	Tuber blight	Protec-tant	Cura-tive	Eradi-cant		
Chlorothalonil	++	0	(+)	0	++	0	0	++(+)	Contact
Copper	+	0	+	+	+(+)	0	0	+	Contact
Cyazofamid	+++	0	+	+++	+++	0	0	+++	Contact
dithiocarbamates [2]	++	0	+	0	++	0	0	+(+)	Contact
famoxadone+cymoxanil	++	0	+(+)[4]	N/A	++	++	+	++(+)	Contact + translaminar
Fluazinam	+++	0	+	++(+)	+++	0	0	++(+)	Contact
zoxamide+mancozeb	+++	0	+ [4]	++	+++	0	0	++(+)	Contact + contact
cymoxanil+mancozeb or metiram	++(+)	0	+(+)	0	++	++	+	++	Translaminar + contact
dimethomorph+mancozeb	++(+)	0	+(+)	++	++(+)	+	++	++(+)	Translaminar + contact
fenamidone+mancozeb	++(+)	0	+(+)[4]	++	++(+)	0	0	++	Translaminar + contact
benalaxyl+mancozeb [3]	++	++	++	N/A	++(+)	++(+)	++(+)	+++	Systemic + contact
metalaxyl-M + mancozeb or fluazinam [3]	+++	++	++	N/A	++(+)	++(+)	++(+)	+++	Systemic + contact
propamocarb-HCl+mancozeb or chlorothalonil	++(+)	+(+)	++	++	++(+)	++	++	+++	Systemic + contact

[1] The scores of individual products are based on the label recommendation and are NOT additive for mixtures of active ingredients. Inclusion of a product in the list is NOT indicative of its registration status either in the EU or elsewhere in Europe.
[2] Includes maneb, mancozeb, propineb and metiram.
[3] See text for comments on phenylamide resistance.
[4] Based on limited data.
Key to ratings: 0 = no effect; + = reasonable effect; ++ = good effect; +++ = very good effect; N/A = not recommended for control of tuber blight.

Table 2. Provisional ratings for the effectiveness of new fungicide products /co-formulations for the control of P. infestans *in Europe. These ratings are the opinion of the Fungicides Sub-Group at the Jersey late blight workshop (2004) and are based on field experiments and (very) limited experience under commercial conditions (Bradshaw, 2004).*

Product [1]	Effectiveness				Mode of action			Rainfastness	Mobility in the plant
	Leaf blight	New growing point	Stem blight	Tuber blight	Protec-tant	Cura-tive	Eradi-cant		
Benthiavalicarb + mancozeb	+++	?	?	+(+)	+++	+(+)	+	++(+)	Translaminar + contact
Fenamidone + propamocarb- HCl	+++	?	?	++(+)	++(+)	?	?	?	Translaminar + systemic

[1] The scores of individual products are based on the label recommendation (to be registered) and are NOT additive for mixtures of active ingredients. Inclusion of a product in the list is NOT indicative of its registration status either in the EU or elsewhere in Europe.

Key to ratings: 0 = no effect; + = reasonable effect; ++ = good effect; +++ = very good effect; ? = no experience in trials and/or field conditions.

http://www.potatocrop.com). These sites provide up to date information on late blight outbreaks, as reported by scouts, plus information on favourable weather conditions for late blight indicated by Smith periods. It is also possible to register for SMS or e-mail late blight incidences updates. Expert comments and recommendations for control are provided every week and help to schedule spray programmes. In Germany also a number of websites is available that provide information on monitoring for late blight and disease pressure. Recommendations are based on the DSS Simphyt (http://www.phytophthora.de, http://www.lfl.bayern.de). In Switzerland an Internet based DSS is available that informs on infected fields. Warnings are sent to growers based on weather information, cultivar resistance and history of fungicide applications (http://www.phytopre.ch). Companies and institutes like Dacom (http://www.dacom.nl), Opticrop (http://www.opticrop.nl), Pro-Plant (http://www.proplant.de) and Arvalis (http://www.mildilis.arvalisinstitutduvegetal.fr/v2/) have developed DSS for control of late blight and offer these services -after registration and payment- also on the Internet. In France, Belgium and Italy also other DSS are used but not available on the Internet. Taking into account all available information to control late blight is of course more complicated than applying fungicides in a regime with fixed intervals. Moreover, factors that can obstruct the implementation of IPM strategies can be the higher costs, the higher perceived risk and the availability during the growing season of fungicides with different action modes and unpredictable weather conditions. The presence of the new, aggressive population of late blight will force growers to take all information into account because the risk of regimes with fixed spray intervals will often be too high (Turkensteen, 2002). An important task for the near future is to update the DSSs with information on the epidemiology of the new aggressive population of *P. infestans*. Issues such as the influence of temperature and relative humidity on the infection process, the role of primary and secundary inoculum sources and the resistance ratings for foliar and tuber blight will have to be addressed. Also the control of early blight caused by *Alternaria* will have to be integrated in control strategies for late blight.

References

Boonekamp, P.M. (2003). The Netherlands Initiative on Late Blight (NILB). Global Initiative on Late Blight, Newsletter, 21, 2-3.

Bradshaw, N.J. (2004). Discussion of potato early and late blight fungicides, their properties & characteristics: report of the fungicide subgroup. In: Westerdijk, C.E. and H.T.A.M. Schepers (editors), Proceedings of the eighth workshop of an European network for development of an integrated control strategy of potato late blight. Jersey, Channel Islands, 31 March-4 April 2004, 151-156.

Hansen, J.G., B. Kleinhenz, E. Jörg, J.G.N. Wander, H.G. Spits, L.J. Dowley, E. Rauscher, D. Michelante, L. Dubois and T. Steenblock (2002). Results of validation trials of Phytophthora DSSs in Europe, 2001. In: Westerdijk, C.E. and H.T.A.M. Schepers (editors), Proceedings of the 6th workshop of an European network for development of an integrated control strategy of potato late blight. Edinburgh, Scotland, 26-30 September 2001, 231-242.

Kessel, G.J.T., W.G. Flier, H.G. Spits, J.G.N. Wander and H.T.A.M. Schepers (2004). Exploitation of cultivar resistance using reduced fungicide dose rates, the Wageningen UR approach. In: Westerdijk, C.E. and H.T.A.M. Schepers (editors), Proceedings of the 8th workshop of an European network for development of an integrated control strategy of potato late blight. Jersey, Channel Islands, 31 March-4 April 2004, 211-214.

Nærstad, R. (2002). Exploitation of cultivar resistance in potato late blight disease management and some aspects of variation in *Phytophthora infestans*. PhD-thesis Agricultural University of Norway, Ås. 112 pp.

Schepers, H.T.A.M., J. Dogterom and J.P. Kloos (2000). The Masterplan Phytophthora: a nationwide approach to late blight. In: Schepers, H.T.A.M. (editor), Proceedings of the 4th workshop of an European network for development of an integrated control strategy of potato late blight. Oostende, Belgium, 29 September-2 October 1999, 131-136.

Spits, H.G., J.G.N. Wander and G.J.T. Kessel (2004). DSS development focussed on variety resistance in the Netherlands, 2003. In: Westerdijk, C.E. and H.T.A.M. Schepers (editors), Proceedings of the 8th workshop of an European network for development of an integrated control strategy of potato late blight. Jersey, Channel Islands, 31 March-4 April 2004, 215-222.

Turkensteen, L.J., A. Mulder, W.G. Flier and J.P. Kloos (2002). Feiten rondom Phytophthora in 2002 (Facts concerning Phytophthora in 2002). Aardappelwereld magazine, 1, 16-25.

Zwankhuizen, M.J., F. Govers and J.C. Zadoks (2000). Inoculum sources and genotypic diversity of *Phytophthora infestans* in Southern Flevoland, the Netherlands. European Journal of Plant Pathology, 106, 667-680.

Curriculum vitae – Dr. H.T.A.M. Schepers

Hubertus Thomas Antonius Maria Schepers was born on 21 June 1956 in Zevenbergsche Hoek, The Netherlands. He received his MSc degree in plant pathology, in 1980 and his Ph.D. in plant pathology in 1985, both from the Agricultural University Wageningen, The Netherlands. In the Extension Service of the Ministry of Agriculture he co-ordinated the advisory team of crop protection in horticultural crops from 1985 to 1987. As a technical manager he was responsible for the Research, Development and Registration of agrochemicals with the company Hoechst Holland (1987-1991). From 1991 to 1992 he was an expert at the Ministry of Social Affairs in the field of exposure to pesticides and risk assessment in registration procedures. He joined the Applied Plant Research (PPO) in Lelystad in 1992. His research interests focus on the epidemiology and integrated control of airborne fungal diseases in arable crops and vegetables such as potato, onion, winter wheat and lettuce. Pathogens of particular interest include *Phytophthora infestans* in potato, Fusarium Head Blight in cereals, *Peronospora destructor* in onion and *Bremia lactucae* in lettuce. He was the co-ordinator of an EU-funded Concerted Action on integrated control of late blight in potato (1996-2000) and is chairman of the European network for integrated control of early and late blight in potato. This network regularly organises workshops and publishes Proceedings in which all aspects of (chemical) control of late and early blight are presented. In The Netherlands Initiative on Late Blight he is co-ordinator of the theme "Phytophthora Toolbox" in which results from all studies are translated into practical solutions resulting in an integrated control strategy for *P. infestans* in potato.

Company profile – Applied Plant Research

Applied Plant Research (PPO) is the Netherlands' organization for applied research and development in agricultural and horticultural production. PPO is an interdisciplinary organization developing and transferring knowledge and innovations for the benefit of producers as well as consumers. We form the link between science and practical application of knowledge and are part of Wageningen University and Research Centre (Wageningen UR), a leading organization in agricultural, environmental and related sciences in the Netherlands. PPO operates in a productive and sustainable horticultural and agricultural environment to

which we make a significant contribution. We have an excellent track record of innovations in all fields of agriculture and horticulture. Applied Plant Research also studies plant-related issues in the direct environment of these production systems through to those in agricultural nature development and our living environment. The ongoing research for innovation is crucial for the competitive position of our clients, and we support them with knowledge, products, production processes, logistical concepts and marketing concepts. Our ambition is to transfer knowledge and innovations worldwide, to local and national governments, and to clients and partners throughout the production chain. These include growers and their cooperatives and organizations, trade, food processing industry, manufacturers and suppliers of agricultural inputs such as crop protection products and fertilizers, and other industries.
All of the above applies to the potato crop. We have strong programmes on controlling pests and diseases in potato, in particular on the control of nematodes and late blight. We also developed a strong position on post harvest management and research as well as on quality and chain management.

Applied Plant Research
Arable Farming, Field Production of Vegetables and Multifunctional Agriculture
P.O. Box 430, 8200 AK Lelystad, , The Netherlands
Infoagv.ppo@wur.nl, www.ppo.wur.nl

Primary outbreaks of late blight and effect on the control strategy

Peter Raatjes
Dacom PLANT-Service BV, PO Box 2243, 7801 CE Emmen, The Netherlands

Abstract

In The Netherlands, Dacom plays an important role in the mapping of disease outbreaks of *Phytophthora infestans* in potatoes. Each year farmers are surprised to find *P. infestans* in their fields so early. Because Dacom is a well known reporting point for Late blight disease mapping, these farmers call the Dacom-office with: "Help, I have early Late Blight in my crop...".

By order of the Dutch Masterplan Phytophthora, Dacom has to investigate first outbreaks of late blight in cooperation with experts from Plant Research International at Wageningen University and Research Centre (PRI), like dr. L.J. Turkensteen. The findings are reported to the Masterplan. In summary, the source of first outbreaks over the years, based on the inspections. From 1995 till 1999 the epidemic typically started on potato dumps. In 1999 a field of early potatoes under plastic caused severe early infections in a wide radius. In 2000 the epidemic was mainly started by infections from seed tubers. Early infection on volunteer potatoes in other crops were the main source of primary infections in 2002. Oospores and infected seed caused large scale outbreaks in 2003. In 2004 and 2005 some early infected potato dumps were reported, followed by in-field infections from seed and oospores.

The size and development of lesions on leaves and stems matches the infection events that the PLANT-Plus disease forecasting tool calculates. It basically shows that infection events are necessary for the late blight to develop and spread in the field. Chemical treatment around infection events will stop the spread of the disease, whatever the primary source would be.

Keywords: outbreaks, primary sources, monitoring, control strategy

Introduction

"Help, I have early infections of late blight in my crop!" Every year, a call like this will come to the Dacom office in the Netherlands. It is not always the farmer himself who calls but also his advisor. Dacom coordinates the research of the first outbreaks in the Netherlands. Looking back over a number of years shows that not one year is the same. This presentation will give a summary of the years before 1999 and detailed information from the years 1999 till 2005. The outbreaks were characterised and the sources of the outbreaks were determined. For any forecasting model it is important to know if the calculated infection events match with the actual outbreaks in the field, so the size and development of the lesions were matched with the infection events for late blight based on the PLANT-Plus model. Furthermore it is important to know if and how the different sources of primary outbreaks influence the control strategy provided by the model. The PLANT-Plus model also uses the reported and mapped outbreaks

as a parameter in the forecasting model to influence the number of *Phytophthora* spores that can potentially infect a field.

Material and methods

Field surveys have been carried out after receiving reports of outbreaks. The field surveys have been carried out by different experts, like dr. ir. L. Turkensteen, dr. ir. G. Kessel (PRI), ing. H. Germs (Dacom) and ing. P. Raatjes (Dacom). Initially the surveys were made to map the outbreaks and to define the type of outbreak, like commercial fields, volunteer potato, dumps, organic fields or home gardens. Since 1999 the surveys have been extended towards determination of the primary source of the outbreaks, divided into: infected seed, oospores, volunteers, nearby source and other.

Part of the survey has been conducted within the Masterplan Phytophthora, a joint action from the potato industry in the Netherlands to contribute to the fight against late blight, including e.g. epidemiology research, breeding, genomics and a national blight warning system.

Results

1995 - 1999

In the information in the disease mapping system of Dacom, not only a spotted infection is recorded but also where the infection source was found. The graph in Figure 1 shows a typical example of how the epidemic developed in those years from the different primary sources.

Generally infections on dumps are the first to be reported (first line to appear). This is followed somewhat later in the season with reports in commercial fields (line going straight up in the middle of the graph). One of the first actions of the Masterplan was to act on prevention of the dumps by putting a fine on having an uncovered dump. This resulted in clearly less reports of dumps from the start in 1999.

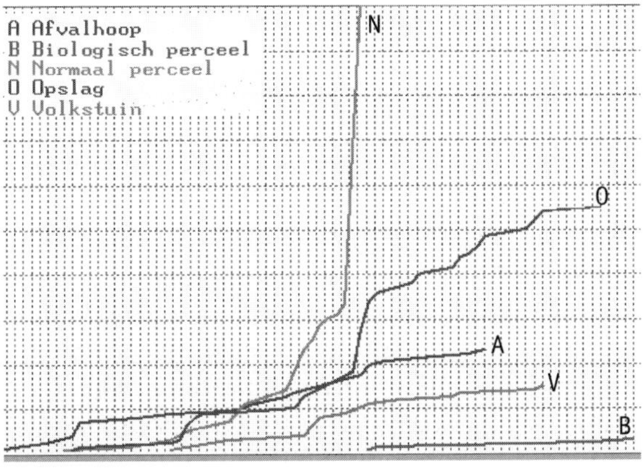

Figure 1. Example of annual epidemic development with types of outbreaks (1995-1999).

1999

In the early part of the 1999 season, just after the commercial fields emerged, a number of infection events took place as shown in dots on the map in Figure 2. Normally, this early in the season, this is not a big problem as the total amount of infected leaf area is still small. In this case, however, many reports of late blight came in from the South Western part of the Netherlands as the map shows.

Looking at the reports in combination with the wind direction, a certain spot in the West showed up as a possible source of infection, marked as the spooky field in Figure 2. Spooky, because it was only confirmed by rumours. From 4 infected fields in that region a total of 18 isolates were taken and the DNA was sampled by PRI. The results showed that all the isolates came from the same primary source. On top of these early infections, another infection event occurred in the middle of June, causing a big alarm and predictions of great losses. However the rest of the season was relative easily with just a small number of infection events. The panic was easily forgotten.

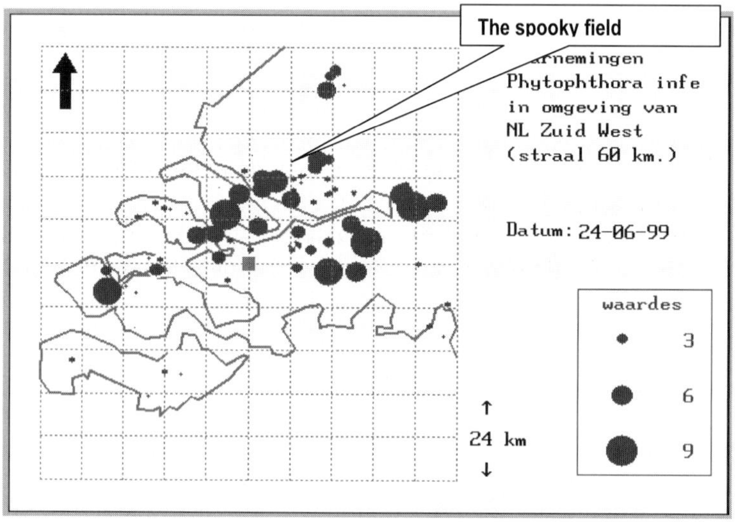

Figure 2. Locations of early outbreaks in the South West of the Netherlands.

2000

In 2000 the first outbreaks could be traced to the seed tuber on a broad scale. On the first report (Figure 3), a dump in the North West of the Netherlands, the seed tuber was infected. Also the following reports in commercial fields were easily traceable to the seed tuber.
In combination with a number of early infection events, the season started off with many reports. This picture continued all summer long as infection events kept re-occurring.

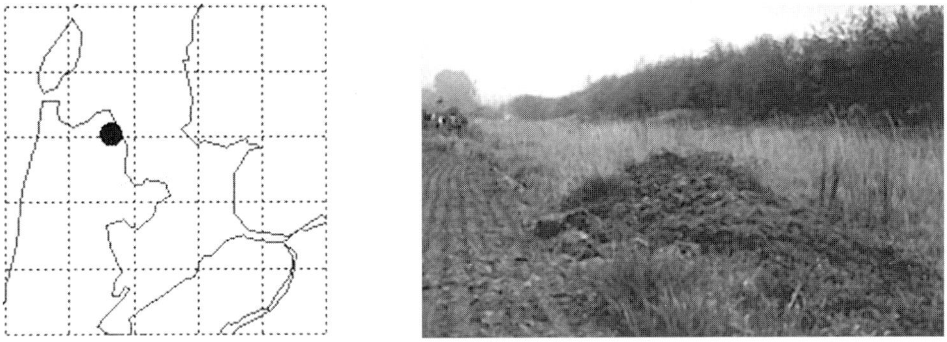

Figure 3. location and view of the first outbreak in 2000.

2001

In 2001 the first reports came in on May 15. In general farmers and field managers were very alert on outbreaks because of the situation at the end of last season. This made determination easy as the infections could be read as the growing rings in a tree. In the first half of the season, just three major infection events took place causing a slow spreading of the disease through the commercial fields. By the end of the year, the panic was forgotten.

2002

The first part of the season was relative dry. The first infections were all caused by infected volunteers. Typically, this first infected plant would be almost totally destroyed before it could infect near-by plants. Especially given the fact there were only few infection events. In other words, the infection will spread in the primary plant itself, but has no opportunities to spread around.

2003

The PLANT-Plus model calculated an infection event based on the weather forecast for May 20 till May 24. Based on the Masterplan telephone alarm system, all farmers in the Netherlands received a message to protect their crops, as emergence was already well on the way. Many farmers decided not to treat as the plants were still small. In the last week of May many outbreaks were reported, especially in the sandy North Eastern part of the Netherlands. Given the relatively dry spring and the sudden infection event it turned out that a lot of oospores managed to infect the crops. An early treatment would probably not have prevented the outbreak, but surely have suppressed it. Oospores are particularly dangerous as the number of primary sources within a field can be extremely high, up to thousands of plants per hectare.

2004

In 2004 the first in-field infections were only found in early June. Unfortunately also some infected dumps were reported again, the first increase since the downward tendency started with the Masterplan measures in 1999. These first infections were caused by infection events on May 9 and 10. Later on there was a significant three day infection event again starting on May 31. In some specific areas an excessive amount of rainfall was measured, causing the farmers not to be able to apply chemical treatments. This resulted in several outbreaks from infected seed, although in some cases it was not possible to determine the primary source as the infection was already too far established in the crop.

2005

In 2005 the first infections were unfortunately again reported on dumps, already on May 2nd. The size of the lesions fitted the infection events from mid April. Although there is a regulation that farmers are obliged to cover dumps with plastic before April 15th to prevent plants from growing, it takes continuous effort to convince farmers to really do it. Given the relatively dry and cold weather conditions, the infection did not really spread into fields. The first large scale in-field infections were found mid May. The size of the initial lesions exactly matched the first large infection event on May 3 and 4 with only very little spread from a small infection event on May 20 and 21. The primary source could be identified as oospores and infected seed. The primary source has a large effect on the potential risk: where infected seed usually leads to dozens of infected plants on a hectare, oospores can lead to thousands of infected plants per hectare.

Discussion

The above résumé indicates that it is very difficult to predict the primary source of infection that will start the epidemic in any given season. Although this starting-point for each new season is cause for concern, the problems will only occur under the right conditions. Given the weather conditions in the spring, sources like oospores and/or infected seed might even not lead to infection of the crop at all. And if the conditions are so that they can infect, infection events for late blight will be necessary to cause spread to other plants. Timely treatment with chemicals before such infection events can prevent that from happening. Analyses of the outbreaks learned that the PLANT-Plus model accurately calculates late blight infection events as the age of leaf and stem lesions matches the days after the infection events. Elimination and prevention of primary sources is of course also important, although not always possible. It goes without saying that removing or covering potato dumps should be the first step.

The proposed strategy in order to control early infections in the crops includes:

- Cover or remove potato dumps.
- Early inspections of the field.
- Keep an eye on volunteers.
- Protect crop before Infection Events with the right product.

Acknowledgements

The project to monitor, inspect and analyse the *Phytophthora* outbreaks has been co-funded by Hoofd Productschap Akkerbouw, Dacom and the Masterplan Phytophthora. We have to thank all the farmers and field-scouts who have reported the Phytophthora infections thus enabling us to inspect, analyse and learn.

Curriculum vitae – Peter Raatjes

Peter Raatjes studied Agronomy at the Agricultural College Prof. H.C. van Hall Instituut in Groningen, the Netherlands from which he got his BSc diploma in 1992. He took special courses in software programming, database management and farm economics. For AVEBE he preformed a study on the factors influencing the quality of starch potatoes. Since 1993 he works for Dacom PLANT-Service, currently as the operations manager. Since 1995 he has been the project leader for the Dacom disease mapping system in the Netherlands. Since 1999 he has been a member of the team of experts that inspects and analyses the early outbreaks in the Netherlands.

Company profile – Dacom PLANT-Service BV

Dacom is a company specialized in the development and marketing of decision support and traceability systems using the latest technology. In the mean time, farmers in most of the European countries, Japan, South Africa, Egypt, the USA and Canada are using the system. Demonstrations are underway in Russia, South America, Australia and New Zealand. In this perspective we can speak of a global system. Advice is not only given for a variety of crops and fungal diseases but also on irrigation management and an insect model is being tested. The PLANT-Plus system consists of integrated modules for data-communication, crop recording, weather data and crop advice. The communication module basically uses the internet but can also interact by phone with a voice response system. Warning messages can be communicated by fax or to a cell phone by text messages. The total concept includes disease mapping and data sharing with crop consultants and partners in the processing chain. The system is coordinated from a central databank in the Dacom office in Emmen.
For potatoes the PLANT-Plus system includes two state-of-the-art disease forecasting models: *Alternaria solani* and *Phytophthora infestans*.

The Netherlands Umbrella Plan Phytophthora in (inter)national perspective

Piet M. Boonekamp
Plant Research International, Wageningen UR, P.O. Box 16, 6700 AA Wageningen, The Netherlands, Email: piet.boonekamp@wur.nl.

Abstract

Late blight is the most important world wide disease of potato, and control of the disease is only possible by frequently spraying fungicides. For a sustainable production of potato the input of fungicides has to be reduced and replaced by integrated crop protection, including resistant varieties. To accomplish this, a strong interaction between integrated research and stakeholders in the potato production chain has to take place, with a joint focus on short- and long term programmes. A few years ago, such a consortium was formed in The Netherlands, the Umbrella Plan Phytophthora. In this paper the research themes and organization of the Umbrella Plan Phytophthora will be shown, as well as some possibilities to use this plan as a platform for initiating new and challenging ways for European and global cooperation in our fight against late blight.

Introduction

Potato, at the global level the fourth food crop after the cereals wheat, rice and maize is becoming increasingly important. The reason is that potatoes are highly nutritious, can be grown in a wide variety of soil- and climatic conditions, and the production/input ratio is high in terms of water, fertilizers, conversion of light to starch, if compared with cereals. In addition potato is very useful for processing in various products which are highly wanted by the modern urban society. There is one major world wide disease, late blight caused by the oomycete *Phytophthora infestans*. Although late blight became a major problem more than 150 years ago causing a famine in Western Europe, and much research has been performed during the last decades, no sustainable solution has been found yet. Optimal potato production is highly dependent on extensive spraying of fungicides against late blight (Figure 1). It is estimated that late blight is responsible for annual pesticide costs and yield losses amounting to at least € 3 billion. Therefore it would be a challenge if forces could be joined to conquer late blight on a global scale. The Global Initiative on Late blight (GILB) is an important initiative to bring world wide research on late blight together. However GILB is rather coordinating present research than setting the agenda for new integrated short- and long term research, which is needed for a sustainable solution. Therefore in this paper we propose that various late blight research programmes should be linked in a network, which can co-ordinate present research, and organize consortia to finance future research as well. We think that the Umbrella Plan Phytophthora, which has been launched two years ago, is a good starting point for world wide extension.

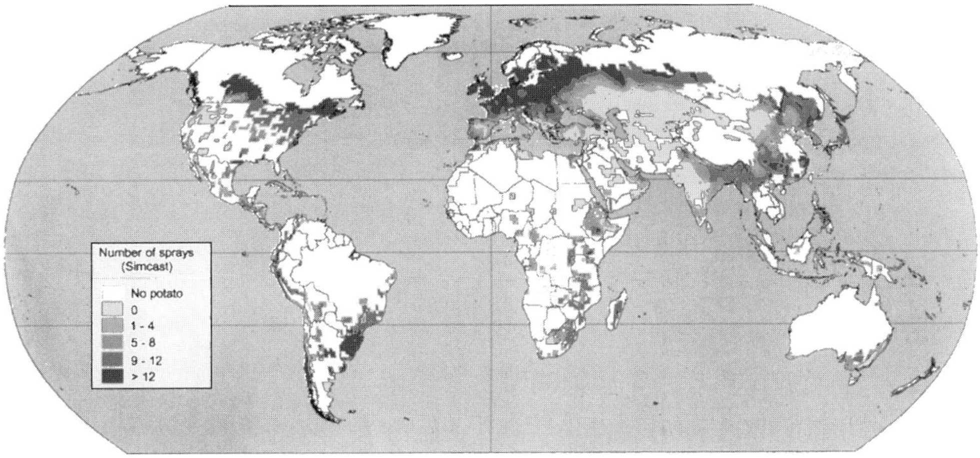

Figure 1. The number of annual sprays against late blight is indicated, if all potato-growing areas in the world bring their yield of potato production at the western European standard (reprinted from R.J. Hijmans et al. *Plant Biology (2000), 49, 697 - 705).*

Production of potatoes is of prime agricultural importance in The Netherlands. Late blight, caused by the oomycete *Phytophthora infestans* is the most important threat to this crop. Control of the disease is only possible by frequently spraying fungicides, which amounts to more than 50% of all fungicides used in The Netherlands. For more than a century the *Phytophthora* population was rather uniform and control was possible by integrated crop protection in combination with more or less resistant potato cultivars. However, due to import of infected potato material, the *Phytophtora* population has changed during the last 20 years, becoming more aggressive, producing persistent oospores, showing more genetic variability, thus enabling adaptation to the present resistant cultivars. The increased aggressiveness, the decrease of resistance and, consequently, the large input of fungicides form a threat to sustainable potato culture in The Netherlands. This threat has led to the initiative to start an Umbrella Plan Phytophthora, 2003-2012.

The aim of the Umbrella Plan Phytophthora is to reduce the negative impact on the environment of the use of fungicides to control *P. infestans* in potato by 75% in 2012 by three strategies. Firstly, to integrate all present and new research, and to focus all research on the reduction strategy. Secondly, to hand over the steering of all research to a board of representatives from the potato sector to ensure commitment to and application of the results of all short term and long term research. Thirdly, to combine the three parties - research (Wageningen UR), policy (Ministry of Agriculture) and potato sector - in one consortium to ensure that each party takes its responsibility for reaching the 2012 aim (Figure 2).

Research themes in the Umbrella Plan Phytophthora

All research is brought together in six themes that form a highly integrated programme, (Figure 3).

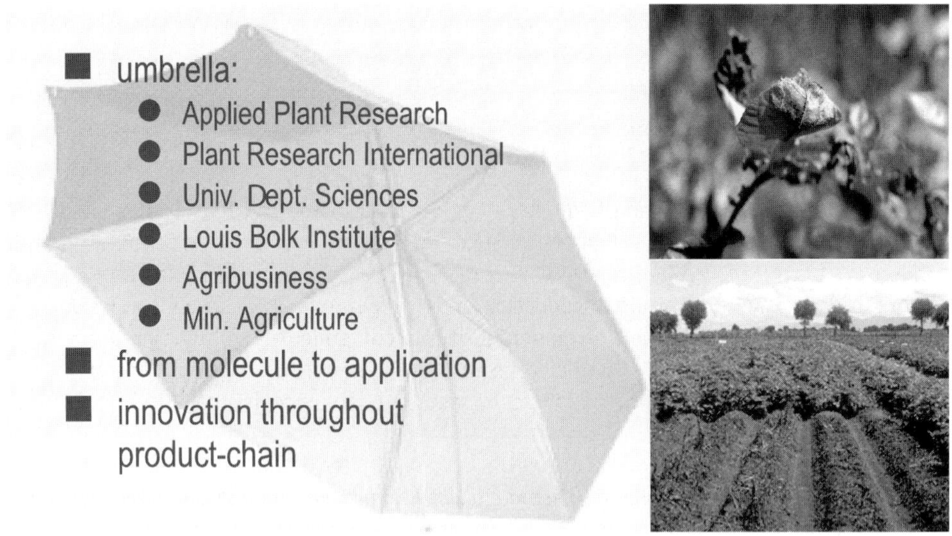

umbrella:
Applied Plant Research
Plant Research International
Univ. Dept. Sciences
Louis Bolk Institute
Agribusiness
Min. Agriculture
from molecule to application
innovation throughout product-chain

Figure 2. Consortium formation of all the Dutch research partners, with all partners in the potato production chain and the Ministry of Agriculture for a unique joint sustainability aim.

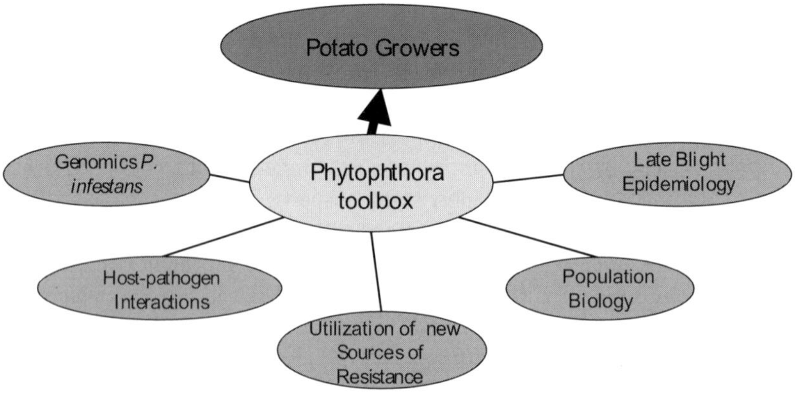

Figure 3. The results from five themes of short- and long-term research are integrated in the Phytophthora *toolbox (theme 6) and translated to be applicable by the potato sector.*

Theme 1: *Phytophthora* Toolbox

Within the Toolbox results from the other five themes will be translated into practical solutions resulting in an integrated control strategy for *P. infestans* with a minimal input of fungicides. The strategy will have to be adapted to the specific Dutch conditions regarding crop intensity, varieties, pathogen population, pathogen epidemiology, available fungicides, efficacy of fungicides during the growth cycle, and weather conditions. Integrated practical knowledge will be generated by experimental applied research, on-farm research using different farming systems and by evaluation of farmers' results after applying the new knowledge. For the first results, see H. Schepers, in these proceedings.

Theme 2: Population Biology

High levels of genetic variation mark *P. infestans* populations in the Netherlands, which may be associated with the presence of sexual reproduction by means of oospores in diseased crops. This theme is aimed at generating knowledge on functional variation in *P. infestans* that can be used to improve late blight management strategies and risk assessments. Research topics are monitoring the various sources of early infections and of specific virulence and genetic diversity, risk assessment of future introduction of new *Phytophthora* pathogens in Europe from the Andean highlands, evaluation of epidemic fitness components, and studies on mutation rates in *P. infestans* that lead to fungicide resistance. Some results of this theme are:

- Monitoring and analysis of fungicide resistance and virulence in the Dutch *P. infestans* population.
- Monitoring and analysis of early late blight foci in the Netherlands.
- Analysis of fitness traits of *P. infestans* genotypes as related to build up of the population during the growing season, survival of the population during the winter and emergence of potato late blight through early foci in the following growing season.
- Monitoring of risk on introduction of new, pathogenic *Phytophthora* spp.

Theme 3: Epidemiology

This theme aims to contribute to improved, reliable, year round control strategies through additional (quantitative) insight in the potato late blight disease cycle and epidemic build up. Projects focus on reduction of tuber-borne primary inoculum, improved estimates for disease pressure as related to timing and input of chemical control measures, improving yield and yield quality of organic potato by advancing yield formation and increasing the partial resistance level through agronomic measures and physiological optimization of seed tubers. Prototype control measures are evaluated in a practical context in the theme 'toolbox'. Some results of this theme are:

- Improved estimation of the infection risk for potato crops incorporating past, present and future weather, cultivar resistance and fungicide protection level. New components include survival of spores during transport in the atmosphere and updated climate relationships for infection parameters such as the minimum leaf wetness period needed for infection at a wide range of temperatures, infection efficiency, sporulation, etc.
- Estimation of the risk on tuber infection. The tuber infection cycle is quantitatively and qualitatively analyzed to arrive at specific warnings for risk on tuber infection and retrospective identification of seed tuber risk - lots based on agronomic history. Both actions reduce the number of tuber infections and thus the primary infection pressure.

Theme 4: New sources for resistance

Development of durable disease resistant potato varieties is the most effective way to control late blight. The major resistance genes in the current potato varieties are no longer effective against *Phytophthora*. In this theme an elaborate and systematic search for new genes is performed in some hundred wild species, related to potato. Whenever new major resistance genes are found, they will be characterized for variation in function and combined by breeding.

After testing for resistance, promising material will be handed over to the breeders for integration in their potato breeding programmes. Some results of this theme are:

- A systematic search for resistance is in progress by evaluating 1000 *Solanum* accessions (5 genotypes per accession) representing 164 *Solanum* species.
- Intra- or interspecific crosses between highly resistant and fully susceptible genotypes are made to generate 1:1 segregating populations for major resistance genes.
- So far 4032 genotypes have been tested for resistance and 1209 highly resistant, 714 intermediately resistant, and 2109 fully susceptible genotypes have been identified.
- Out of these 4032 genotypes a subset of 855 genotypes is selected for in vitro maintenance and crossing purposes.

Theme 5: Genomics potato-*Phytophthora* interaction

It is anticipated that breeding for resistance in potato, aiming at substantially contributing to disease management, requires, by one way or another, the deployment of known and yet unknown *R* genes. Combining *R* genes is a strategy that could lead to durable resistance, if genes with complementary mechanisms, and recognizing pathogen factors that are essential for pathogen fitness, are combined in one potato clone. However, our understanding of resistance mechanisms is still insufficient to be able to predict which *R* genes should be combined to obtain durable resistance. The aim of this theme is to increase the molecular understanding of the interaction process by identifying genes of potato and *Phytophthora*, which in the initial stages of an infection determine the outcome of the interaction, both at the host and non-host level. The generation and analysis of defence-related mutants and the implementation of transient silencing techniques like virus induced gene silencing (VIGS) in *Solanum* spp. enable the functional analyses of key regulators of defence in late blight resistance. Recent results include:

- the identification of genes differentially expressed during compatible and incompatible interactions between potato and *Phytophthora* using cDNA micro-arrays;
- the identification of *Arabidopsis* deletion lines that display varying levels of susceptibility to *Phytophthora*;
- over-expression of candidate key regulators with defending properties and their phenotypic analyses;
- the implementation of VIGS in numerous *Solanum* species and the transient silencing of *R* genes and other defence related genes.

Theme 6: Genomics *Phytophthora infestans*

In this theme current genomic and EST sequence data of *P. infestans* will be screened as efficiently as possible to identify candidate genes that possibly encode for pathogenicity and (a)virulence factors and to identify variable regions in the genome that can be used for the efficient development of molecular markers. These markers can subsequently be used to study the variation and dynamics of natural *P. infestans* populations. The final goal is to identify factors that are essential for the overall fitness of the pathogen. The targeting of such factors for chemical control will lead to the development of innovative control strategies. For the first results, see F. Govers in these proceedings.

Organization and cooperation

The organization of the Umbrella Plan is outlined in Figure 4. Central is the Steering Committee with representatives from the stakeholders. Although the Dutch Ministry of Agriculture is financing the major part of the research at Wageningen UR, their role in the Steering Committee is modest. The potato sector (breeders, growers and trade) has the most prominent role in the Steering Committee, as this sector is able to keep research on the applied track and to communicate and implement the results into practice. As shown in Figure 4 "steering" occurs in two annual loops: one is steering the Wageningen UR research to fill the *Phytophthora* 'Toolbox', another is implementing the toolbox in practice. The implementation is monitored in different farming systems with respect to environmental effects, and problems will be reported to Wageningen UR for feed-back. Whenever a tool is found to be useful and robust, the Steering Committee can make the application of it mandatory for all farmers.

The Umbrella Plan Phytophthora might be an excellent platform for further international collaboration as indicated in Figure 5.

Although a substantial research programme was launched in the Umbrella Plan, it was realized that much more research input would be needed to conquer this very complex disease. Therefore from the start onwards, the aim of the Umbrella Plan was to start international collaboration. Similar to the Umbrella Plan, research groups in the Nordic countries (Sweden, Norway, Denmark and Finland) have joint forces in NORPHYT, a research programme on potato late blight epidemiology and warning systems. Cooperating institutes and universities are:

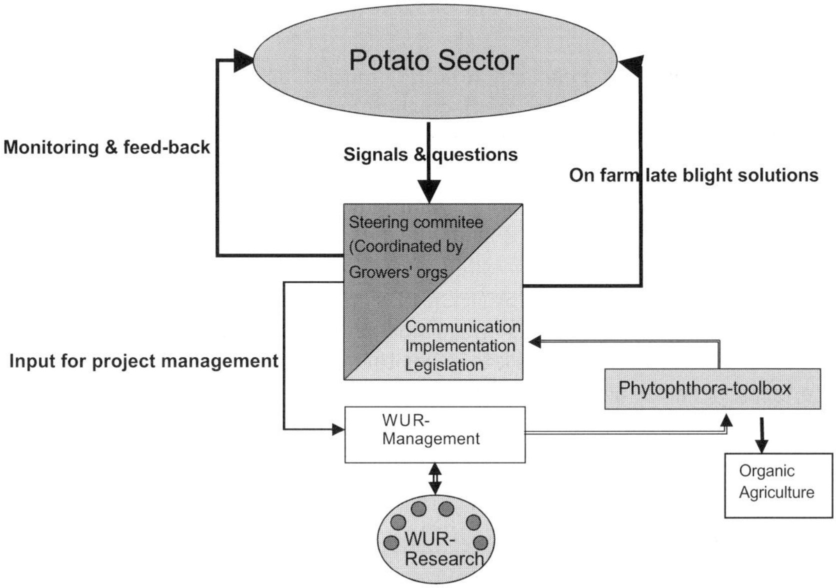

Figure 4. Organization-scheme of the Umbrella Plan Phytophthora. The Steering Committee is steering Wageningen UR research as well as communication and implementation of the results.

Towards global sustainable potato production

Building a case through 4 levels:

1) National (Umbrella Plan Phytophthora)
↓
2) International (WUR – Norphyt cooperation potato late blight)
↓
3) European (NoE ENDURE)
↓
4) Global, the ultimate aim

Figure 5. An ambitious scheme to establish a global initiative on integrated new research and implementation in world wide practice.

- the Danish Institute of Agricultural Sciences (DIAS)
- Norwegian Crop Research Institute (NCRI)
- Swedish University of Agricultural Sciences (SLU)
- Agrifood Research Finland (MTT)

NORPHYT is coordinated by Professor Jonathan Yuen at SLU, and the objectives are to:
- Evaluate and quantify epidemiological parameters of the new population of *P. infestans* in the Nordic countries;
- Implement these results in existing and new forecasting- and DSS;
- Develop a common framework for a core Internet based late blight DSS.

As the objectives of Umbrella Plan and NORPHYT turned out to be complementary, both initiatives joint forces. This cooperation is effectuated through a memorandum of understanding and effectively doubles the research capability on epidemiology and warning systems through coordination of research goals and experiments.

Similar to the Umbrella Plan, NORPHYT is also strongly linked to EUCABLIGHT, an EU concerted action. This collaboration will be further extended in a new initiative, a Network of Excellence on Crop Protection, (NoE ENDURE). INRA (France) is coordinating the NoE, and 10 other EU-countries participate, including some Central European countries. An NoE is a relatively new instrument of the EU commission with the objective to establish an EU-wide network aimed at 1) a durable restructuring of European R&D on the use of pesticides, and 2) integrating and improving expertise across disciplines to develop durable disease management strategies. The NoE programme will be submitted in the fall 2005, and it is anticipated that it can start early 2006.

One of the pilots will be potato, with *Phytophthora* as the main target to coordinate forces. Therefore the alliance of the Umbrella Plan and NORPHYT will be extended with other late blight research programmes in other European countries. A pragmatic approach will be chosen. In a first stage the critical factors for controlling late blight will be evaluated, 'best practices' of control systems in the different countries will be compared, and it will be discussed if 'best practice tools' can be transferred to another farming system in another country. This stage will be completed in one year with an inventory of tools that can be transferred from one country to another, and a 'white paper' describing new research activities which need to be performed for a durable solution of the late blight problem. This 'white paper' will be used for agenda setting of the FameWork 7 call, which will start in 2007. It is anticipated that the FW7 will contain ample possibilities to include new research on late blight, as 'sustainability' and 'agriculture' will apparently be higher ranking topics in FW7 than in the present FW6 EU-programmes.

Global need for new late blight research

As already mentioned the Umbrella Plan and NORPHYT participate in the Global Initiative of Late Blight (GILB). It would be important to go one step further than GILB, rather than coordination of research, starting new strategic and long term research as outlined in the Umbrella Plan would be essential. These research fields have to include functional genomics of *Phytophthora infestans* (gene functions that are essential and are therefore targets for new pesticides and resistance breeding), new resistance genes and breeding (genes from hosts and non-hosts and new breeding technologies), population genetics (to study population changes world wide and methods to avoid it) and epidemiology (nano-chips and GPS for early warning and robotized control measures).

A second approach might be the export of a 'sustainability toolbox' together with the seed potatoes. Western Europe is an important exporter of seed potatoes. After export these potatoes are grown under various conditions in the importing countries as climatic and logistic conditions are very different. If the knowledge we have on sustainable growing of potatoes could be optimized for the situation in an importing country and could be co-exported together with the seed potatoes, the world wide sustainability of potato growing will be improved, and the competition power of potato over other starch crops will be reinforced.

It might be a challenging result of Potato2005 if all stakeholders of the potato-network, including breeders, growers, trade, processors and research could join, to start this world wide initiative.

Eucablight: a late blight network for Europe

L.T. Colon[1] , D.E.L. Cooke[2], J. Grønbech-Hansen[3], P. Lassen[3], D. Andrivon[4], A. Hermansen[5], E. Zimnoch-Guzowska[6] and A.K. Lees[2]
[1]Plant Research International, PO Box 16, 6700 AA Wageningen, the Netherlands
[2]Scottish Crop Research Institute, Invergowrie, Dundee DD2 5DA, United Kingdom
[3]Danish Institute of Agricultural Sciences, Department of Agroecology, Research Centre Foulum, 8830 Tjele, Denmark
[4]UMR INRA-Agrocampus Rennes BiO3P, INRA Centre de Rennes, Domaine de la Motte, BP 35327, F-35653 Le Rheu Cedex, France
[5]Norwegian Crop Research Institute, Plant Protection Centre, Dept. of Plant Pathology, Hoegskoleveien 7, 1432 Ås-NLH, Norway
[6]Plant Breeding & Acclimatization Institute, Mlochow Research Center, 19 Platanowa street, 05-831 Młochow, Poland

Abstract

The pan-European Concerted Action on late blight 'Eucablight' was set up and launched in 2003, with the aim of providing tools for investigating variation in both the host and the pathogen. Objectives include the construction of a database (www.Eucablight.org) containing information on past and current potato cultivars and late blight populations available in the participating countries, and the design, testing and recommendation of protocols for testing host resistance and pathogen diversity. The database is structured, and made accessible, in such a way that many different target groups can use the data for their own purpose. Farmers will be able to find out about the characteristics of the pathogen and the best performing cultivars in their region. DSS advisers will be able to access the model parameters they need to build locally adapted forecasting systems and scientists will be able to employ the data to study host and pathogen interactions, and to place this in an historical perspective. Once the database is operational, maintenance and the entry of new data should take minimal effort. Submission of data is open to anybody who wants to contribute and who follows the standard protocols. We intend the database to generate enough interest to be kept up to date by members of the research community who will continue to submit their results. If this can be achieved, the database will be an important instrument for potato breeders, scientists, advisors and policy makers to follow the co-evolution of host and pathogen in Europe and inform the use of appropriate resistance genes and control measures.

Keywords: *Phytophthora infestans*, late blight, resistance, pathogen population

Introduction

Late blight (caused by the oomycete pathogen *Phytophthora infestans*) first occurred in European potatoes in 1845, when *P. infestans* was imported from the Toluca Valley in Mexico which is thought to be its centre of origin. The disease was first noted in Belgium, and spread rapidly to The Netherlands, France, Britain and Ireland, where it was responsible for causing famine in 1845-1848 (Bourke, 1964). In the 160 years since this introduction, *P. infestans* has remained one of the most devastating potato pathogens in Europe.

Almost immediately after the appearance of late blight, attempts were made to solve the problem using host resistance in the form of *Solanum demissum*, a Mexican relative of the potato *(S. tuberosum)*. This wild species was used as alternative crop (Lindley, 1848; Jühlke, 1849) and, without much success, as a source of major resistance genes in potato breeding programmes around 1850 (Focke, 1881). It was used again between 1909 and 1950, with the first resistant cultivars appearing on the market around 1930 (Müller and Black, 1952). Unfortunately, virulent strains of the pathogen overcame this resistance as early as 1849 (Jühlke, 1849) and have continued to do so. Virulence is usually a recessive trait, and European *P. infestans* strains are often polyploid, which allows them to store large numbers of unexpressed virulence genes in their genetic background without fitness costs. Virulent strains originating from this hidden reservoir have an enormous capacity to multiply rapidly and spread over long distances. Strains with virulence for all 11 R genes from *S. demissum* occur worldwide, although some virulence factors, such as Avr8 and Avr5, are less frequent, or, in the case of Avr9, rare. As a result, breeders have turned to new sources of major genes such as the Mexican species *S. bulbocastanum* or the South American species *S. microdontum*, using both direct gene transfer and traditional breeding methods.

Integrated strategies for the control of potato late blight using reduced fungicide inputs would benefit from the increased availability of durable blight resistance in commercial potato

Figure 1. Symptoms of late blight, caused by Phytophthora infestans, *on a potato leaf. The necrotic lesions (arrows), brown in colour, are surrounded by white halos consisting of sporangiophores carrying numerous sporangia.*

cultivars. Durable resistance, often termed field resistance, or race-non-specific (or horizontal) resistance, exists. However, in potato it is usually associated with late maturity and is selected against in breeding because breeders strongly favour early genotypes. Many sources of late blight resistance exist in wild and primitive potatoes and developed cultivars, but the nature of that resistance is often poorly understood. The current European data on host resistance are fragmented and the methods used to collect this data are often not well documented.

P. infestans is also a tuber pathogen, and therefore both foliage and tuber resistance should be considered.

The effective deployment of resistant commercial cultivars creates a "moving target" for *P. infestans*, but such a strategy can only be effective if we understand the existing pathogen population structure and can predict its ongoing evolution. *Phytophthora infestans* populations have undergone many significant changes since 1845, with several major migration events (Fry *et al.*, 1992), adaptive response to new resistance genes or agrochemicals introduced to combat the pathogen (Wastie, 1991), and the recent occurrence of sexual reproduction in several parts of the European continent (Drenth *et al.*, 1994). Many studies on the structure and evolution of these populations have been undertaken, but there is still a lack of general understanding of the mechanisms involved, including for instance, the extent of metapopulation structures and the impact of local and long scale migrations.

The Eucablight project, a concerted action funded by the European Union, is undertaking to improve this situation. The collation of the available data into a harmonised and readily accessible database will allow breeders and geneticists to compare or exploit sources of resistance in their breeding programmes. Other target groups are farmers and advisers using Decision Support Systems, who need reliable information about the aggressiveness of the late blight population and the level and stability of resistance in the potato cultivars in their region. The Eucablight project started in February 2003 and will run for three years. Eucablight includes 24 member institutions from EU countries and representatives from several non-EU countries. It is coordinated by Alison Lees, and operates on both a thematic (host or pathogen) and a regional basis. This dual structure favours interactions between members. Objectives include the construction of a database containing reliable information available in participating countries on past and current potato cultivars and late blight populations, and the design, testing and recommendation of protocols for testing host resistance and pathogen diversity.

The database

The database (www.Eucablight.org) is accessed and structured in such a way that many different target groups can use the data for their own purpose. Farmers will be able to find out about the type of pathogen and the best performing cultivars in their region. DSS advisers will be able to access model parameters they need to build locally adapted forecasting systems and scientists will be able to employ the data to study host and pathogen interactions and to place this in an historical perspective.

The database has been created, and is hosted, by the Danish Institute of Agricultural Sciences (DIAS) at Foulum, Denmark. The technical details of the database are described by Lassen and

Hansen (2005). The information in the database consists of phenotypic data from a number of potato cultivars and late blight strains. All the information is associated with specific years and geographic areas, allowing direct comparisons of specific host events (e.g. breaking of a new resistance gene) to specific pathogen events (introduction of a new genotype of the pathogen).

The data can readily be enriched by new entries using two specially designed software programmes, Cultivar.exe and Phytophthora.exe, developed by the project.

An important characteristic of the database is that the data contributed (primary data) can only be viewed directly, or retrieved by the contributors themselves. This will ensure that the copyright of the original data remains with the contributor. Both the host and pathogen data will be processed into secondary variables and displayed on the website as tables, graphs and maps. Examples are AUDPC or frequency of metalaxyl sensitivity. This will provide synthetic overviews of the resistance of the cultivars, the diversity of the pathogen population and their distribution in time and space across Europe. The displays, automatically updated when new data are added to the database, will be made available to Eucablight members and non members through the Eucablight website.

Of course, the database is open to data from Eucablight non-members, provided that they follow the protocols defined or agreed by Eucablight.

Protocols and training

Protocols for assessing biological, pathogenic and genetic diversity vary between laboratories engaged in host resistance and population analyses of *P. infestans* worldwide, but also in Europe. These differences make the comparison of data collected by different teams difficult, and sometimes downright impossible. Furthermore, technologies for the analysis of genetic and phenotypic diversity evolve rapidly. Therefore, one of the main objectives of Eucablight is to collate, formalise, assess and recommend the most suitable protocols, through the collective work of its members. These protocols are available through a dedicated section of the Eucablight website, and formed the basis of training courses available to both members and non-members. The overall aim is to provide reliable, standardised methodologies generating readily comparable data, and to facilitate the use and adoption of improved protocols for future work.

The protocols were compiled by members of the 'host' and 'pathogen' technical groups of Eucablight, and discussed during project meetings before validation as 'Eucablight Recommended protocols'.

Links to other databases

The host data in the Eucablight database will complement the European Cultivated Potato Database (www.Europotato.org), currently the best source of information on late blight resistance in potato, which collates all the available information on a large number of European potato cultivars. This information includes blight resistance ratings, which have been taken from existing national lists. These provide some insight into the resistance level of the cultivars, but the range of classification that is given to a single cultivar usually shows a large variation because data originate from trials conducted in different years and with different pathogen populations, using different methods (Zimnoch-Guzowska and Flis, 2002; Hansen *et al.*, 2005).

Host data

The Eucablight database contains resistance data from trials that include seven standard cultivars and which have been scored using standard disease assessment methods. The standard cultivars have been chosen on the basis of their availability and the expected durability of the resistance, and represent the extremes of susceptibility and resistance in the three maturity classes 'first early' (Eersteling and Gloria), 'maincrop' (Bintje and Escort) and 'late' (Alpha and Robijn). Sarpo Mira, was added as a representative of the highest resistance class, although the durability of its resistance has yet to be established. To facilitate the dissemination of the seven standards, they have been distributed as seed potatoes and are also being made available from the Scottish Agricultural Science Agency (SASA) in East Craigs as *in vitro* cultures. In addition to these standards, the inclusion of SASA's 11 single R gene differentials R1 - R11 in the trials is also recommended to ensure that high resistance scores are not due to a simple pathogen race structure.

A requirement of the trials is the use of standard protocols to carry out resistance assessments on foliage and tubers. Protocols for foliage and tuber resistance include a field trial, a detached leaflet assay and a whole plant glasshouse assay, and a field trial, a whole tuber lab test and tuber slice lab test, respectively. A protocol relating to plant maturity is also available, since this trait is highly correlated with resistance to late blight in the foliage. All protocols and other information are available as PDF documents at www.Eucablight.org.

Trial data [Name: 2004_DK_09_01]

Print this page

Show graphics | Results for all trials | Documentation

Year: 2004 Country: Denmark Region: Vestsjælland Number of cultivars: 46 Number of dates: 14 Number of replicates: 3 Responsibilty: Bent J. Nielsen

Cultivar/Genotype	AUDPC		RAUDPC		AIR (Logistic)		Goodness of fit (R^2)		AIR (Gompertz)		Goodness of fit (R^2)		Delay, 1% disease (days)		Delay, 5% disease (days)	
	mean	SE	mean	SE	mean	SE	min	max	mean	SE	min	max	mean	SE	mean	SE
Sarpo Mira	212,7	67,6	0,03	0,01	*	*	*	*	*	*	*	*	*	*	*	*
Kuras	1.821,6	74,3	0,26	0,01	0,15	0,01	0,94	0,98	0,09	0,00	0,89	0,97	19,4	1,5	30,7	1,3
Valiant	1.850,3	94,2	0,27	0,01	0,15	0,01	0,94	0,94	0,10	0,01	0,89	0,90	18,4	0,8	29,7	0,5
Mercury	2.403,2	181,5	0,35	0,03	0,21	0,01	0,95	0,97	0,15	0,01	0,89	0,92	22,4	1,7	30,2	1,6
Aviala	2.800,1	90,0	0,41	0,01	0,20	0,02	0,94	0,96	0,15	0,01	0,90	0,95	17,5	1,9	25,8	1,3
Kardent	2.831,5	98,8	0,41	0,01	0,19	0,01	0,90	0,97	0,13	0,01	0,86	0,95	16,6	1,7	25,1	1,4
Robijn	2.941,2	154,3	0,43	0,02	0,14	0,01	0,87	0,91	0,10	0,01	0,92	0,96	7,3	4,4	19,2	3,2
Canasta	3.037,4	22,2	0,44	0,00	0,24	0,01	0,92	0,96	0,16	0,01	0,89	0,96	19,4	0,7	26,3	0,5
Escort	3.318,4	110,4	0,48	0,02	0,35	0,04	0,94	0,98	0,23	0,02	0,89	0,96	22,2	3,2	27,1	2,5
Kardal	3.321,8	127,4	0,48	0,02	0,20	0,01	0,93	0,95	0,15	0,00	0,95	1,00	13,3	1,7	21,6	1,4
Bilbo	3.374,0	142,1	0,49	0,02	0,17	0,01	0,87	0,93	0,13	0,01	0,89	0,98	8,8	1,9	18,4	1,7
Producent	3.409,0	55,9	0,49	0,01	0,18	0,00	0,89	0,91	0,13	0,00	0,98	0,98	11,6	0,8	20,7	0,7
Tivoli	3.441,0	104,1	0,50	0,02	0,35	0,03	0,97	0,99	0,24	0,00	0,92	0,98	21,4	0,3	26,1	0,3
Starter	3.724,7	111,1	0,54	0,02	0,30	0,02	0,93	0,95	0,22	0,02	0,89	0,99	16,7	1,6	22,3	1,3

Figure 2. Example of table of results generated by the Eucablight database from disease observations in an inoculated field trial in Denmark, 2004.

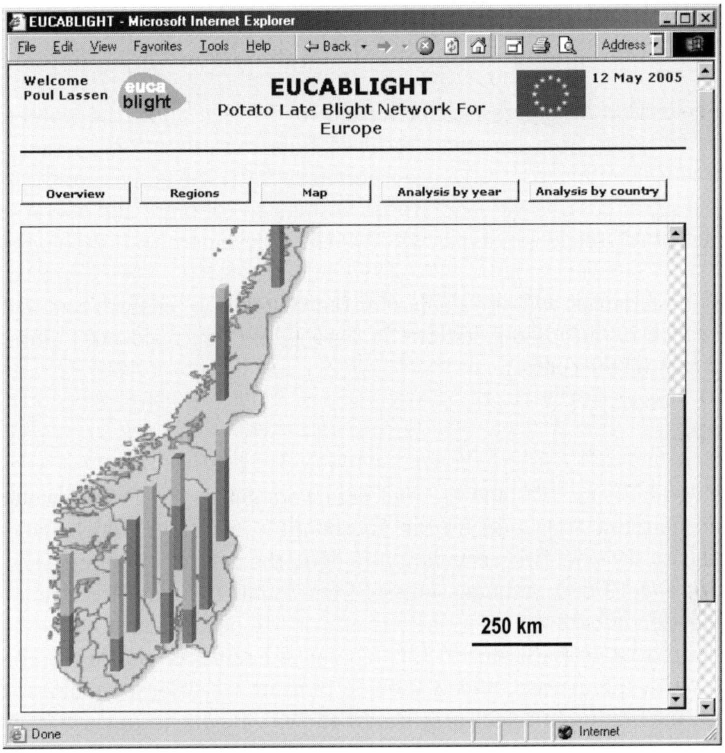

Figure 3. Example of map with overview of selected pathogen data, in this case mating type distribution in Norway.

In order to allow monitoring of the resistance of a cultivar from the time of first selection, the database contains data from both cultivars and breeding lines. A database containing details of more than 4000 cultivars will help to link the identities of breeding lines to cultivars, and to avoid duplicated entries due to misspellings. This section of the Eucablight database also contains data relating to presence of R genes, fertility, use (ware, processing or starch) and ploidy level of the germplasm, collated from published and unpublished sources.

Pathogen data

The pathogen data in the Eucablight database mainly represent isolates collected since 1990, and contain both phenotypic (e.g. mating type, phenylamide resistance, virulence) and genotypic (e.g. isozymes, mtDNA fingerprints, SSR alleles) information. The database is designed to accommodate new markers as these become widely available.

The reliability of virulence tests has been questioned repeatedly, and Eucablight has therefore designed experimental validation trials in the form of blind ring tests. The virulence ring test involves 10-12 laboratories within the Eucablight Consortium who, in 2005, tested ten coded isolates according to an agreed protocol and on the same set of differential clones (provided in 2004 by SASA).

A small core collection of isolates, characterised for most or all of the traits entered in the database, was compiled from donations received from the existing collections of Eucablight participants and is maintained as reference material by SCRI. The purpose of this core collection is to serve as a set of controls for laboratories wishing to calibrate and operate a technology, as well as a set of reference genotypes with which to develop new technologies.

These Eucablight activities will be valuable for the description, analysis and understanding of the evolution of *P. infestans* populations in Europe, and thus aid both current and future strategies for late blight control.

Future steps

The database was built in 2003 and the first data were entered in 2004. Encouragingly, there were also contributions from outside the Eucablight consortium, which indicates that the concept appeals to the scientific community. To date, the database contains over 4 000 potato cultivar entries and 10 000 pathogen entries from countries across the whole of Europe. In 2005, the collection of data will continue and efforts will be concentrated on the outputs that will be made available on the website. This requires a number of choices to be made relating to statistics and to the disease models used to transfer the disease readings into secondary variables. The next step is to select graphics to allow comprehensive viewing of the database across space and time.

The database should also be a means to link important information on resistance phenotype to rapidly emerging knowledge about the corresponding potato gene sequences. This should help molecular scientists to decide which genes are worth attention.

Conclusion

A particular advantage of the Eucablight project has been to facilitate intensive and wide-ranging discussions on host resistance testing methods and to use our accumulated expertise to disseminate standardised protocols, information and training across Europe and internationally. Once the database is operational, maintenance and entry of new data should take minimal effort. Submission of data is open to anybody who wants to contribute and who follows the standard protocols. We intend the database to generate enough interest to be kept up to date by members of the research community who continue to submit their results. If this can be achieved, the database will be an important instrument for potato breeders, scientists, advisors and policy makers to follow the co-evolution of host and pathogen in Europe and inform the use of appropriate resistance genes and control measures.

New contributors are welcome and can contact Alison Lees at alees@scri.ac.uk.

Acknowledgements

The 'Eucablight - A Late Blight Network for Europe' project (QLK5-CT-2002-00971) is supported by the European Commission under the Fifth Framework Programme.

References

Bourke, P.M.A., 1964. Emergence of potato blight, 1843-46. Nature, 203, 805-808

Drenth A., I.C.Q. Tas and F. Govers, 1994. DNA fingerprinting uncovers a new sexually reproducing population of *Phytophthora infestans* in the Netherlands. European Journal of Plant Pathology, 100, 97-107.

Focke, W.O. 1881. Die Pflanzenmischlinge. Ein Beitrag zur Biologie der Gewächse. (Contribution to the Biology of Crop Plants) Verlag von Gebrüder Borntraeger, Berlin, 567 pp.

Fry W.E., S.B. Goodwin, J.M. Matuszak, L.J. Spielman, M.G.Milgroom and A. Drenth, 1992. Population genetics and intercontinental migrations of *Phytophthora infestans*. Annual Review of Phytopathology, 30, 107-129.

Hansen, J.G., M. Koppel, A. Valskyte, I. Turka and J. Kapsa, 2005. Evaluation of foliar resistance in potato to *Phytophthora infestans* based on an international field trial network. Plant Pathology, 54, 169-179

Jühlke, 1849. Beitrag zu dem *Solanum utile* von Klotsch. (Contribution to Klotsch's *Solanum utile*) Allgemeine Gartenzeitung, 17, 356.

Lassen, P. and J.G. Hansen, in press. EUCABLIGHT database: collecting, storing and analysing data related to potato late blight. Proceedings of the Joint 5th Conference of the European Federation for Information Technology in Agriculture, Food and Environment and the 3rd World Congress on Computers in Agriculture and Natural Resources, July 25-28, 2005, Vila Real, Portugal.

Lindley, J., 1848. Notes on the wild potato. Journal of the Royal Horticultural Society, 3, 65-72.

Müller, K.O. and W. Black, 1952. Potato breeding for resistance to blight and virus diseases during the last hundred years. Zeitschrift für Pflanzenzüchtung, 31, 305-318

Wastie R.L., 1991. Breeding for resistance. In '*Phytophthora infestans*, the cause of late blight of potato' (D.S. Ingram, P.H. Williams, eds). Advances in Plant Pathology, 7, 193-224. Academic Press, London.

Zimnoch-Guzowska, E. and B. Flis, 2002. Evaluation of resistance to *Phytophthora infestans*: A survey. In: Lizarraga, C. (ed.) Proceedings of the Global Initiative on Late Blight (GILB) Conference, 11-13 July 2002, Hamburg, Germany. CIP Lima, Peru: 37-47.

Curriculum vitae – Leontine T. Colon

Leontine T. Colon studied Phytopathology and Plant Breeding at Wageningen University, attaining the MSc degree in 1986. In 1986 she joined the research staff of Plant Research International, where she began her work on resistance to *Phytophthora infestans* (late blight) in potato. In 1994 she received her PhD degree in Plant Sciences defending a thesis on sources of durable resistance in the cultivated potato and in wild potato species. She continued her work on potato late blight concentrating on genetics and mechanisms of resistance, the relation with maturity and the practical use of durable late blight resistance with special emphasis on organic farming. She was one of the initiators of the European potato late blight database 'Eucablight' and manages the host section of this database.

Company profile – Plant Research international B.V.

Plant Research international B.V. is part of the Plant Sciences Group, a partnership with Wageningen University, Plant Sciences and Applied Plant Research (PPO).

Within Wageningen UR the Plant Sciences Group brings together scientific education and fundamental, strategic and applied research in the fields of biology, of plants in relation to their environment, of plant-related organisms and of plant production.

Plant Research International B.V. specialises in strategic and applied research. Thanks to a sophisticated combination of knowledge and experience in genetics and reproduction, genomics, proteomics, metabolomics, bioinformatics, crop protection, crop ecology and agrosystems, Plant Research International offers a unique range of perspectives for government and industry. It services the entire agro-production chain with scientific products, from the DNA level to production system concepts. Plant Research International takes a business-like approach and delivers exceptional quality results. The institute regularly has articles in the leading scientific publications and has a superb infrastructure.

Please contact us through info.plant@wur.nl or consult our website www.plant.wur.nl for further information.

Potato blight populations in Ireland and beyond

L.R. Cooke[1,2] and K.L. Deahl[3]
[1]Department of Applied Plant Science, Faculty of Science and Agriculture, The Queen's University of Belfast
[2]Applied Plant Science Division, Department of Agriculture and Rural Development, Newforge Lane, Belfast BT9 5PX, UK
[3]USDA, ARS, PSI, Vegetable Laboratory, Beltsville, MD 20705, USA

Abstract

Phytophthora infestans in Northern Ireland was studied using isolates collected mainly from commercial potato crops between 1995 and 2002. Phenotypic diversity was assessed using mating type and metalaxyl resistance; genotypic diversity was assessed using two allozyme loci (glucose-6-phosphate isomerase, *Gpi,* and peptidase, *Pep*), mitochondrial DNA haplotype and the multilocus RFLP probe RG57. Based on published data, the *P. infestans* population in Northern Ireland was compared with that in the Republic of Ireland. In the island of Ireland, the population appears to consist of a limited number of clones of the A1 mating type, A2 types are very infrequent, if present at all, and there is no evidence of sexual recombination. Comparisons with populations in Great Britain and mainland Europe indicate that these are more diverse than the population in Ireland. In some regions of Europe, sexual recombination may occasionally contribute to this diversity, while in others it appears to occur frequently, but it is unclear to what extent it influences the epidemiology of the pathogen. *P. infestans* populations in Europe are compared with those in N. America. Differences in *P. infestans* populations within Europe and between Europe and N. America suggest a number of questions, which will only be answered by international co-operation.

Keywords: *Phytophthora infestans*, allozymes, mating type, mitochondrial DNA haplotype, RG57 analysis

Introduction

In the past 25 years, world-wide changes in populations of *Phytophthora infestans* have been associated with increased problems in controlling late blight on both potatoes and tomatoes. These changes most probably occurred as a result of pathogen migrations brought about by introductions of new strains of the pathogen from its presumed centre of origin in the central highlands of Mexico (Niederhauser, 1991).

Before the 1970s, European populations of this pathogen appear to have consisted exclusively of a single clonal lineage of the A1 mating type, known as US-1, which has mitochondrial DNA (mtDNA) haplotype Ib, an allozyme genotype (based on glucose-6-phosphate isomerase, *Gpi* and peptidase, *Pep* allozyme loci), *Gpi 86/100, Pep 92/100* and a characteristic fingerprint based on the multilocus RFLP probe RG57 (Goodwin *et al.*, 1994). In recent years, analyses of *P. infestans* isolates from many European countries have generally failed to detect this "old" clonal lineage of *P. infestans*, and have shown the presence of more diverse, new populations often containing both A1 and A2 mating type strains.

When A1 and A2 mating type strains occur together, *P. infestans* can undergo sexual recombination, providing the potential for greater variation and thus more rapid adaptation to control measures such as fungicides or cultivars. The sexually-produced oospores also allow the pathogen to survive over winter in soil and may potentially act as an early infection source for crops planted in contaminated land. Although the introduction of sexually recombinant populations has been suggested to be the major problem posed by the late 20th century *P. infestans* migrations, the influence of sexual reproduction on the epidemiology of the pathogen remains unclear. The introduction of new and, presumably, fitter strains, has, however, apparently increased the severity of late blight in many regions even where there is no evidence for sexual recombination (e.g. Deahl *et al.*, 2002).

World-wide, the annual cost of late blight is estimated to be US$3 billion including an annual fungicide cost of US$750 million; crop losses due to late blight are estimated to account for 10-15% of annual global potato production (Turkensteen and Flier, 2002). In the Republic of Ireland, trials over a 20 year period (1983-2002) have shown an average loss in marketable yield of *c.* 23% when untreated plots were compared with those protected by fungicides (Dowley, L.J., personal communication). In Northern Ireland, in the early 1990s the annual crop loss due to blight was estimated to be 8%. Losses from blight in Northern Ireland have been reduced over the past ten years by adoption of more effective control measures advocated by the Department of Agriculture & Rural Development (DARD) and are now estimated to be the order of *c.* 5%, but this has been achieved largely by increased usage of blight fungicides or by use of more effective, but more expensive products. In the last five years, most seed crops have received 8-9 applications of foliar fungicides and some as many as 14 sprays to protect them from foliar and tuber infection by *P. infestans* (L. R. Cooke, unpublished data). Fungicides to a value approaching £1 million (US$1.8 million) are applied to the Northern Ireland potato crop annually.

This paper will consider the structure of the *P. infestans* population in the island of Ireland and compare it with populations elsewhere in the British Isles, mainland Europe and beyond.

Materials and methods

In 1995 and 1996, nine and 20 potato crops, respectively, were sampled; single infected leaves or tubers were collected and 223 single lesion isolates obtained (Carlisle *et al.*, 2001) of which 161 were fully characterised. 1995-1997: 87 isolates for mating type. During 1995-2002, members of the Department of Agriculture and Rural Development's Potato Inspection Service collected blighted potato material from commercial potato crops in the major production areas of Northern Ireland. Each sample consisted of infected leaves and stems (and infected tubers if present) from a single crop and isolates were obtained by bulking together the sporangia from each sample. Some additional isolates were obtained from infected potato dumps, volunteers, garden crops (including one tomato), breeder's clones, woody nightshade (*Solanum dulcamara*) and naturally-infected field trial sites.

Mating type was ascertained by growing the test isolates on agar plates with known reference isolates of the A1 or A2 mating types (Cooke *et al.*, 1995). Metalaxyl sensitivity was determined using the floating potato leaf disc method (Carlisle *et al.*, 2001). Isolates were tested on 2 and 100 mg metalaxyl L^{-1}; those unable to sporulate on leaf discs on either concentration (but sporulating on untreated discs) were designated sensitive; isolates sporulating on discs on both metalaxyl concentrations were designated resistant and isolates sporulating on discs on 2, but not on 100 mg metalaxyl L^{-1} were designated intermediate.

Genotypes at two polymorphic allozyme loci, *Gpi-1* (glucose-6-phosphate isomerase) and *Pep-1* (peptidase), were determined using cellulose acetate electrophoresis (CAE) using the protocols of Goodwin *et al.* (1995a) as modified by Carlisle *et al.* (2001). The genotypes of unknown isolates were determined by comparison with reference isolates. Mitochondrial DNA (mtDNA) haplotypes of isolates were determined by PCR-RFLP using the method of Griffith and Shaw (1998) as modified by Carlisle *et al.* (2001). DNA fingerprinting using the moderately repetitive probe RG57 was carried out on a sub-set of 91 isolates from 1998-2002 using a modification of the method described by Goodwin *et al.* (1992).

Results

All of the *P. infestans* isolates collected between 1995 and 2002 were of the A1 mating type, except for one isolate from Co. Antrim in 1995 which was A2 (Table 1). Over all years, 69% of the isolates tested were sensitive to metalaxyl, 30% contained resistant strains and 1% were classified as intermediate (Table 1); within years the percentage of resistant isolates ranged from 11% in 1997 to 68% in 2001.

All isolates tested were homozygous (*100/100*) at the *Gpi* locus by CAE (Table 2). The vast majority of isolates were also homozygous at the *Pep* locus (*100/100*), but four isolates had the *Pep 83/100* genotype and six had the *Pep 96/100* genotype. The four *Pep 83/100* isolates

Table 1. Mating type and metalaxyl resistance of Northern Ireland isolates of Phytophthora infestans, *1995-2002.*

Year	Metalaxyl resistance*				Mating type	
	number of isolates			(%) R+I	Number A1	Number A2
	S	I	R			
1995	79	0	25	24	103	1
1996	152	0	31	20	183	0
1997	16	0	2	11	18	0
1998	32	0	15	32	47	0
1999	22	4	9	37	35	0
2000	12	0	14	54	26	0
2001	11	1	26	71	38	0
2002	25	1	32	57	58	0
Total	349	6	154	31	508	1

* S = metalaxyl-sensitive, I = metalaxyl-intermediate, R = metalaxyl-resistant.

Table 2. Characteristics of Northern Ireland isolates of Phytophthora infestans, *1995-2002.*

No. of isolates*	mtDNA haplotype	Allozyme genotype		Metalaxyl sensitivity			Years found
		Gpi	*Pep*	S	I	R	
61	Ia	*100/100*	*100/100*	11	0	50	1996-2002
6	Ia	*100/100*	*96/100*	0	0	6	1999-2001
4	Ia	*100/100*	*83/100*	0	0	4	1998, 2002
344	IIa	*100/100*	*100/100*	274	6	64	1995-2002
5	IIb	*100/100*	*100/100*	2	0	3	1996, 1998
420	Total			287	6	127	

* all isolates were A1 mating type.

were all recovered from naturally-infected plants at the Agriculture & Food Science Centre, Belfast (one in 1998, three in 2002). The six *Pep 96/100* isolates were obtained from widely disparate commercial crops in 1999, 2000 and 2001.

Three mtDNA haplotypes were detected among the isolates analysed (Table 2). No isolates with the old Ib haplotype were identified. Five isolates with the IIb haplotype (rare in Europe) were identified from 1996 and 1998, but this haplotype was not found in subsequent years. The majority (82%) of isolates were haplotype IIa, the remainder (17%) being Ia. Haplotype IIa predominated in each year except 2001, but the percentage of IIa isolates declined from 100% in 1995 to 45% in 2001 with a recovery to 61% in 2002.

RG57-fingerprinting of 91 Northern Ireland isolates from 1998-2002, identified eleven polymorphic bands from the 25 bands detected by this probe. Each type was designated by a code, using NI for Northern Ireland followed by a number. Types differing from the most common types by only one band were designated as sub-types indicated by lower case suffixes. Nine different RG57 fingerprints were distinguished, of which one (designated NI-1) occurred in just over 50% of isolates and the second most common (NI-2) in 29% of isolates (Table 3). Of the other seven fingerprints, two were found in 8% and 4% of isolates, respectively, while the remainder were only represented by one or two isolates.

All RG57 types were associated with a single mtDNA haplotype (Table 3). For example, all isolates with the RG57 fingerprint NI-1 had the mtDNA haplotype IIa, while all NI-2 isolates possessed mtDNA haplotype Ia. Four isolates with the *Pep 96/100* genotype were fingerprinted; all had the same distinctive RG57 fingerprint and were mtDNA Ia. Three isolates with the *Pep 83/100* genotypes were fingerprinted and these had two distinct and unique fingerprints and were mtDNA Ia.

Metalaxyl resistance tended to be associated with mtDNA haplotype Ia, 85% of Ia isolates being metalaxyl-resistant. Metalaxyl-sensitive isolates with haplotype Ia were relatively infrequent (15% of Ia isolates); eight were fingerprinted, seven had the same the RG57 fingerprint not found among other isolates and the eighth had a unique fingerprint. In contrast, the majority (80%) of haplotype IIa isolates were metalaxyl-sensitive, but there was no

Table 3. Genotypes, phenotypes and sources of Northern Ireland isolates of Phytophthora infestans, *1998-2002.*

RG57 type	mtDNA haplotype	Allozyme genotype *Gpi*	*Pep*	Metalaxyl sensitivity	Number of isolates	Year found
NI-1	IIa	*100/100*	*100/100*	S, I, R	46	1998-2002
NI-1a	IIa	*100/100*	*100/100*	S, R	2	1999, 2000
NI-2	Ia	*100/100*	*100/100*	R	26	1999-2002
NI-2a	IIa	*100/100*	*100/100*	S, I	2	1999, 2002
NI-2b	Ia	*100/100*	*100/100*	S	1	1998
NI-3	Ia	*100/100*	*100/100*	S	7	1998, 1999, 2002
NI-4	Ia	*100/100*	*96/100*	R	4	1999, 2001
NI-5	Ia	*100/100*	*83/100*	R	2	2002
NI-5a	Ia	*100/100*	*83/100*	R	1	2002

association between metalaxyl sensitivity and specific genotypes. Each of the three RG57 mtDNA IIa genotypes included both metalaxyl-resistant and -sensitive isolates.

Discussion

The Northern Ireland *Phytophthora infestans* population

All *P. infestans* isolates from Northern Ireland crops collected in 1995-96 (Carlisle *et al.*, 2001) and more recently belonged to the "new population" (as defined by allozyme genotypes and mtDNA haplotypes). However, although A2 mating type strains had been found in Northern Ireland between 1987 and 1995 (Cooke *et al.* 1995), since 1995 no A2 isolates have been identified. The 1995-96 study of 29 crops (Carlisle *et al.*, 2001) indicated relatively little diversity among the isolates examined and a preponderance of *Gpi 100/100, Pep 100/100* mtDNA IIa isolates. This was supported by results of RAPD analysis of selected isolates, all mtDNA IIa. The more extensive survey of isolates up to 2002 indicated somewhat greater diversity; the proportion of mtDNA Ia isolates was greater and two additional *Pep* genotypes, *Pep 83/100* and *Pep 96/100*, were detected. The RG57 analysis revealed relatively few fingerprints and particular fingerprints were strongly associated with other genotypic characters. Thus, over 90% of mtDNA IIa isolates had the same fingerprint (NI-1). Similarly, all metalaxyl-resistant *Gpi 100/100, Pep 100/100* mtDNA Ia isolates had the same fingerprint (NI-2). These results suggest that the Northern Ireland *P. infestans* is highly clonal and exclusively asexually reproducing, although more detailed sampling and use of additional markers would be required to exclude the possibility of very infrequent sexual recombination.

The *Phytophthora infestans* population on the island of Ireland

Although in the Republic of Ireland, the A2 mating type was first isolated in 1988 and 35% of isolates from 1988-89 were A2, A2 isolates were only obtained from a limited number of cultivars (O'Sullivan *et al.*, 1995), the majority from tubers of cv. Cara. Subsequently, Griffin *et al.* (1998) failed to detect any A2 isolates in 1995 and only four isolates (3%) from 1996 were A2 mating type; these had all been obtained from a trial at the Crops Research Centre, Carlow. Griffin *et al.* (2002) characterised *P. infestans* isolates collected between 1995 and

1999 from the Republic of Ireland using mating type, metalaxyl resistance, mtDNA haplotyping and RG57 fingerprinting (Goodwin *et al.*, 1992). Their study indicated a low level of diversity in the Irish population of *P. infestans*; only 12 genotypes were detected using RG57 and the four A2 mating type isolates from 1996 (the only A2 isolates included) all had the same RG57 fingerprint. They concluded that their results strongly suggested that sexual recombination of *P. infestans* does not occur in Ireland.

Comparison of multilocus genotypes identified in Northern Ireland with those detected in the Republic of Ireland indicated that the commonest types occur throughout Ireland. Thus Griffin *et al.* (2002) found their type IE-1 to be the commonest among Irish isolates (65%); this is identical with the commonest Northern Ireland RG57 type, NI-1 (51% of isolates), in terms of RG57 fingerprint, mating type and mtDNA haplotype (IIa). The second most frequent type in Northern Ireland, NI-2 (29% of isolates), was identical to the second most frequent (11% of isolates) in the Republic of Ireland (IE-2) identified by Griffin *et al.* (2002). As with NI-2, all IE-2 isolates were metalaxyl-resistant and mtDNA Ia. Throughout Ireland, *P. infestans* strains with the mtDNA IIa haplotype predominated, representing 67% of Northern Ireland isolates from 1998-2002, 96% of Northern Ireland isolates from 1995-96 (Carlisle *et al.*, 2001) and 78% of isolates characterised by Griffin *et al.* (2002). The *P. infestans* population in Ireland appears to be clonal, almost exclusively A1 mating type and predominantly haplotype IIa; there is no evidence of sexual recombination.

Comparison of the *Phytophthora infestans* populations in Great Britain and Ireland

The *P. infestans* population in Ireland is differentiated from that of mainland Britain in terms of mating type and multilocus genotype frequencies, but genotypes are shared between the populations. In Scotland, characterisation of nearly 500 *P. infestans* isolates collected between 1995 and 1997 indicated considerable diversity (Cooke *et al.*, 2003). Approx. 20% of isolates were A2 mating type, although very few of these were metalaxyl-resistant; spread of the A2 population was possibly restricted by its sensitivity to phenylamide fungicides. Among a sub-sample of 292 isolates examined by AFLP, 56% of isolates had unique fingerprints, but the majority of A2 isolates belonged to a single AFLP group. Whereas in Northern Ireland, only few isolates were classified as intermediate in sensitivity to metalaxyl, in Scotland the proportion of intermediate isolates increased from 2% in 1995 to 29% in 1997 and Cooke *et al.* (2003) suggested that they arose from recombination between resistant and sensitive strains. The presence of both mating types within sites, an increasing frequency of isolates with intermediate sensitivity to metalaxyl and the extent of AFLP diversity, all suggested that sexual recombination occurred occasionally within the Scottish *P. infestans* population.

Day *et al.* (2004) characterised 2,691 *P. infestans* isolates collected from England, Scotland and Wales in 1995-1998 and found that the A2 mating type isolates were rare (3% of isolates) and all were metalaxyl-sensitive. These authors determined the mtDNA haplotype and RG57 fingerprint of 1,459 of their isolates. In contrast to Ireland, haplotype Ia isolates predominated (91% of isolates). Thirty fingerprints were detected by RG57, of which four were frequent (77% of isolates), widespread and found in all years. Of these four, the most frequent (RF039, 47% of isolates) has rarely been found in Ireland; it was not detected in Northern Ireland and occurred in only one isolate from the Republic of Ireland found by Griffin *et al.* (2002). However, the next two most frequent fingerprints (RF002 and RF006) were also common in

Ireland (as IE-2/NI-2 and IE-1/NI-1, respectively), while the last, RF040, was associated with the A2 mating type and was identical to the fingerprint associated with A2 isolates from Ireland by Griffin *et al.* (2002). Day *et al.* (2004) concluded that, while circumstantial evidence suggested that sexual recombination occurred at some sites in Great Britain, it was not possible to determine its extent or its contribution to the evolving gene pool.

The *Phytophthora infestans* populations in mainland Europe

In mainland Europe, as in Ireland, the great majority of isolates characterised since 1990 have proved to be *Gpi 100/100, Pep 100/100*. Starch gel electrophoresis identified *Gpi 90/100* in Europe, but this is not distinguished from *Gpi 100/100* by CAE, used for most recent studies. The 'old' *Gpi* genotype *Gpi 90/100* was seldom found. The *Pep* genotype *83/100*, which occurred rarely (0.9% of isolates) in Northern Ireland, has been reported elsewhere in Europe, in Great Britain (Carlisle *et al.*, 2001), Poland (Sujkowski *et al.*, 1994) and France, where it was the commonest *Pep* genotype among potato isolates (Lebreton and Andrivon, 1998). Similarly, the *Pep 96/100* genotype which occurred in 1.4% of isolates from Northern Ireland has been reported from Poland (Sujkowski *et al.*, 1994) and Hungary (Bakonyi *et al.*, 2002b), but is rare in western Europe (Bakonyi *et al.*, 2002a).

In contrast to the situation in Ireland, most *P. infestans* isolates characterised from mainland Europe possessed mtDNA haplotype Ia. For example, studies of French *P. infestans* populations indicated that haplotype Ia predominated (Lebreton and Andrivon, 1998). The old mtDNA Ib haplotype has very rarely been found in recent years.

Studies using RG57 fingerprinting indicate similarities in genotypes across Europe. The commonest fingerprint in England, Scotland and Wales (Day *et al.*, 2004), which is rare in Ireland, was the most frequent (18%) among 1,048 isolates from the Netherlands characterised by Zwankhuizen *et al.* (2000). Similarly, the commonest fingerprint in Ireland (IE-1/NI-1), which was also widespread in Great Britain (Day *et al.*, 2004), was common in Norway and Finland (Brurberg *et al.* 1999) and is probably also the same as the most frequent among isolates from Poland (Sujkowski *et al.*, 1994). While convergent evolution can result in isolates having the same RG57 fingerprint on differing genetic backgrounds (Purvis *et al.*, 2001), the widespread occurrence of these fingerprints in northern European *P. infestans* populations, suggests that they may represent widely distributed clonal lineages. RG57 fingerprinting has also confirmed that displacement of the old clonal US-1 *P. infestans* population from Europe is virtually complete.

Nonetheless, despite this evidence of common *P. infestans* genotypes across Europe, populations differ very markedly in the relative frequencies of the two mating types and the likelihood of sexual recombination. In the Netherlands (Turkensteen *et al.*, 2000) and Scandinavia (Hermansen *et al.*, 2000; Stromberg *et al.*, 2001; Lehtinen and Hannukkala, 2004), the A2 mating type is common and widespread and oospores have been observed in naturally-infected potato leaves in the field. In these countries, there is strong, although largely circumstantial, evidence for sexual recombination contributing to pathogen diversity.

Elsewhere in Europe, the situation is less clear. In France, Lebreton and Andrivon (1998) found that the A2 mating type was rare on potato, but occurred more often in isolates from tomato;

while isolates from potato tended to be very similar, those from tomato were more diverse and these authors suggested that sexual reproduction of *P. infestans* might occur on the latter host. Knapova and Gisi (2002), who studied isolates collected from potato and tomato in France and Switzerland in 1996 and 1997, also considered that sexual recombination was more likely to occur in the tomato population. In Great Britain, as already noted, sexual recombination may occur within *P. infestans*. In contrast in Ireland, the population appears to be clonal and not sexually reproducing. The differentiation of the Irish *P. infestans* population from other European populations may be a result of clonal selection mediated by restricted introduction of new *P. infestans* genotypes due to controls on importation of seed tubers (only import of high-grade seed is permitted). However, since *P. infestans* in Ireland, as elsewhere in Europe, belongs to the new population, it is clear that new strains must be introduced periodically (as indeed occurred when the pathogen arrived in 1845).

Comparison with *Phytophthora infestans* populations in the USA and Canada

How do *P. infestans* populations in Europe compare with those in the USA and Canada? Current US and Canadian genotypes differ from those now established in Europe in terms of allozyme genotypes, RG57 fingerprints and mating type, indicating that they resulted from separate pathogen migrations. In the US, introductions of new strains of *P. infestans* from Mexico probably occurred during the late 1980s and early 1990s as a consequence of increased cross-border trade between the US and Mexico. The multilocus genotype most prevalent in the US is US-8 (Fry and Goodwin, 1997; Douches *et al.* 2004) ; this is A2 and metalaxyl-resistant (a combination rare in Europe) and possesses the allozyme genotype *Gpi 100/111/122, Pep 100/100*, which has never been reported from Europe. Although numerous clonal lineages of *P. infestans* have been identified in N. America (Goodwin *et al.*, 1998), relatively few have become established. In most locations in the US and Canada, epidemic populations have been composed of a single clonal lineage (Fry and Goodwin, 1997) and sexual reproduction appears to have contributed to diversity in only a few areas (Goodwin *et al.*, 1995b; Goodwin *et al.*, 1998). The dominance of US-8 in the eastern US may be a result of its aggressiveness compared with other US genotypes. In Michigan it out-competed other US genotypes in field trials in 2003-04 (Young *et al.*, 2004). The domination of *P. infestans* populations in N. America by a limited number of genotypes seems to be greater than has occurred in Europe and nowhere has frequent formation of oospores in the field, as occurs in the Netherlands and Scandinavia, been reported. Although aggressive genotypes of both A1 and A2 mating types co-exist, sexual reproduction resulting in fit progeny seems to be the exception rather than the rule.

Unanswered questions

Why have differences developed in *P. infestans* populations across Europe and why is the population structure so different from that in N. America? The contrast with the global distribution of the old US-1 clonal lineage of the pathogen up to the 1970s is striking. Although separate migrations were involved, in both cases new strains were apparently introduced from Mexico, the pathogen's centre of diversity. But there are many unanswered questions. Why was the displacement so complete? Why have A2 types failed to establish in some regions, but thrive in others? Why are A2 strains in Europe seldom metalaxyl-resistant whereas in the US the metalaxyl-resistant A2 US-8 is the commonest genotype on potato. Why do strains with mtDNA haplotype IIa predominate in Ireland, but nowhere else in Europe? How much is random, due to founder effects and how much due to differential fitness? Even in regions where oospores are formed in the field, our understanding of their role in late blight epidemiology is very limited. What factors restrict the ability of strains to produce viable progeny? How often do oospores germinate in the field and cause late blight?

In terms of practical disease control, has the role of sexually recombinant populations been over-rated in terms of the contribution of oospores to initiating infection? The importance of the introduction of fitter, more aggressive strains into populations, by migration or *in situ* sexual recombination, is paramount. Characteristics of new strains such as short latent periods and wider temperature tolerances may have a more drastic effect on blight epidemics than oospores acting as a reservoir of soil-borne infection or initiators of early outbreaks. Even if new strains occur only very sporadically, they can quickly dominate populations, making late blight control much harder. Understanding the drivers of the selection process remains a challenge. International co-operation through initiatives such as the EU-funded EU.NET.ICP and Eucablight projects and GILB is vital.

Acknowledgements

This paper draws on data produced by Diane J Carlisle (during her doctoral research funded by Department of Education for Northern Ireland Quota Award), Michele Quinn and Cathy Donaghy of Applied Plant Science Division, DARD and by Frances Perez of the ARS Vegetable Laboratory, USDA, all of whom are thanked. A further publication dealing with the Northern Ireland *P. infestans* population 1998-2002 is in press (Cooke *et al.*, 2005). Many discussions with colleagues in EU.NET.ICP and Eucablight have also been invaluable.

References

Bakonyi, J., B. Heremans and G. Jamart (2002a). Characterization of *Phytophthora infestans* isolates collected from potato in Flanders, Belgium. *Journal of Phytopathology,* 150, 512-516.

Bakonyi, J., M. Láday, T. Dula and T. Érsek (2002b). Characterisation of isolates of *Phytophthora infestans* from Hungary. *European Journal of Plant Pathology,* 108, 139-146.

Bruberg, M.B., A. Hannukkala and A. Hermansen (1999). Genetic variability of *Phytophthora infestans* in Norway and Finland as revealed by mating type and fingerprint probe RG57. *Mycological Research,* 12, 1609-1615.

Carlisle, D.J., L.R. Cooke and A.E. Brown (2001). Phenotypic and genotypic characterisation of Northern Ireland isolates of *Phytophthora infestans*. *European Journal of Plant Pathology,* 107, 292-303.

Cooke, L.R., D.J. Carlisle, C. Donaghy, M. Quinn, F.M. Perez and K.L. Deahl (2005). The Northern Ireland *Phytophthora infestans* population 1998-2002 characterised by genotypic and phenotypic markers. *Plant Pathology,* in press.

Cooke, D.E.L., V. Young, P.R.J. Birch, R. Toth, F. Gourlay, J.P. Day, S.F. Carnegie and J.M. Duncan (2003). Phenotypic and genotypic diversity of *Phytophthora infestans* populations in Scotland (1995-97). *Plant Pathology,* 52, 181-192.

Cooke, L.R., R.E. Swan and T.S. Currie (1995). Incidence of the A2 mating type of *Phytophthora infestans* on potato crops in Northern Ireland. *Potato Research,* 38, 23-29.

Day, J.P., R.A.M. Wattier, D.S. Shaw and R.C. Shattock (2004). Phenotypic and genotypic diversity in *Phytophthora infestans* on potato in Great Britain, 1995-98. *Plant Pathology,* 53, 303-315.

Deahl, K.L., L.R. Cooke, L.L. Black, T.C. Wang, F.M. Perez, B.C. Moravec, M. Quinn and R.W. Jones (2002). Population changes in *Phytophthora infestans* in Taiwan associated with the appearance of resistance to metalaxyl. *Pest Management Science,* 58, 951-958.

Douches, D., J. Coombs, K. Felcher and W. Kirk (2004). Foliar reaction to *Phytophthora infestans* in inoculated potato fields in Michigan. *American Journal of Potato Research,* 81, 443-448.

Fry, W.E. and S.B. Goodwin (1997). Re-emergence of potato and tomato late blight in the United States. *Plant Disease,* 81, 1349-1357.

Goodwin, S.B., B.A. Cohen and W.E. Fry (1994). Panglobal distribution of a single clonal lineage of the Irish potato famine fungus. *Proceedings of the National Academy of Science, USA,* 91, 11591-11595.

Goodwin, S.B. A. Drenth and W.E. Fry (1992). Cloning and genetic analyses of two highly polymorphic moderately repetitive nuclear DNAs from *Phytophthora infestans. Current Genetics,* 22, 107-115.

Goodwin, S.B., R.E. Schneider and W.E. Fry (1995a). Use of cellulose-acetate electrophoresis for rapid identification of allozyme genotypes of *Phytophthora infestans. Plant Disease,* 79, 1181-1185.

Goodwin, S.B., C.D. Smart, R.W. Sandrock, K.L. Deahl, Z.K. Punja and W.E. Fry (1998). Genetic change within populations of *Phytophthora infestans* in the United States and Canada during 1994 to 1996: role of migration and recombination. *Phytopathology,* 88, 939-949.

Goodwin, S.B., L.S. Sujkowski, A.T. Dyer, B.A. Fry and W.E. Fry (1995b). Direct detection of gene flow and probable sexual reproduction of *Phytophthora infestans* in northern North America. *Phytopathology,* 85, 473-479.

Griffin, D., E. O'Sullivan, M.A. Harmey and L.J. Dowley (1998). Distribution of metalaxyl resistance, mating type and physiological races in Irish populations of *Phytophthora infestans*; a preliminary report. *PAV-Special Report number 3.* Lelystad, the Netherlands: Applied Research for Arable Farming and Field Production of Vegetables, 153-159.

Griffin, D., E. O'Sullivan, M.A. Harmey and L.J. Dowley (2002). DNA fingerprinting, metalaxyl resistance and mating type determination of the *Phytophthora infestans* population in the Republic of Ireland. *Potato Research,* 45, 25-36.

Griffith, G.W. and D.S. Shaw (1998). Polymorphisms in *Phytophthora infestans*: Four mitochondrial DNA haplotypes are detected after PCR amplification from pure cultures or from host lesions. *Applied and Environmental Microbiology,* 64, 4007-4014.

Hermansen, A., A. Hannukkala, R.H. Naestad and M.B. Brurberg (2000). Variation in populations of *Phytophthora infestans* in Finland and Norway: mating type, metalaxyl resistance and virulence phenotype. *Plant Pathology,* 49, 11-22.

Knapova, G. and U. Gisi (2002). Phenotypic and genotypic structure of *Phytophthora infestans* populations on potato and tomato in France and Switzerland. *Plant Pathology,* 51, 641-653.

Lebreton, L. and D. Andrivon (1998). French isolates of *Phytophthora infestans* from tomato and potato differ in genotype and phenotype. *European Journal of Plant Pathology,* 104, 583-594.

Lehtinen, A. and A. Hannukkala (2004). Oospores of *Phytophthora infestans* in soil provide an important new source of primary inoculum in Finland. *Agricultural and Food Science,* 13, 399-410.

Niederhauser, J.S. (1991). *Phytophthora infestans* - the Mexican connection. In: J.A. Lucas, R.C. Shattock, D.S. Shaw and L.R. Cooke (eds.) *Phytophthora*. Cambridge, UK: Cambridge University Press., 272-294.

O'Sullivan, E., L.R. Cooke, L.J. Dowley and D.J. Carlisle (1995). Distribution and significance of the A2 mating type of *Phytophthora infestans* in Ireland. In: L.J. Dowley, E. Bannon, L.R. Cooke, T. Keane, E. O'Sullivan (eds.) *Phytophthora 150*. Dublin, Ireland: Boole Press, 232-239.

Purvis, A.I., N.D. Pipe, J.P. Day, R.C. Shattock, D.S. Shaw and S.J. Assinder (2001). AFLP and RFLP (RG57) fingerprints can give conflicting evidence about the relatedness of isolates of *Phytophthora infestans. Mycological Research,* 105, 1321-1330.

Stromberg, A., U. Bostrom and N. Hallenburg (2001). Oospore germination and formation by the late blight pathogen *Phytophthora infestans in vitro* and under field conditions. *Journal of Phytopathology,* 149, 659-664.

Sujkowski, L.S., S.B. Goodwin, A.T. Dyer and W.E. Fry (1994). Increased genotypic diversity via migration and possible occurrence of sexual reproduction of *Phytophthora infestans* in Poland. *Phytopathology,* 84, 201-207.

Turkensteen, L.J. and W.G. Flier (2002). Late blight: its global status in 2002 and beyond. In: 'Late blight: Managing the global threat'. *Proceedings of the Global Initiative on Late Blight Conference*, Hamburg, Germany, 11-13 July 2003, pp. 1-9.

Turkensteen, L.J., W.G. Flier, R. Wanningen and A. Mulder (2002). Production, survival and infectivity of oospores of *Phytophthora infestans. Plant Pathology,* 49, 688-696.

Young, G.K., L.R. Cooke, W.W. Kirk and P. Tumbalam (2004). Competitive selection of *Phytophthora infestans* in the US and Northern Ireland. Eighth Workshop on the European Network for Development of an Integrated Control Strategy of potato late blight. Jersey, 31 March - 4 April 2004. *PPO-Special Report No., 10,* 271-275.

Zwankhuizen, M.J., F. Govers and J.C. Zadoks (2000). Inoculum sources and genotypic diversity of Phytophthora infestans in Southern Flevoland, the Netherlands. *European Journal of Plant Pathology,* 106, 667-680.

Curriculum vitae – Louise R. Cooke

Louise R. Cooke gained a B.Sc. in biochemistry from Bristol University in 1974, then pursued her interest in plants at Long Ashton Research Station working on the chemical control of Dutch elm disease, research for which she was awarded a Ph.D. from Bristol University in 1978. She spent three more years at Long Ashton funded by the British Potato Marketing Board to investigate the control of fungal potato diseases, including late blight. In 1981, she moved to Belfast to work as a plant pathologist for the Department of Agriculture for Northern Ireland (now the Department of Agriculture & Rural Development) and is now a principal scientific officer there. She holds a dual appointment as a lecturer in Applied Plant Science of Queen's University, Belfast, teaching and supervising undergraduates and postgraduates. Potato blight is her major research interest, but she is also involved in projects on other fungal potato diseases, on fungicide resistance in cereal pathogens and the epidemiology and control of apple canker. She served as President of the Society of Irish Plant Pathologists 2000-2003 and is a member of the editorial board of *Plant Pathology*.

Curriculum vitae – Kenneth L. Deahl

Kenneth L. Deahl is a Lead Scientist in the Vegetable Laboratory, Plant Sciences Institute at Beltsville Agricultural Research Center Beltsville, Maryland which is a part of the USDA's Agricultural Research Service. He conducts research on the genetics and molecular biology of plant pathogens, mainly *Phytophthora infestans*. He focuses on understanding its evolution, using the methods of population genetics to find out how variation is generated and maintained, inheritance of molecular markers and phenotypic traits, the epidemiology, and management of late blight.

Company profile – The Applied Plant Science Division

The Applied Plant Science Division (APSD) of the Department of Agriculture & Rural Development (DARD) in Northern Ireland is part of DARD's Science Service, due to become the Agri-Food and Biosciences Institute in 2006. The Science Service currently undertakes programmes that integrate research, statutory testing, analytical and diagnostic work, technology transfer and degree level education. The Applied Plant Science Division carries out programmes to promote DARD's policy of improving the economic performance and competitiveness of the Northern Ireland agri-food industry and supporting its pursuit of quality, while conserving the rural environment and protecting the public. Research areas include plant breeding, horticulture, agronomy, crop protection, ecology and management of environmentally sensitive farming practices, the biology of alternative land use and diversification opportunities. Scientists in APSD are involved in many collaborative projects within the UK and internationally and seek external funding through these opportunities and activities such as fungicide evaluation.

Company profile – The Vegetable Laboratory

The Vegetable Laboratory is based at the Plant Sciences Institute, Beltsville Agricultural Research Center, Maryland which is a part of the USDA's (United States Department of Agriculture) Agricultural Research Service. The mission of the Vegetable Lab is several-fold: to unravel fundamental principles of biology, biochemistry, biocontrol agents, genetics and molecular biology of horticultural crops; to develop state-of-the-art technologies for production of high quality vegetables and for reducing losses due to diseases, pests and physiological disorders; and to test and optimize biocontrol agents and genetically-enhanced vegetables in environmentally-friendly sustainable production systems. Specific crops targeted for improvement and quality enhancement are potatoes, tomatoes, beans, peppers, and other vegetables.

Late blight resistance in Sárpo clones: an update

D.S. Shaw and D.T. Kiezebrink
Sárvári Research Trust, Henfaes Research Centre, Abergwyngregyn, Llanfairfechan, LL33 0LB, UK

Abstract

Sárpo clones, bred in Hungary, have been assessed for their resistance to late blight disease in many trials in western Europe over the last 10 years. Several clones showing high levels of rate-limiting, non-isolate-specific resistance have been selected for commercialisation. These show other characteristics, including weed suppression, long dormancy and resistance to virus, slugs and wireworm, that suit them to low-input growing conditions.

Keywords: Sárpo, late blight, resistance, low-input, *Phytophthora infestans*

Introduction

There is increasing evidence that late blight disease is evolving rapidly following the migration of new strains from Mexico to most parts of the world over the last 30 years (Day *et al.*, 2004). The pathogen can now survive for many years in the soil as resistant oospores and is able to recombine during its sexual phase to increase its diversity and become a more formidable pathogen. Late blight control using fungicides is costly and not always effective and there is increasing pressure from the supermarkets and the consumer for pesticide-free food and less pollution of the environment. Use of copper by organic growers is being phased out in the UK and already prohibited in other EU member states. Clearly the solution is to grow varieties of potato which can survive high pressures of late blight disease without protection.

Breeding for late blight resistance

Genes for late blight resistance have been identified in many wild *Solanum* spp. (Müller and Black, 1952). Those from *S. demissum* were deployed in the breeding of new cultivars more than 50 years ago but this did not provide permanent protection as the pathogen evolved strains which could overcome these R-genes. Later attempts to make use of rate-limiting genes have been more successful as the resistance has proved to be more durable and largely isolate-non-specific (Wastie, 1991).

High levels of resistance in Sárpo clones

The Sárvári family have been breeding disease resistant potatoes in Hungary since 1945, initially at the Georgikon Institute, Keszthely and latterly privately, at Zirc. Eight *Solanum* spp. have been mined for a variety of genes and many of their clones have been shown to survive high late blight pressures in the field. Very large numbers of progeny from each cross were screened for foliage- and tuber-blight resistance in their first year and the most highly resistant clones were selected for further late blight screening the following year.

How does the resistance stand up?

Following successful trials in Scotland in the 1990s, Sárpo Kft. was formed to support and upgrade the breeding effort at Zirc. More advanced clones have since been selected and trialled in the United Kingdom, Denmark and North Africa (e.g. Shaw and Johnson, 2004). By exposing promising clones to diverse pathogen populations at many sites, it has been possible to eliminate clones showing isolate-specific resistance. Evidence for labile R-genes in these clones has been confirmed in detached leaflet and whole plant tests in the laboratory, using inoculum isolated from Sárpo clones and from standard cultivars (e.g. Carnegie and Cameron, 2005).

Field trials over several seasons have been conducted at many sites within the UK, in Denmark and elsewhere in which high and sustained late blight pressure has occurred naturally or artificially. The progression of the disease has been scored against standard, reference cultivars, well known for their relatively durable resistance (e.g. Cara, Valor and Stirling). In season 2004, one trial included late blight-resistance standards agreed by the Concerted Action, EUCABLIGHT; these are now being used for late blight assessment throughout Europe so that results from trials at different sites will be more comparable. Results have shown that many Sárpo clones have a higher level of resistance than any of the standard cultivars but they are not immune to attack. Infected Sárpo clones have shown a phenotype characterised by low infection efficiency, slow lesion growth and low sporulation. These are well known features of cultivars with horizontal, partial resistance which has proved to be durable. In 2004, four advanced Sárpo clones were planted at INIFAP, Toluca, Mexico. These were inspected in late August; typical symptoms of slow-blighting were observed and tuberisation was well advanced.

Resistance to tuber blight has been tested in field trials and in inoculation experiments in the laboratory. Results have shown that tuber resistance is also high. Early breeding efforts, necessarily accomplished in the presence of high population densities of aphid vectors, resulted in resistance to virus, particularly PLRV and PVY^{NTN}; resistances to PVX and PVA have also been selected.

How near-market are they?

Progress is now being made in the selection of clones with higher market potential as some of the first resistant clones had serious defects. The well known association between late blight resistance and late-maturity (Visker, 2005) also applies to some of the Sárpo material. Careful seed management to advance the crop can allow later clones to bulk and mature satisfactorily. Some clones with earlier maturity and substantial late blight resistance, able to bulk under heavy late blight pressures are now being evaluated. Useful characteristics of most Sárpo clones growing in organic or low-input systems include: a vigorous haulm which out-competes weeds; a long, natural dormancy and slug and wireworm resistance. High dry-matter contents in most clones tend to make them less suitable for use in pre-packs but assessment for baking and French-fry processing have been more successful. Two clones, Sárpo Mira and Axona have been Nationally Listed in the UK and Denmark. Over 17000 gardeners throughout the UK will grow these varieties in 2005. Sárpo Mira has been trialled in Morocco, Algeria and Senegal by Danespo where it has shown useful heat and drought tolerance.

Discussion

Much effort is now spent on using genetic modification and marker-assisted-selection techniques to accumulate genes for late blight resistance. It will be most interesting to compare products of these methods with the conventionally-bred Sárpo clones to see how the resistance phenotypes compare and how they survive over time in the field. Now that *P. infestans* has the advantage of recombination as well as mutation it can be expected to develop new ways around even the most promising resistance eventually. The arms race will continue and the potato breeder must stay out in front.

Conclusions

Sárpo clones have shown high levels of rate-limiting resistance in field trials in many countries. They possess many characteristics, including virus resistance, which make them suitable for growing in organic and low-input systems.

Acknowledgements

The work of the Sárvári Research Trust is part-funded by the European Regional Development Fund, the Welsh Development Agency and the British Potato Council. Thanks are due to Mr Adam Anderson for his continuing inspiration and support.

References

Carnegie, S.F. and A.M. Cameron (2005). Evaluation of Late Blight Resistance in Sárpo Varieties. In: R.K.M. Hay and A.J. Milne (eds) Scientific Review 2000-2003 Scottish Agricultural Science Agency. Scottish Executive, Edinburgh: pp. 32-34.

Day, J.P., R.A.M. Wattier, D.S. Shaw and R.C. Shattock (2004). Phenotypic and genotypic diversity in *Phytophthora infestans* on potato in Great Britain, 1995-98. Plant Pathology, 53, 303-315.

Müller K.O. and W. Black (1952). Potato breeding for resistance to blight and virus diseases during the last hundred years. Z. Pflanzenzüchtung, 31, 305-318.

Shaw, D. and L. Johnson (2004). Progress in the selection of cultivars with resistance to late-blight disease. PPO-Special Report, 10, 203-209.

Visker, M. (2005). Association between late blight resistance and foliage maturity type in potato. PhD Thesis, Wageningen University, The Netherlands: pp. 160.

Wastie, R.L. (1991). Breeding for resistance. In: D.S. Ingram and P.H. Williams (eds.) Advances in Plant Pathology. Volume 7, Academic Press, London, pp. 193-224.

Curriculum vitae – David Shaw

David Shaw graduated BSc and then PhD from University of Glasgow. His research was on the genetics of *Phytophthora cactorum*. After a Post-doctoral Fellowship at the Dept of Botany and Plant Pathology, Michigan State University, 1965-67, he was appointed Lecturer and then Senior Lecturer at University of Wales, Bangor. He continued to work on the biology of *Phytophthora* and especially on the molecular diversity of populations of *P. infestans*.

In 2002, Dr Shaw retired from University of Wales to direct research for the Sárvári Research Trust.

Curriculum vitae – Daan Kiezebrink

Daan Kiezebrink graduated from Wageningen Agricultural University in 1993. He gained a PhD from Dundee University for his work on the interaction between nematodes and soil structure carried out at the Scottish Crop Research Institute. From 2000 - 2004 he was employed on a British ministry of Agriculture (DEFRA) grant working on the effects of growing genetically modified crops on soil ecosystems at Leeds University.

Currently he is trials manager for the Sárvári Research Trust.

Company profile – The Sárvári Research Trust

The Sárvári Research Trust was registered in 2002 and is researching late blight disease of potato. Our current work concerns the development of improved methods of assessing late blight resistance and the evaluation of new clones of potato with high late blight resistance, bred in Hungary by the Sárvári family. The potential of new clones for low-input and organic growing is being investigated. For further details regarding our work or to discuss collaboration please contact us via info@sarvari-trust.org.

Infinito®: a novel fungicide for long-lasting control of late blight in potato

S. Tafforeau[1], M.P. Latorse[2], P. Duvert[2], E. Bardsley[3], T. Wegmann[1] and A. Schirring[1]
[1]Bayer CropScience AG, Alfred-Nobel-Straße 50, 40789 Monheim, Germany
[2]Bayer CropScience S.A., 16 rue Jean-Marie Leclair, 69266 Lyon, France
[3]Bayer CropScience, Hauxton, CB2 5HU, Cambridge, UK

Abstract

Infinito® is a new generation fungicide for control of potato late blight. It is a combination of fluopicolide with propamocarb-HCl. Fluopicolide is an innovative active ingredient from the new chemical class acylpicolides, which has a novel mode of action and impressive fungicidal properties.

In a large number of field trials conducted since 2002, Infinito applied in preventive spray programmes consistently provided outstanding control of late blight on leaves and stems of potatoes. Long persistence and anti-sporulant properties resulted in the combination of high yields and excellent protection from tuber blight.

Combining the different modes of action of two fungicides, Infinito is a powerful new tool in the management of resistance of *Phytophthora infestans* shown by existing fungicides.

Infinito is under development in most potato growing countries and the first registrations are expected at the beginning of 2006.

Keywords: fluopicolide, propamocarb-HCl, *Phytophthora infestans*, late blight, tuber blight

Introduction

Late blight is the most important fungal disease in potato in Northern European countries. During the potato growing season, in humid climatic conditions, the disease may spread very quickly over the fields. In the absence of adequate fungicide protection, the disease can lead to a complete destruction of the crop, thus reducing dramatically the yield. Tuber infections also affect the quality during storage and can result in the rejection of an entire batch of potatoes.

Further to the introduction of new populations of *Phytophthora infestans* in the 1970s (Flier *et al.*, 2004) and the overall spreading of the A2 mating type in Europe in the 1980s, the disease epidemiology has significantly changed in the recent years. Shorter life-cycles of the fungi and earlier primary infections result in increased aggressiveness of the pathogen with more severe damage to leaves and stems. In these conditions, farmers need more effective and longer lasting products to maintain good protection against potato late blight without increasing the overall number of applications.

Of the combinations tested during the development of fluopicolide for control of potato late blight, Infinito, a liquid formulation of fluopicolide and propamocarb-HCl, showed the greatest benefits. This suspension concentrate (SC) contains 62.5 + 625 g a.i./litre of the respective active ingredients and has been specifically developed for effective and long lasting control of potato late blight.

This paper describes the fungicidal and agronomic characteristics of Infinito and gives a summary of three years of field trials, demonstrating the superiority of Infinito to commercial fungicides for control potato late blight. The following features will be discussed in detail:

- fluopicolide has a novel mode of action;
- solid pro-active anti-resistance management;
- synergistic effects and enhanced performance;
- (e.g. translaminar and anti-sporulant properties);
- favourable toxicological, ecotoxicological and residue profile;
- consistent high level of control of late blight on leaves and stems;
- protection against tuber blight;
- long lasting activity;
- yield improvement.

Materials and methods

The mode of action of fluopicolide has been investigated in laboratory and greenhouse studies conducted by Bayer CropScience from 1998. The characterisation of the biochemical activity of fluopicolide was done in standard *in vitro*-tests on different organisms such as *Pythium ultimum* and *P. aphanidermatum*. Microscopic observations of different stages of development of *P. infestans* were conducted *in vitro*, after addition of different compounds in the medium, in order to determine the fungicidal characteristics of fluopicolide. From 2002, Infinito and various commercial fungicides were tested by spray applications on potato plants artificially inoculated with *P. infestans* under greenhouse. Efficacy assessments combined with laboratory and microscopic observations were performed in order to elucidate the fungicidal characteristics of the combination.

Field experiments reported in this paper were usually conducted in 2002-2004 for registration of Infinito in Germany, France, United Kingdom and the Netherlands under good experimental practice (GEP), with randomised block designs, a minimum of 4 replicates and minimum plot size of 20 m^2. Efficacy of Infinito against potato late blight was assessed in comparison with different commercial fungicides applied at uniform dose rates (Table 1). Foliar applications of the products were made at 7 day intervals in preventative programmes (except in one specific trial to assess long lasting efficacy). Disease assessments and yield determination were performed following EPPO guidelines. Disease progression on leaves for each treatment in individual trials is summarised using the RAUDPC calculation (Relative Area Under the Disease Progress Curve). Efficacy was calculated where necessary (% Abbott).

Commercial fungicides used in comparison with Infinito were not always the same in all field trials. The standard fungicides present in more than 10 trials are reported in this paper to compare late blight efficacy. In the case of tuber blight, only the trials with high infection levels (> 10% in untreated control) are presented.

Table 1. Dose rates of Infinito and commercial fungicides applied in field trials.

Treatments	Dose rates	
	(L or Kg product / ha)	(g a.i./ha)
Infinito (fluopicolide + propamocarb-HCl)	1.2	75 + 750
Infinito (fluopicolide + propamocarb-HCl)	1.6	100 + 1000
propamocarb-HCl + chlorothalonil	2.0	750 + 750
dimethomorph + mancozeb	2.0	180 + 1200
cymoxanil + mancozeb	2.5	113 + 1700
fluazinam	0.4	200
cyazofamid (+ 150 ml adjuvant)	0.2	80
mancozeb	2.1	1575
untreated control	-	-

Results from laboratory and greenhouse tests

Fluopicolide has a novel mode of action

Although the biochemical target site of fluopicolide is not yet known, the mode of action of fluopicolide is novel: it is not active at known target sites of oomycete fungicides such as rRNA synthesis (metalaxyl/mefenoxam), oxidative phosphorylation (fluazinam), inhibition of electron transport in complex III (QoI: famoxadone, fenamidone; QiI: cyazofamid). In addition, the overall biochemical and biological characteristics of fluopicolide are different from other known fungicides.

Fluopicolide is effective at all stages of the life-cycle of *P. infestans*: sporulation, zoospore and cyst formation, zoospore mobility, cyst germination, penetration into the plant tissues and mycelium growth. It is active on both the direct and the indirect germination of sporangia of *P. infestans*, providing a strong and reliable activity against this disease whatever the temperature.

The rapid effect of fluopicolide on zoospore mobility at very low dose rates is remarkable (LC_{90} < 0.05 ppm on *P. infestans* - see Table 2). Microscopic observations show that fluopicolide application immediately stops the movement of zoospores, and makes them burst within a few minutes.

Table 2. In vitro activity of fluopicolide on mobility of P. infestans *zoospores.*

Fungicides	LC_{90} (mg a.i./litre) Inhibition of zoospore mobility (*P. infestans*)
fluopicolide	< 0.05
fenamidone	0.25
fluazinam	0.25
mancozeb	> 10

Solid pro-active anti-resistance management

As *P. infestans* is considered a high risk pathogen with regard to resistance development to fungicides (Brent and Hollomon, 1998), fluopicolide has been exclusively developed in combination with fungicides with different modes of action such as propamocarb-HCl in potato.

After more than 15 years of successful use in potato, *P. infestans* has not shown any resistance to propamocarb-HCl. This fungicide has a very low risk of resistance. As Infinito proved to control both A1 and A2 mating types of *P. infestans* in field trials, as well as strains resistant to phenylamide fungicides in laboratory tests, it can be included in spray programmes for control of potato late blight, giving farmers a new and powerful tool to support a strong long term anti-resistance strategy.

Synergistic effects and enhanced performances

The synergism of fluopicolide and propamocarb-HCl for potato late blight control has been demonstrated in 1999 in laboratory. Infinito offers enhanced biological activity, optimum uptake and distribution of the compounds in the plant.

Propamocarb-HCl quickly penetrates into the potato leaves and moves through the plant tissues where it inhibits mycelial growth and development of sporangia. After application, fluopicolide is distributed on the leaf surface, providing protection from the pathogen. Some fluopicolide is also taken up by the leaves, redistributed *via* the xylem and translocated into the leaf tissues, thus providing translaminar activity. In laboratory conditions the penetration of fluopicolide in potato leaves increases two fold when formulated with propamocarb-HCl.

Infinito has a very strong translaminar effect against *P. infestans* based on the movement of both a.i.'s from the upper surface to the lower surface of the leaflets. Infinito applied at dose rates equivalent to 1.2 and 1.6 litre/ha efficiently controlled late blight development (Figure 1).

These properties are particularly important in practical field situations where spray coverage may not be optimal.

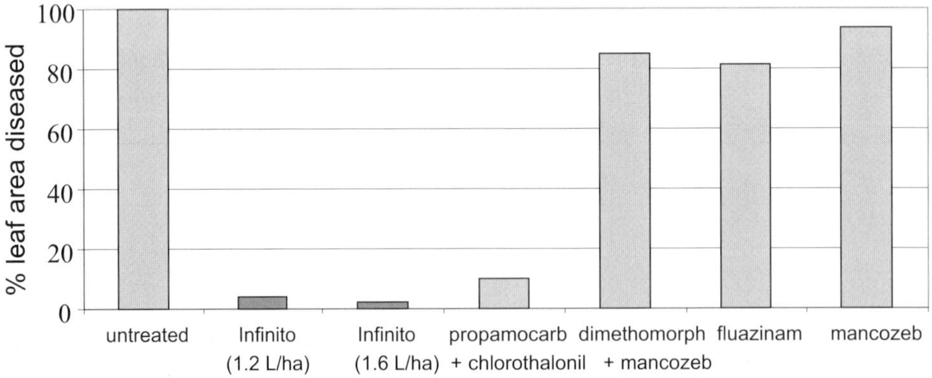

Figure 1. Translaminar activity of Infinito (greenhouse trial, application at the upper leaf surface, inoculation to the lower leaf surface 24 hours later, assessment 7 days after inoculation).

Table 3. Antisporulant effect of Infinito on P. infestans *in potato (application 24 hours after inoculation, counting of sporangia washed off from leaflets 7 days after inoculation).*

Treatments	Dose rate (g a.i./ha)	Anti-sporulant effect	
		Sporangia / ml	% inhibition
Untreated	-	165 900	-
Infinito (fluopicolide + propamocarb-HCl)	75 + 750	0	100
propamocarb + chlorothalonil	750 + 750	6 700	96.0
dimethomorph + mancozeb	180 + 1200	22 500	86.4
fluazinam	200	63 200	61.9

The anti-sporulant effects of Infinito observed under field conditions was confirmed in one glasshouse trial conducted in 2003 (Table 3). Infinito exhibited strong anti-sporulant activity.

Toxicological and ecotoxicological properties of Infinito

Overall, Infinito has a very favourable toxicological, ecotoxicological and residue profile.

No residues of fluopicolide or propamocarb-HCl have been detected above the limit of quantification (LOQ = 0.01 mg/kg) in potato tubers and processed potatoes, following applications of Infinito according to the recommended uses. Infinito is suitable for use in integrated pest management systems (IPM) in potato crop.

Table 4. Toxicological and ecotoxicological properties of Infinito.

Toxicology	Acute oral rat	LD50 > 2500 mg/kg
	Acute dermal rat	LD50 > 4000 mg/kg
	Acute inhalation rat	LD50 > 3195 mg / m3 air
	Skin irritation	Negative
	Eye irritation	Negative
	Skin sensitisation (Buehler)	Negative
Ecotoxicology	Acute trout	LC50 6.6 mg/L
	Acute daphnia	LC50 > 100 mg/L
	Acute algae (green)	EC50 > 100 mg/L
	Beneficials / pollinators / soil organisms	Safe profile

Field trial results

Consistently high level of control of late blight on leaves and stems

Infinito has been widely tested in field conditions in comparison to commercial fungicides since 2002. Infinito demonstrated excellent crop safety in all field trials carried out on a wide range of potato varieties. The preventive activity of Infinito was assessed when fungicides were applied throughout the season in a 7 day spray programme (Figure 2). The results clearly demonstrate that Infinito provided excellent control of the disease on leaves with a shallow

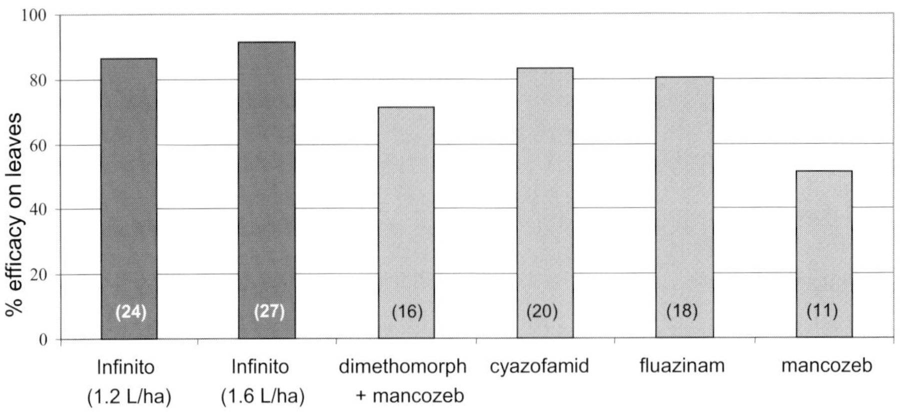

Figure 2. Efficacy of Infinito for control of P. infestans *on leaves (source: 29 field trials in Europe from 2002 to 2004, numbers of trials per product are mentioned in brackets).*

dose response at rates from 1.2 to 1.6 litre/ha. Even at 1.2 litre/ha, Infinito performed better than the commercial standards.

In three trials Infinito also reduced the disease severity of stem blight by 89% on average, performing better than the reference fluazinam.

Consistency in the efficacy of the fungicides in these trials was statistically evaluated using the Boxplot chart method (Tukey, 1977). The graph (Figure 3) shows that Infinito has a lower variability than the commercial standards. This confirms the consistency and reliability of Infinito treatments for maintaining a healthy crop.

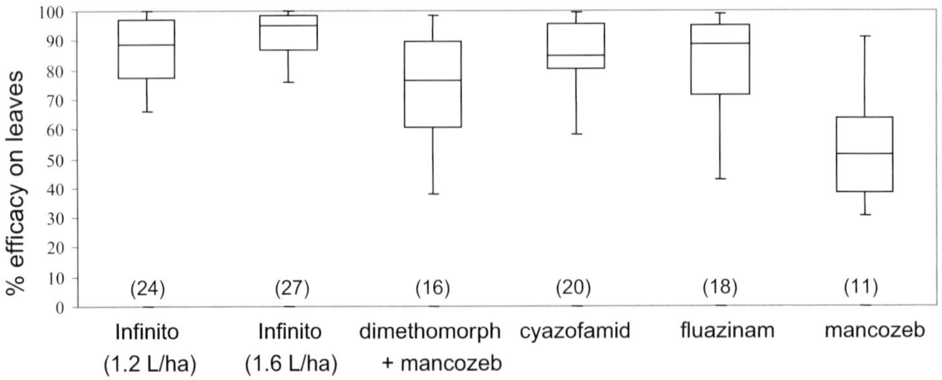

Boxplot chart: black lines across the boxes represent the medians, boxes include 50% of the values and the vertical lines represent the minimum and maximum values of a treatment excluding the outliers

Figure 3. Consistency of Infinito for control of P. infestans *(source: 29 field trials in Europe from 2002 to 2004, number of trials per product are mentioned in brackets).*

Long lasting activity

Infinito demonstrated remarkable long lasting activity in field trials. Specific trials were conducted in 2004 to confirm its long lasting potential. Fungicides were applied only twice at disease onset (full flowering). Disease progress curves in Figure 4 show that Infinito delayed the development of *P. infestans* by more than one week during the main period of rapid disease development.

When applied at 7 to 10 day intervals in a spray programme, Infinito provided long lasting blight control, thus offering the farmer better flexibility in use and greater security.

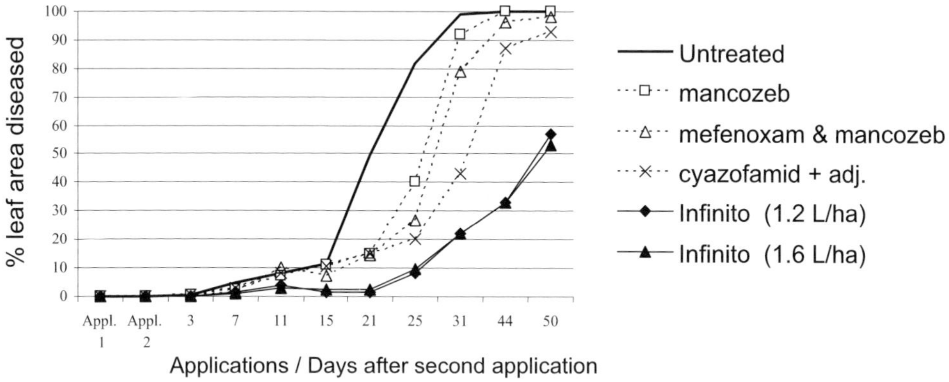

Figure 4. Long lasting activity of Infinito against P. infestans *(field trial on cv. 'Bintje', France, 2004).*

Protection against tuber blight

In four trials carried out in the Netherlands in 2004, *P. infestans* on tubers was assessed at harvest. The results summarised in Figure 5 show that Infinito applications throughout the entire season provided excellent protection against tuber blight in comparison to the commercial standards fluazinam and cymoxanil + mancozeb.

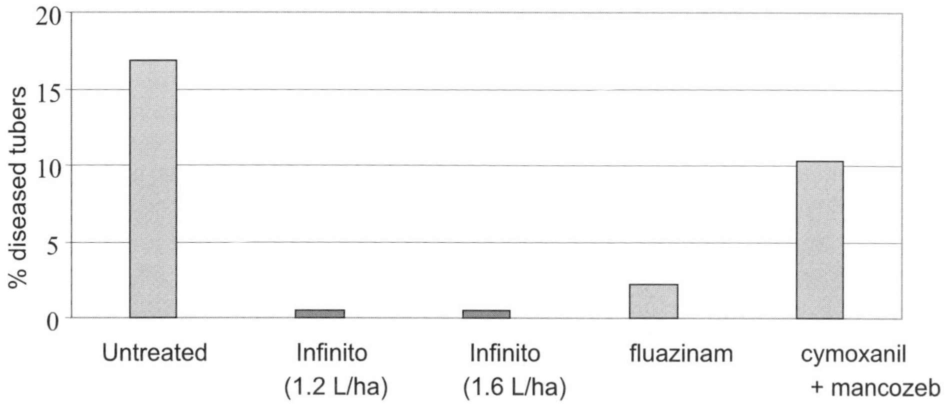

Figure 5. Efficacy of Infinito against tuber blight (means in 4 trials, Netherlands, 2004).

Yield improvement

Yield was assessed in trials conducted in France and results were expressed as percentage of yield relative to untreated (Figure 6). Both dose rates of Infinito gave higher yields than standard products. Infinito moreover exhibited a clear dose response as a consequence of better leaf, stem and tuber disease control with excellent plant compatibility.

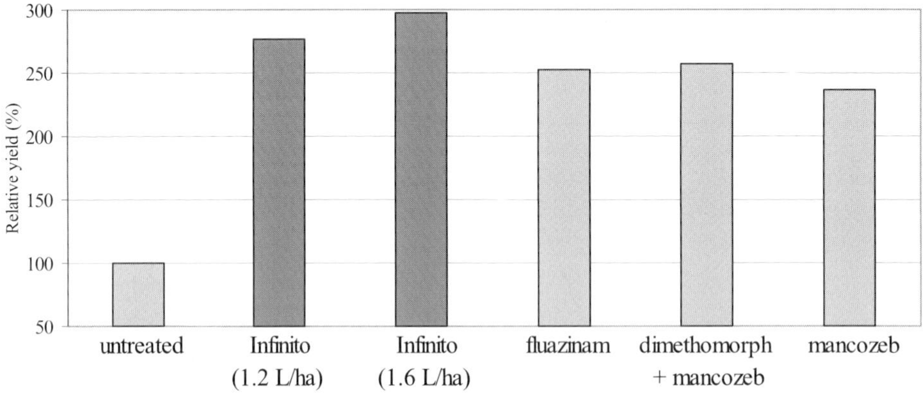

Figure 6. Yield enhancement with Infinito compared to commercial standards (7 registration trials in France in 2002 and 2003, preventive applications at 7 day spray intervals).

Discussion

Mid-season application and tuber blight protection

It is essential for potato growers to prevent the development of *P. infestans* in their crops. The key to achieve both optimal yield and tuber quality suitable for storage is successful protection of leaves and stems and therefore prevention from sporangia dissemination in the crop. Sporangia which are washed down to the soil may be viable for several weeks and tubers are susceptible to *P. infestans* infections from the tuber initiation phase onwards. Consequently it is important to protect crops from late blight from before the start of tuber initiation (Andrivon, 1994). In that respect, intensive observations of *P. infestans* dynamics in Europe during three years of field trials indicate that applications of Infinito in mid-season efficiently reduce tuber infection. During the period when the disease may develop very quickly with intensive emissions of sporangia, Infinito provides the robust foliar protection and strong anti-sporulant activity needed to achieve optimal yield and tuber blight protection at harvest.

Conclusions

Infinito is a new fungicide combining two different active ingredients: fluopicolide, having a novel mode of action, and propamocarb-HCl, a well established potato fungicide. Infinito is a robust quick-acting late blight product. It has strong anti-sporulant activity and long lasting efficacy, resulting in excellent and consistent control of leaf, stem and tuber blight in late blight control programmes.

Infinito has a good environmental profile and is suitable for use in integrated pest management systems in potato crop. Infinito sets a new standard for potato late blight control.

Acknowledgements

The authors would like to thank all colleagues who contributed to the development of Infinito.

References

Andrivon, D. (1994). Dynamics of the survival and infectivity to potato tubers of sporangia of *Phytophthora infestans* in three different soils. Soil Biology and Biochemistry, 26(8), 945-952.

Brent, K.J. and D.W. Hollomon (1998). Fungicide resistance: the assessment of risk. FRAC monograph No. 2, 48 pp.

Flier, W.G., G.J.T. Kessel and H.T.A.M. Schepers (2004). The impact of oospores of *Phytophthora infestans* on late blight epidemics. Plant Breeding and Seed Science, 50, 5-13.

Tukey, J.W. (1977). Exploratory Data Analysis. Addison-Wesley, Reading, Massachusetts.

Curriculum vitae – Sylvain Tafforeau

Sylvain Tafforeau is responsible for the development of potato and vegetable fungicides at Bayer CropScience. Based in the company headquarters in Monheim (Germany) since 2002, he is taking care of the new oomycete fungicide, fluopicolide, which is expected to be launched in the main markets from 2006 onwards. Sylvain Tafforeau joined the company in 1996 in Côte d'Ivoire where he took over different managing positions in field research, development, marketing and sales. He studied Plant Biology & Physiology at the University of Angers (France) from which he received the Master degree in 1989. He fulfilled his skills in Agronomy and in Plant Protection at the High School of Horticulture (ENSH) in Versailles (France) and received the diploma of Engineer in Agronomy in 1991.

sylvain.tafforeau@bayercropscience.com

Company profile – Bayer CropScience

Bayer CropScience, a subsidiary of Bayer AG with annual sales of about EUR 6 billion, is one of the world's leading innovative crop science companies in the areas of crop protection, non-agricultural pest control, seeds and plant biotechnology. The company offers an outstanding range of products and extensive service backup for modern, sustainable agriculture and for non-agricultural applications. Bayer CropScience has a global workforce of about 19,000 and is represented in more than 120 countries.

Further information is available at: www.bayercropscience.com

The role of spray technology to control late blight in potato

J.C. van de Zande[1], J.M.G.P. Michielsen[1], H. Stallinga[1], R. Meier[2] and H.T.A.M. Schepers[2]
[1]Wageningen UR - Agrotechnology and Food Innovations, P. O. Box 17, 6700 AA Wageningen, The Netherlands
[2]Applied Plant Research (PPO), P.O. Box 430, 8200 AK Lelystad, The Netherlands

Abstract

During the growing season of a potato crop late blight is controlled with approximately 15 sprays. Spray technology affects spray deposition and biological efficacy. In a series of field experiments the effects of spray volume, nozzle type, spray quality, dose and air assistance were examined. In addition the effect of sprayer boom movement on variation in spray deposition and protection against late blight was also determined. Spray dose was varied between full dose, 75% and 50% of the recommended rate, including non-treated plots. Spray interval was fixed either at a weekly or a fortnight interval to both mimic normal commercial practice and to stress fungicide performance. Spray deposit was measured in a quantitative and a qualitative way by, respectively using, a fluorescent tracer and water sensitive papers. Foliar late blight infestation was evaluated at a weekly interval. Results show that air assistance (5/8 of full capacity) increased spray deposits on the leaves of the whole potato crop by 8% more than that gained with conventional practice. Spray deposition patterns within the crop also changed with the use of air assistance; quantities retained at the top, middle and bottom leaf levels were different from those attained conventionally with more spray being able to penetrate through the canopy to lower leaf areas and more being retained on lower leaf surfaces. Ground deposits did increase in some experiments by the use of air assistance. Late blight control was improved by using air assistance when spraying at weekly intervals. In general, increasing the length of this spray interval to two weeks - and/or reducing dose rates - caused the level of late blight control to be decreased. In a separate series of experiments, sprayer boom movement - an effect also known to have a high influence on deposit variability - was measured for a sprayer passing along the same track within the potato field. Boom movements were affected with growing season since track conditions changed as a consequence of the repeated passage over them. Spray deposition was measured over a 10 m crop row length and patterns of variation related to sprayer boom movement explain the interaction. Sampled potato leaves from along this 10 m length, having had their spray deposits recorded, established the correlation between spray deposition patterns, efficacy and subsequent level of protection against late blight.

Keywords: spray technique, nozzle type, air assistance, spray deposition, sprayer boom movement, biological efficacy

Introduction

The reduction of the emission of plant protection products to the environment is for a long period now an issue in the Netherlands (Van de Zande *et al.*, 2000a). Spray free and crop free buffer zones were introduced, to minimise the risk of mainly spray drift (Water Pollution Act, Plant Protection Act) for intensively sprayed crops such as potato. The Dutch crop protection policy plan however also aims to reduce the use of agrochemicals by 90% in the year 2010 (LNV, 2004). Increased efficiency in the application of crop protection products should lead directly to scopes for a reduction in their applied doses and therefore the total emission into the environment. New spray application techniques like an air-assisted field sprayer, using a fan on the sprayer to produce air to convey the spray towards the crop canopy, do reduce spray drift significantly (Van de Zande *et al.*, 2000c). Whether air-assisted spraying changes the spray deposition in the crop canopy and whether this change has its effect on biological efficacy was questioned. In a series of experiments, spray-deposition and biological effect were determined in a potato crop using a commercially made sprayer equipped with an air-assisted system (Van de Zande *et al.*, 1999; 2000b; 2002). Spray interval, the rate of active ingredient and the use of air assistance were compared.

A better placement on the target and a more even distribution are ways to minimise the required dose. As dose is reduced concentration of actives in spray drift are reduced also and therefore the risk for the environment. This requires however more advanced spray techniques. Especially the effect of sprayer boom movements was seen as a source for variation of spray deposit (De Jong *et al.*, 2000a; Ooms *et al.*, 2003) and biological efficacy in the field. It was assumed that during season, as sprayers move repeatedly through the same spray track, boom movements increase as the spray track becomes more pronounced. This can cause peaks and gaps in the spray applied and, in the context of late blight control in potato, possibly result in those areas being the sources of late blight infections that are first seen in the field. This research is one in a series of experiments designed to evaluate whether spray deposit forms can be preferentially enhanced through application technique to thereby reach the intended dose reduction levels that have been set (Van de Zande *et al.*, 2004). A view is also given on the likely contribution of other developments in spray technique that are relevant to potato disease protection and which may further support this goal.

Materials and methods

Field trials have been performed to assess spray deposition in potato (Van de Zande *et al.*, 1999, 2000b) and the role that spray boom movement (Van de Zande *et al.*, 2004) may contribute to biological efficacy.

Spray deposition

During four growing seasons (1991-1994) field trials were established in a potato crop (cv. Bintje) at the Kollumerwaard experimental farm (Van de Zande *et al.*, 2000b). Measurements on deposition and biological effect were carried out, comparing spray intervals, rate of active ingredient and the use of air assistance. A fungicide treatment, 2.5 kg/ha cymoxanil/mancozeb 4.5 / 65% (Turbat), was applied using doses that were 100%, 75% to 50% of that recommended on the label. A trailed "Hardi Twin" sprayer with a sleeve boom capable of treating a swath of 18 m wide was used. When using air assistance, the sprayer was operated at 6/8 of its maximum airflow, an air speed/volume that was judged appropriate for the range of crop growth stages to be treated. Nozzles were vertical, as in conventional spraying practice, but the air curtain could then be angled forward - having entrained the sprayed drops - as the swath progressed down the experimental plots. The sprayer boom was equipped with Hardi 4110-18 nozzles operating at 2.5 bar pressure, producing a medium spray quality (Southcombe *et al.*, 1997). The forward speed was 7.2 km/h to apply a sprayed volume of 200 l/ha. The sprayer boom height was approximately 0.50 m above crop canopy. The trials to measure the deposition were conducted separately from the trials monitoring biological efficacy although both studies were in the same potato crop. At different times during the growing season, spray deposition measurements were carried out by adding the fluorescent dye (Brilliant Sulfo Flavine BSF) to the spray liquid at a concentration of 0.5 g litre^{-1}. The non-ionic surfactant (Agral N) was added at a concentration of 1 g litre^{-1} to mimic a pesticide formulation. After spraying the target crop, the dye was extracted from spray collectors made from chromatography paper strips (Whatman no.1; 20 cm long x 2 cm wide), which were folded around and attached to the leaves with paper clips in such a way as to predict spray quantities that are available to the upper and lower surfaces of the leaves. Four collectors were placed systematically at three leaf heights (top, middle and bottom) within the vertical depth of the developing crop. Further spray deposit collectors (1.00 x 0.08 m filter tissues; Camfil CM360), were placed on the soil surface on and between the ridges to judge the risk of exposure of the soil surface to the fungicide. Deposition measurements were carried out at three points across the swath that has been treated such that these values could be related to a known point on the sprayer boom. All such tests were repeated at three different growth stages during the growing season. In each growing season the number of infected plants, the level of infected leaf-area of the plants and the number of infected tubers with *Phytophthora infestans* were measured. Levels of infection in the crop foliage were recorded at weekly intervals from the first time of application until harvest. From this data, the area under the disease progress curve (AUDPC) was calculated and analysis of variance was performed.

Sprayer boom movement

During the 2002 growing season a second field trial was established in a potato crop (cv. Agria) at the Oostwaardhoeve experimental farm (Van de Zande *et al.*, 2004). A trailed 24 m "Hardi Commander Twin Force" sleeve boom sprayer was used to permit two experimental treatments to be applied; each treatment having a working width of 24 m and being 200 m long and having a 3 m bare soil surface [uncropped] path in between. The sprayer was only used in the conventional, non air-assisted, way. The sprayer boom was equipped with TeeJet DG11004 nozzles operating at 3.0 bar pressure, producing a coarse spray quality (Southcombe *et al.*, 1997). The forward speed was 6 km/h, resulting in a sprayed volume of 300 liter ha^{-1}. The sprayer boom height was set to 0.50 m above crop canopy. The trials to measure the deposition were conducted simultaneously with the trials of the biological efficacy measurements at the same plots in the potato crop. Potatoes were protected against late blight using the fungicide fluazinam (Shirlan Flow). Two strips received a full practical dose (0.3 liter ha^{-1} Shirlan; maximum label dose is 0.4 litre ha^{-1}) treatment and two strips a 50% dose. Fungicide treatments took place between 31-5-2002 and 3-9-2002 on a weekly basis (13 times). Throughout the season measurements of sprayer boom movement were recorded (De Jong *et al.*, 2000b) as the sprayer moved through the same track in the field. Spray deposition measurements took place on 10 m long arrays of collectors (Technofil TF-290; 0.5 m long and 0.10 m wide) whilst potato leaves were picked to evaluate protection against late blight (*Phytophthora infestans*) repeatedly through the season.

Results

Spray deposition

Van de Zande *et al.* (2000b) assessed spray deposition in the potato crop for conventional and air-assisted spraying (Figure 1). The deposit of spray liquid on the upper part of the plant canopy was significantly higher than that measured at the middle and lower leaf levels. In general, deposit on the topside of leaves was higher than on the bottom side of leaves. With the additional use of air assistance on the field sprayer, deposition on the middle and lower leaf levels were higher for both the top and bottom side of the leaves compared to that gained with the conventional application. With air assistance, deposition on the under surfaces of middle and lower leaf levels was higher than with conventional application. Deposition on the lowest leaf level for air-assisted spraying was as high as on the middle leaf level for the conventional spraying.

The total spray volume predicted to be deposited on the crop was computed by multiplying the individual leaf deposits, expressed as a volume flux, normalised by the estimated leaf surface areas at the appropriate heights. Typical mean leaf surface area measurements were 2.3 m^2 per m^2 of soil surface (i.e. Leaf Area Index = 2.3). The total deposit on plant leaf tissue (Figure 1) was significantly higher for the 200 litres ha^{-1} spray application when air assistance was used (air-assisted 48.6%; conventional 42.7%).

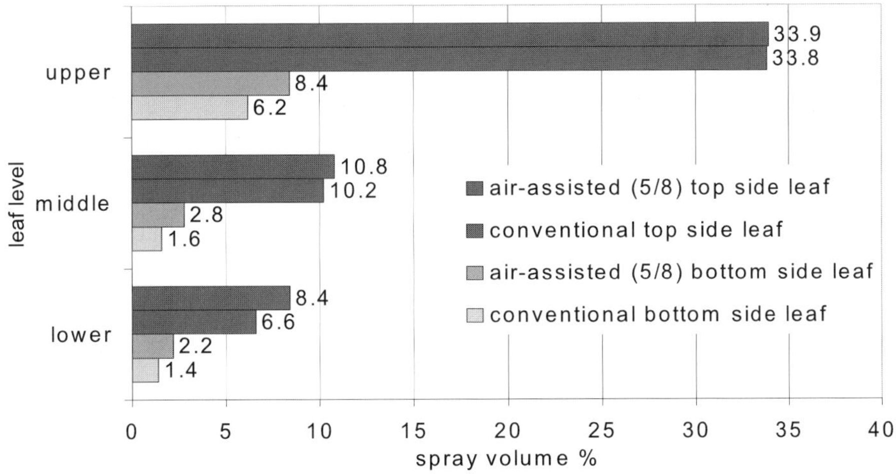

Figure 1. Spray deposition (% of sprayed volume) on upper and lower leaf surfaces at three canopy levels within a potato crop from a 200 l/ha spray volume having a medium spray quality and the use of a conventional or an air-assisted application (after Van de Zande et al.*, 2000b).*

Spray deposition measurements are representative of just that specified time period during the growing season and the extrapolation of such data to other crop stages must be treated with care. In the period when fungicides are applied during the growing season, there are, in general three distinct stages in crop growth to be distinguished:

A. where the leaf canopy is in distinct rows and does not yet touch adjacent rows;
B. where the leaf canopy covers the soil surface completely;
C. where the leaf canopy is decreasing through senescence because of ageing.

There can be changes in coverage of the soil by the crop during each of these periods. From the example above - and other studies - it follows that this affects spray deposition patterns both on the leaf canopy and on the soil beneath the crop (Van de Zande *et al.*, 2003). This is shown in Figures 2 for conventional hydraulic spraying and air-assisted spraying.

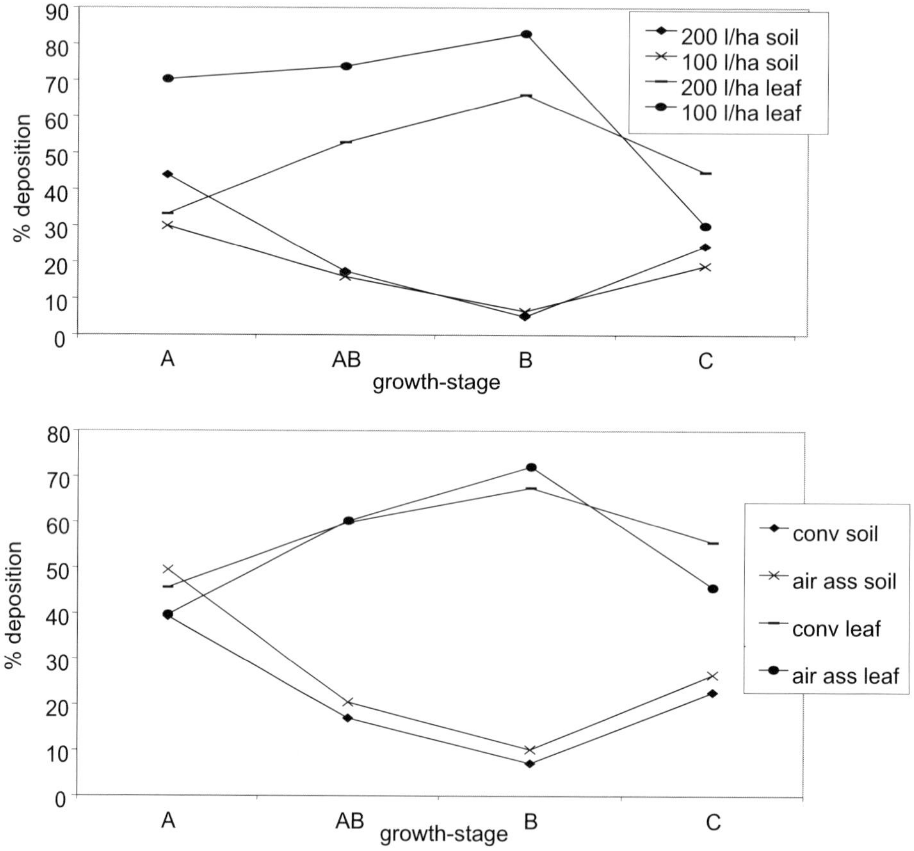

Figure 2. Total spray deposit (% of sprayed volume) on the potato plants and the ground underneath plants during the growing season at different growth stages: effects of spray volumes (top) and air assistance (bottom) (after Van de Zande et al., *2003).*

At growth stage *A* LAI is between 1 and 2 and soil coverage is 20-50%. Deposition in the crop canopy at growth stage A is on average 46% (Table 1). At growth stage *B*, where soil coverage is complete and the crop is growing vigorously, LAI rises to 5.1. Average deposition of spray in the crop canopy at this stage B is 68%. Later in the growing season, when plant stems loose their more vertical stance and may lie between the ridges (stage *C*) LAI reduces again to 1-2. Deposition on the potato plant also reduces and, for conventional spraying, it is around 56%.

Using air assistance on field sprayers changes the deposition pattern within the potato crop canopy. Penetration of the spray into the canopy is increased and by using air assistance more spray is deposited at the middle and lower levels (Figure 1). At early and late growth stages (*A* and *C*) deposition in the crop using air assistance is about 6-10% lower than with conventional techniques. With a fully mature canopy (*B*) the use of air assistance increases spray deposition in the crop by an average of 4%. Improvements in deposition with air assistance are especially increased with fine sprays and low volumes (100 litres ha^{-1}). The effect of spray volume is presented in Figure 2.

The effect of air assistance on soil exposure underneath the crop canopy was significant (Van de Zande *et al.*, 1999) and is shown in Figure 3. Deposition on the soil surface on top of the potato ridges and between them was significantly higher for air-assisted spraying at both volumes of 150 litres ha^{-1} and 300 litres ha^{-1}. Spray deposition on the soil surface underneath the potato crop did not differ between both volumes. Deposition on top of the potato-ridges was higher than between ridges. However, Van de Zande *et al.* (2000b) found no significant effect of air assistance on soil exposure. Deposition on the soil surface underneath the potato

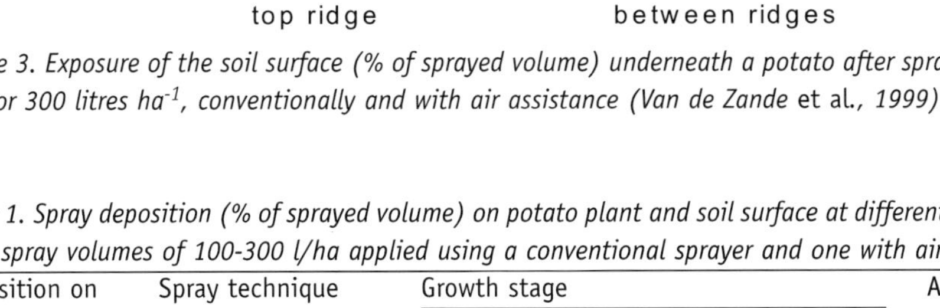

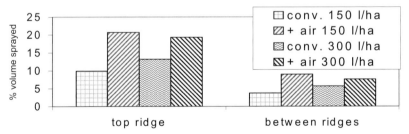

Figure 3. Exposure of the soil surface (% of sprayed volume) underneath a potato after spraying 150 litres ha^{-1} or 300 litres ha^{-1}, conventionally and with air assistance (Van de Zande et al.*, 1999).*

Table 1. Spray deposition (% of sprayed volume) on potato plant and soil surface at different growth stages from spray volumes of 100-300 l/ha applied using a conventional sprayer and one with air assistance.

Deposition on	Spray technique	Growth stage				Avg. all season
		A	AB	B	C	
Potato plant	Conventional	46	60	68	56	57
	Air assistance	40	60	72	46	54
Soil underneath	Conventional	39	17	7	23	22
	Air assistance	50	21	10	27	27

crop was 24% of the applied volume both for conventional and air-assisted spraying, whereas deposition on top of the potato ridges was also higher than between ridges.

During the growing season of the potato, as leaf coverage changes and spray deposit on the crop varies accordingly (Table 1), a change in deposition on the ground also occurs (Van de Zande *et al.*, 2003). In Table 1 spray deposit is presented for the soil surface underneath the potato crop. At growth stage *A* deposition on the soil surface between the potato rows is at full dose (100%). However, this total 'non- target' surface area when averaged with the deposit underneath the plant rows on top of the ridges deposition of the spray, remains at 39%. On a completely covered soil surface in stage B spray deposition on the ground decreased to 7%. However, at growth stage C the average soil deposition is 23%. The use of air assistance increased spray deposit on the soil surface in all growth stages (Figure 2), resulting in an all-season average increase in soil deposition from 22 to 27%. It is expected that at the start of the season more spray will be deposited on the soil surface than on the crop. Since most of the spray deposits on the crop when the maximum LAI occurs and this diminishes towards the end of the season, we can expect that deposition on the soil will show the inverse of this (Van de Zande *et al.*, 2003). In addition, experimental design did not permit changes in air volumes and air speeds from the one that was used. However, a more optimised configuration for use at each - now recognisable - growth stage may be possible in future commercial use.

Biological efficacy

The results of the spray deposition pattern (Figure 1) on leaf late blight infection (AUDPC) (Van de Zande *et al.*, 2000b) are summarised for the weekly spray interval for the research conducted in the three years of experiments (1992-1994) (Figure 4). At the time of first infection, the level of late blight on plots sprayed with air assistance was significantly lower. At the end of the growing season, after repeatedly spraying, no difference could be found. However, on individual recording dates the effect of air assistance on leaf late blight control was significant.

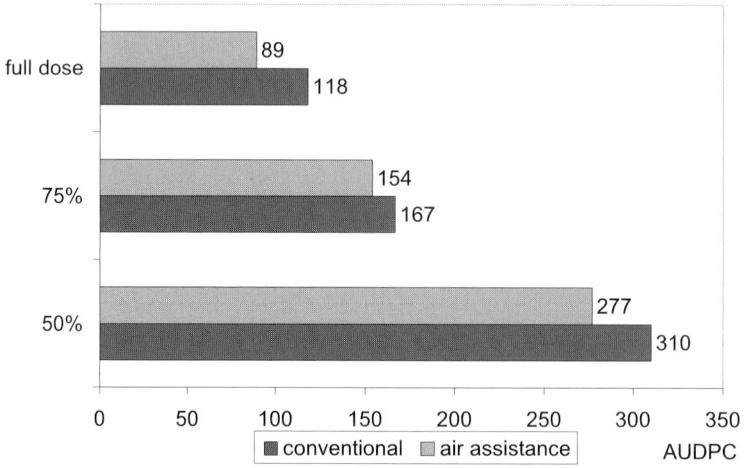

Figure 4. Averaged (1992-1994) foliar infection of late blight (AUDPC values) in potato for conventional and air-assisted spraying, applying 200 l/ha spray volume (100%= 2.5 kg/ha cymoxanil/mancozeb 4.5 / 65% as Turbat) in a 7-days spray interval.

When spraying at weekly intervals, the effect of air assistance on leaf late blight control was significant. Lowering the dosage to 75% and 50% of the recommended dose resulted in lower protection of late blight in the leaf canopy. Tuber late blight control was not different for the recommended and 75% doses. However, tuber infestation was higher with a 50% dose.

Spraying at a fortnight interval resulted in no differences in late blight control between conventional and air-assisted spraying. A dose effect however was clear. Full-recommended dose protected the leaf canopy better against late blight than 75% and 50% doses, with no difference between the latter two. No differences were found in tuber blight control.

Sprayer boom movement

During spray operations the vertical and horizontal position of the spray boom was measured (Van de Zande *et al.*, 2004). Early in the season boom movements were within a range of 5 cm but at the end of the season movements up to 50 cm did occur. Specific movements occurring on a certain point in the track on one date do occur again on the next day in the same direction and at the same site. It is as if there exists an amplifying effect that increases throughout the season. Clearly the rut pattern in the track becomes more pronounced after every pass of a spray application. In particular after a rain shower, these moist conditions increase bumpiness in the track. The standard deviation in horizontal boom tip speed early in the season was around 0.15 m s^{-1} whereas at the end of the season it increased to 0.45 m s^{-1}. The variation in spray deposition underneath the boom tip was closely related to variation in boom tip speed. An example is given (Figure 5) for a measurement early in the season (11-7) in the half dose field. Higher boom tip speeds resulted in lower levels of spray deposition, as can be seen from the area 54-55.5 m and 57-58 m. Lower speeds resulted in higher levels of spray deposition as shown on the 53-54 m area (Figure 5).

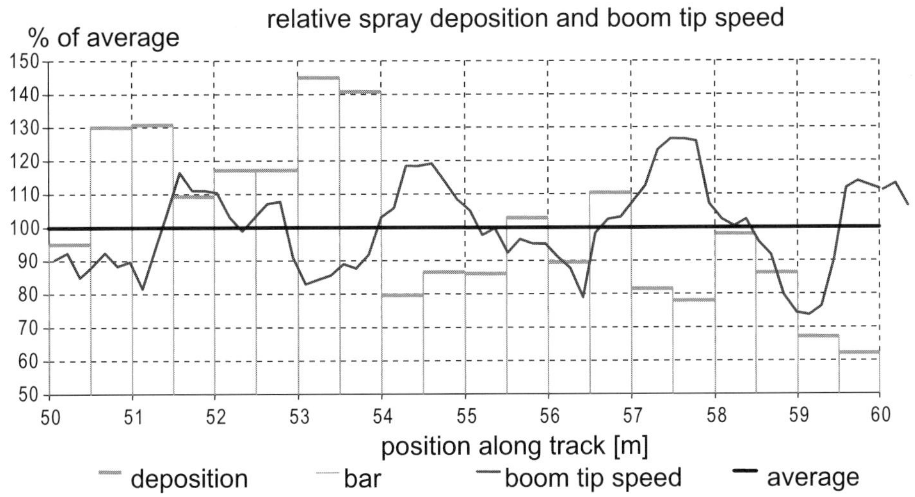

Figure 5. Relation between the variation in sprayer horizontal boom tip speed around average (driving speed) and the variation in spray deposition underneath on top of canopy over a length of 10 m (50-60 m) along the spray track when spraying potato.

On three dates spray deposition was measured (Figure 6). The standard deviation in spray deposition increased for these dates respectively with 19%, 32% and 35%, thereby influencing the coefficient of variation in spray deposition for these dates; it changed from 24% to 31%. The area underneath the boom with a deviation in spray deposit of at least 10% of the emitted dose was 45%, for these measurements, throughout the season. Spray deposit on 20% of the area was lower than 90% of the overall emitted dose. For 35% of the area sprayed, deposit was higher than 110% of this dose. Lowest values of deposit can be near 60%, and highest values up to 180% of intended spray volume.

Spray deposits sampled from exposed 0.50 m collectors within a 10 m length was 104% of sprayed volume, with a coefficient of variation of 31%, suggesting precise application and calibration. Average late blight infection on top leaves over the same 10 m length was 22% with a coefficient of variation of 39% (Figure 7).

These first measurements show that the variation in boom movement and therefore the spray deposition over canopy can be traced back in the level of protection against late blight in the potato leaves. More work is however needed before we can come up with a clear relation as, e.g. a dose-effect curve, to predict the effect of sprayer boom movement on efficacy.

Outlook

Air assistance

Leonard *et al.* (2000) also found that spray penetrated crop foliage deeper using air assistance and that deposits on the under side of leaves were increased with air assistance. But as Jeffrey and Taylor (1991) already indicated, actual deposits on under surfaces were still small compared to those on the upper side of the leaves. However, ground deposits were increased by the use of air assistance when adjusted in the manner described. When spraying a full dose at weekly

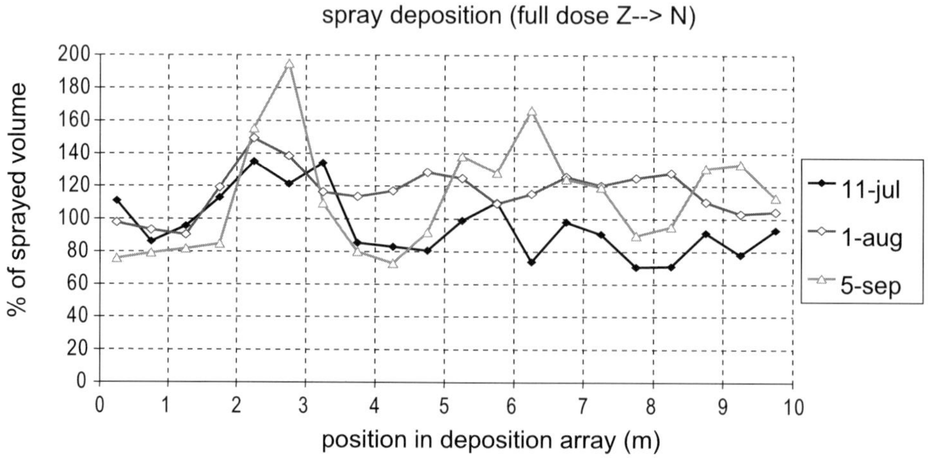

Figure 6. Spray deposition (% of sprayed volume) measured over the same track length of 10 m (50-60 m) underneath the boom tip on different dates during the growing season when spraying potato.

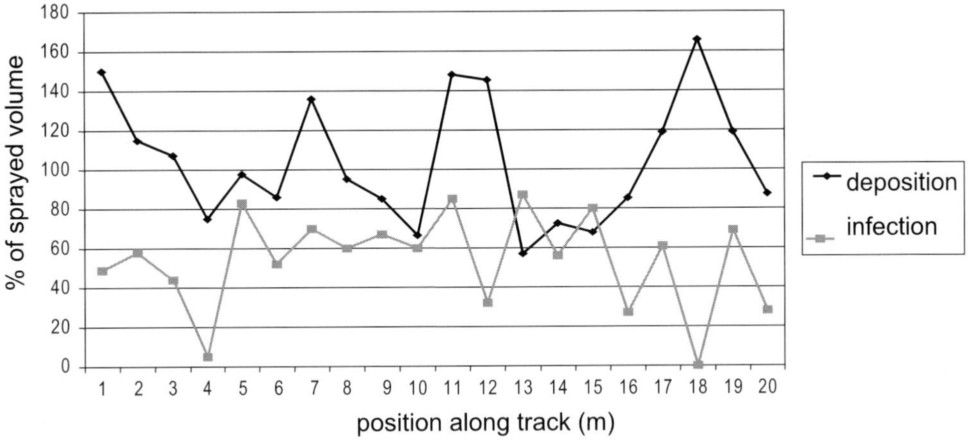

Figure 7. Spray deposition pattern (% of sprayed volume) on top of canopy and infection of late blight in a bio-assay on leaves of 20 potato plants (% of leaves infected) taken over a length of 10 m along the spray track (50-60 m) spraying potato with half (50%) dose of fluazinam (100%= Shirlan 0.3 liter ha^{-1}) with a spray volume of 300 liter ha^{-1}.

intervals, the effect of air assistance on leaf late blight control was significant. Lowering the dosage to 75% and 50% of the recommended dose resulted in lower protection of late blight in the leaf canopy, better for air assistance but not significantly different. When increasing the length of the spray interval, the level of blight control decreased. Reducing spray dose also decreased late blight control. These relations were different, however, for conventional and air-assisted spraying.

Sprayer boom movement

Repeatedly driving through the spray track of a potato crop creates a bumpy track that increased in severity during the season. As a consequence boom movement increased. Typical boom movements seem to occur in the same way at the same site along the track. Repeatedly over- and under dosing of crop protection products are measured at the same place. It is shown that boom movement, especially the variation in horizontal speed of the tip, is correlated to the spray deposition underneath the boom. Increases in boom tip speed lowered spray deposit. The variation shown in spray deposit underneath the boom was found also in the protection against late blight in the potato leaves.

Spray nozzles

The risk of spray drift has led to much attention on low-drift nozzle types especially those generating more coarse spray qualities. Venturi nozzles have been developed to spray low volumes with coarse sprays. Compared to flat fan nozzles Schepers and Meier (2001) found no difference in late blight control when venturi type nozzles were used at spray volumes of 95, 150 and 300 l/ha. Spraying at 300-400 l/ha Wachowiak (1999) also found no differences between leaf late blight infestation and potato tuber yield using venturi type nozzles and standard flat fan nozzles. On vertical targets Wolf and Caldwell (2004) found that a combination of angled double venturi nozzles, and faster travel speed increased (doubled) spray deposition. 60° angles between nozzles increased spray deposition over 30° angles. This was clearer for coarse sprays than for fine sprays. Robinson *et al.* (2000) found increases in spray deposition on potato with angled nozzles, especially in the middle and lower leaf levels. A single nozzle set-up angled 30° backwards had the highest overall spray deposition and gave the best protection against late blight. This has led to the development of the 'potato nozzle' (Syngenta, 2005).

Sprayer speed

Increasing forward speed (8 to 16 km/h) increased the deposition onto vertical surfaces (Webb *et al.*, 2004). Variability of the deposits found on vertical targets was reduced using venturi nozzle types; however, deposition was also lower than for standard flat fan nozzles. Vertically angled air assistance increased deposition onto horizontal surfaces and decreased deposition on vertical surfaces with standard flat fan nozzles and even more with the coarser venturi type nozzles. Vehicle wake influenced deposition of spray in the middle section of the sprayer boom, especially with conventional flat fan nozzles. This led to higher deposits on both horizontal and vertical targets close to the sprayer track (Webb *et al.*, 2002). Nilars (2002) found also that crosswinds increased non-uniformity in spray deposition across the sprayer working width, especially with smaller sized standard flat fan nozzles and higher sprayer speeds. Using venturi type nozzles of the same size reduced this cross-distribution effect but did not meet the standards required (CV < 10%).

Future developments

Giles *et al.* (2002) described the use of sensors to detect the target areas and spraying exclusively to those detected areas. Significant reductions in pesticides doses, spray drift and runoff could be achieved whilst maintaining biological efficacy. Real-time sensing and spraying utilizes a computer based integration model where target position may be tracked only for the duration of the treatment. Map-based systems allow the sensing and treatment processes to be temporarily de-coupled and use spatial coordinates, such as GPS positioning, to ensure that treatment is focused exclusively on the target areas. Robotic micro spraying is possible and even individual plant mapping is within reach. Targeted spraying requires however more precise placement of the crop protection product. Adaptation of sprayer dose can be done by either changing spray volume emitted by one spray solution concentration or by agrochemical injection systems maintaining the same flow rate in the nozzles and with pulse width modulated (PWM) nozzles. PWM nozzles on a sprayer boom can assure a fast response to sensors detecting weeds or diseases and can apply the right dose (Ramon *et al.*, 2002). In fact, new developments such as micro sprayers (Giles *et al.*, 2002) open new perspectives on crop protection possibilities as targets as large as individual leaves can be treated separately. This gives extra reasons to work on boom stabilization (Ramon *et al.*, 2002), especially further work on horizontal boom movements, and a greater emphasis placed on development of active suspension systems.

The concept of Minimum Lethal Herbicide Dosage (MLHD): applying the smallest lethal dose of herbicide is currently developed for potato haulm killing (Kempenaar *et al.*, 2004). In this development, spectral reflections of the crop determine the appropriate dose to be adopted. Initial trials at field level showed that a reduction in agrochemical use of 30% was achievable whilst maintaining efficacy. The use of a YARA N-sensor for crop reflection to steer the sprayer is under development (Achten, 2005). Further developments towards canopy (density) adapted spraying are foreseen (Van de Zande *et al.*, 2005), especially for bed grown crops such as potato but eventually also for other arable crops.

Acknowledgements

This work was funded by the Dutch Ministry of Agriculture, Nature and Food Quality. The spraying equipment was supplied by Hardi International A/S. We also wish to acknowledge the efforts of the people of the Kollumerwaard and Oostwaardhoeve experimental farms for practical help carrying out these experiments, especially Gerrit Goedbloed. Thanks to Bill Taylor (Hardi International) and Tom Robinson (Syngenta) for critically reading of the manuscript.

References

Achten, VTJM (2005). Technology developments in potato quality management. In: A.J. Haverkort and P.C. Struik (ed), Potato in progress, 368pp.

De Jong, A., J.M.G.P. Michielsen, H. Stallinga and J.C. van de Zande (2000a). Effect of sprayer boom height on spray drift. Mededelingen Faculteit Landbouwwetenschappen Rijksuniversiteit Gent, 65/2b, 919-930.

De Jong, A., J.C. van de Zande and H. Stallinga, (2000b). The effects of vertical and horizontal boom movements on the uniformity of spray distribution. Paper 00-PM-015 presented at EurAgEng Warwick 2000, 9 pp.

Giles, D.K., D.C. Slaughter and S.K. Upadhyaya (2002). Biological sensing and sprayer control. Aspects of Applied Biology, 66, International advances in pesticide application, pp. 129-138.

Jeffrey, W. and W.A. Taylor (1991). Foliar distribution of air assisted spray deposits in a potato canopy. In: A Lavers (ed), Air assisted spraying in crop protection. BCPC Monograph no. 46, BCPC, Bracknell, UK, 273 pp.

Kempenaar, C., R.M. Groeneveld and D. Uenk (2004). An innovative dosing system for potato haulm killing herbicides. XIIeme colloque international sur la biologie des mauvaises herbes, Dijon, France, 11 pp.

Leonard, R., B. Rice, L.J. Dowley and S. Ward (2000). The effect of air assistance on spray deposition and biological effect in the control of Phytophthora infestans in potatoes. Aspects of Applied Biolog 57, Pesticide application, pp. 243-250.

LNV (2004). Sustainable Crop Protection. Crop protection policy towards 2010. Ministry of Agriculture Nature Management and Food Quality, The Hague, 2004, 44 pp. (in Dutch)

Nilars, M.S. (2002). Some nozzle performance considerations when using wide booms at higher speeds. Aspects of Applied Biology 66, International advances in pesticide application, pp. 95-106.

Ooms D., R. Ruter, F. Lebeau and M.-F. Destain (2003). Impact of the horizontal movements of a sprayer boom on the longitudinal spray distribution in field conditions. Crop Protection, 22, 813-820.

Ramon, H., J. Anthonis, E. Vrindt, R. Delen, J. Reumers, D. Moshou, K. Deprez, J. DeBaerdemaeker, F. Feyaerts, L. van Gool, R. De Winne and R. van den Bulcke (2002). Development of a weed activated spraying machine for targeted application of herbicides. Aspects of Applied Biology 66, International advances in pesticide application, pp. 147-162.

Robinson, T.H., T.A.J. Scott and D.J. Marshall (2000). The effect of nozzle angle and nozzle types on the deposition and biological performance of potato blight fungicides. Aspects of Applied Biology 57, Pesticide application, pp. 267-272.

Schepers, H.T.A.M. and R. Meier (2001). Effect of an organosilicone adjuvant on the biological efficacy of fungicides applied with low-drift air induction nozzles in potato and onion. In: H. de Ruiter (ed) Proceedings of the 6th International Symposium on Adjuvants for Agrochemicals ISAA 2001, Amsterdam, 13-17 August 2001, pp. 569-574.

Southcombe E.S.E., P.C.H. Miller, H. Ganzelmeier, J.C. van de Zande, A. Miralles and A.J. Hewitt (1997). The international (BCPC) spray classification system including a drift potential factor. Proceedings of the Brighton Crop Protection Conference - Weeds, 1997. November 1997. Brighton, UK, pp. 371-380.

Syngenta (2005). website syngenta crop protection. http://www.syngenta.com

Van de Zande, J.C., H.A.J. Porskamp, J.M.G.P. Michielsen, M.T. van IJzendoorn and R. Meier (1999). Spray deposition and biological efficacy in potatoes. In: J. Pabis and R.S. Rowinski. Techniques in Plant protection, International Conference - Warsaw Poland, 23-26.5.1999, pp. 117-125.

Van de Zande, J.C., H.A.J. Porskamp, J.M.G.P. Michielsen, H.J. Holterman and J.F.M Huijsmans (2000a). Classification of spray applications for driftability, to protect surface water. Aspects of Applied Biology 57, Pesticide application, pp. 57-65.

Van de Zande, J.C., M.T. van IJzendoorn and R. Meier (2000b). The effect of air assistance, dose and spray interval on late blight control *Phytophthora infestans* in potatoes. Proceedings of the Brighton Crop Protection Conference - Pests & diseases, 2000. November 2000. Brighton. UK, pp. 1087-1092.

Van de Zande, J.C., J.M.G.P. Michielsen, H. Stallinga and A. de Jong (2000c). Ranking of low-drift nozzles and air assistance for spray drift. Paper 00-PM-066 presented at AgEng 2000, Warwick, UK, 2000, 9 pp.

Van de Zande, J.C., J.M.G.P. Michielsen, H. Stallinga, H.A.J. Porskamp, H.J. Holterman and J.F.M. Huijsmans (2002). Spray distribution when spraying potatoes with a conventional or an air-assisted field boom sprayer. ASAE Meeting Paper no. 021003, 2002 ASAE Annual International Meeting/ CIGR XVth World Congress, Hyatt Regency Chicago Ill., USA, July 28-31 2002. St. Joseph, Michigan, ASAE, 10 pp.

Van de Zande, J.C., C.S. Parkin and A.J. Gilbert (2003). Optimising pesticide application. In: M.F. Wilson (ed.), 2003. Optimising pesticide use. Chapter 3. John Wiley, UK, pp. 23-44.

Van de Zande, J.C., J.M.G.P. Michielsen, H. Stallinga, R. Meier and H.T.A.M. Schepers (2004). Effect of sprayer boom movement on spray deposition and biological efficacy. Aspects of Applied Biology 71, International advances in pesticide application, pp. 91-98.

Van de Zande, J.C. and V.T.J.M. Achten (2005). Precision agriculture in plant protection technique. Annual Review of Agricultural Engineering (2005): 10 pp.

Wachowiak, M. (1999). The influence of treatment technique on the effectiveness control of *Phytophtora infestans* De Bary. In: J Pabis and RS Rowinski. Techniques in Plant protection, International Conference - Warsaw Poland, 23-26.5.1999, pp. 101-104.

Webb, D.A., C.S. Parkin and P.G. Andersen (2002). Uniformity of the spray flux under arable boom sprayers. Aspects of Applied Biology 66, International advances in pesticide application, pp. 87-94.

Webb, D.A., C.S. Parkin, W.A. Taylor and B.P. Taylor (2004). Influence of nozzle design, application speed and air-assistance on the variability of spray deposits under arable boom sprayers. Aspects of Applied Biology 71, International advances in pesticide application, pp.83-90.

Wolf, T.M. and B.C. Caldwell (2004). Evaluation of double nozzle spray deposits on vertical targets. Aspects of Applied Biology 71, International advances in pesticide application, pp. 99-106.

Curriculum vitae – Jan van de Zande

Jan van de Zande has an education in Agricultural Engineering from Wageningen Agricultural University (1985) and is involved in crop protection research since 1989. In the period 1989-1993 he worked at the Institute for Applied Research for Vegetables and Arable Crops and emphasis was on mechanical weed control and spray technology. Main subjects were the effect of spray techniques (nozzle types and air assistance) on the biological efficacy in fungicide application for disease control and herbicide application for weed control in different crops.

From 1993 onwards he worked at the Institute for Agricultural and Environmental Research (IMAG), which merged to Agrotechnology & Food Innovations in 2003. Main subjects of research are the quantification of spray drift from different spray techniques through field measurements, modelling spray drift and scenario research for environmental risk assessments. Variability of spray distribution at field scale as well as on drop scale are subjects of research, especially related to biological efficacy. Recently he coordinated part of an EU-project (PreciSpray) which led to the development of a precision orchard sprayer making use of canopy density information of individual trees detected from stereoscopic aerial photogrammetry. He is a project leader of numerous projects funded as well as by governmental departments, institutions and industry. He is involved in different working groups on international standardization of spray techniques (ISO, CEN) and the implementation of a classification system for drift reducing techniques in the EU (FOCUS).

Company profile – Agrotechnology and Food Innovations B.V. (A&F)

Agrotechnology and Food Innovations B.V. (A&F) is a research institute part of Wageningen University and Research Centre (WUR). Within A&F (www.agrotechnologyandfood.nl) the research of the Field Technology Innovations (FTI) group is focussed on sustainable crop protection and care for the environment. Special attention is paid to reduction of spray drift and effectiveness of application techniques. Likewise, research is performed on innovative spray techniques. Research is embedded in the Pesticide Act, Water Pollution Act and the theme Sustainable Crop Protection.

Field research is performed on drift reducing spray techniques and techniques leading to a more efficient spraying operation in terms of saving crop protection product use. Both for the agricultural machinery and crop protection industries as for policy makers, effects of crop- or spray free buffer zone width, nozzle type, use of air assisted spraying techniques, sprayer boom movements, spray deposition and biological efficacy are investigated. For manufacturers of spray nozzles and interest groups, measurements are performed to classify nozzles for drift sensitivity. Research is performed on sensors quantifying the crop development stage. These crop sensors are the basis for Canopy Density Spraying. Also, GPS is used to spray crops precisely. The distribution of the spray and its effectiveness is tested also for biological agents. End results are mainly obtained through field research, data analysis, desk studies, feasibility studies, scenario research, and prototyping. The research is executed on experimental farms, in agricultural practice, and indoors in atmosphere-conditioned rooms. The analysis of samples is done in house.

Trade

General trends in the European potato trade

Jörg Renatus
EUROPLANT Pflanzenzucht - Varieties, Competence & Service, Germany

The potato market in the European Union (EU) is generally divided into three segments:
1. Starch potatoes.
2. Processed potatoes for human consumption.
3. Fresh / ware potatoes.

Starch potatoes

The market segment of starch potatoes is ruled by special conditions due to the regulations about subsidies paid by the European Union. As also the varieties used in this market segment are usually not used for other purposes, this market follows its own rules which are predominantly determined by political decisions and not by market forces. Since the profitability for the growers of starch potatoes is depending to a large part on the amount of subsidies paid by the EU and since it is already evident that the amount of subsidies paid will be reduced in the coming years, the pressure on the starch factories and growers to optimize their production will steadily increase. The threat of cheaper maize starch and also tapioca starch from outside the protected EU borders is increasing. In this respect we foresee a further specialization of the growers. The production technique has to be improved by all means. But we will also see a concentration on certain growing areas. The production will be more concentrated close to the factories in order to save freight costs. This will lead to a tighter crop rotation which consequently results in a higher pest and disease pressure, especially from white potato cyst nematode (*Globodera pallida*) and potato wart disease (caused by *Synchytrium endobioticum*). Accordingly, breeders have to offer varieties with higher resistances especially against these diseases and possibly combined with an increased starch yield. Also special purpose varieties such as amylose-free varieties will take more and more place in the factories' production programmes in order to either generate better earnings through higher value products or to save costs in the production.

Processed potatoes for human consumption

This market segment again is divided mainly into the following parts:
a. French fries.
b. Chips or crisps.
c. Prefabricated convenience products.

a. The consumption of french fries is stagnating in the Western European countries over the last years. The reasons are various. One important factor has been the acrylamide discussion, which has made the producers of french fries and chips suffer until 2004 when the consumption slightly started increasing. In other EU countries the discussion has not been as intensive, but it seems to get more onto the agenda. Another main reason is that consumers are becoming more and more health-conscious and prefer to buy light food and snacks with less carbohydrates and fat. Public initiatives are supporting the consciousness about obesity and vascular diseases. After many years of burgers and fries consumers are looking for a variation of products and when we look at the sales strategy of the large fast food chains we can observe that their promotion nowadays is focusing more on salads, pasta and fruit.

b. Also the consumption of chips (respectively crisps) did not show a positive development (see Figure 1). The acrylamide discussion two years ago has led to a heavy drop in consumption of chips and only slowly the marked is recovering. It is doubtful that we will have a further large increase in the consumption. In Europe we are far away from per capita consumptions of more than 2.3 kg as for example in the USA. In general we can say that french fries as well as chips are having an image problem regarding their negative contribution to health and are under careful investigation by the consumers.

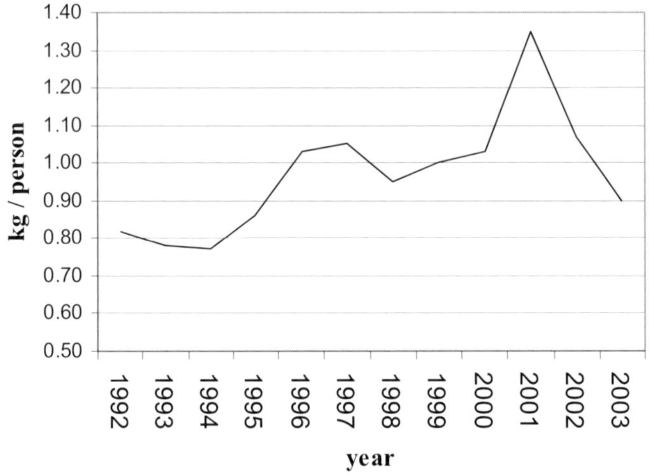

Figure 1. Chips consumption in Germany per capita in kg (ZMP).

c. The third segment, the prefabricated convenience products, are gaining market share steadily over the last years. These products have a good image regarding their nutrition facts and they provide to the consumers what they are looking for: an easy and quick meal. The industry is launching various new variations of hash browns/röstis, potato pockets, dumplings or roasted potatoes to attract the customers. Especially the consumption of products classified as "other products" like precooked meals, potato salad or Gnocchi have developed significantly from 2003 to 2004 (+19,82%).

Since these product are very much in line with the needs and preferences of modern households which have the necessary income to buy these products since they are either single households or double income households, it can be predicted that the market volume of this kind of products will continue to grow.

What all three segments have in common is that for the supply of their raw material they are in a heavy competition to get the lowest prices for processing potatoes. Only for certain variety or quality characteristics that are needed for special products, for example varieties which enable low percentages of acrylamide in fried products, higher prices are paid. Since the input of raw material in relation to the ready product is relatively high with ratios of about 2:1 in french fries and 4:1 in chips and also 2:1 in wet convenience the purchase price of the raw material plays an important role in the product calculation of the industry. So as a further trend for marketers of seed, ware and processing potatoes it can be estimated that the pressure on the development and introduction of new varieties with better characteristics regarding quality, yield and disease resistances will continue. The pressure to optimize the production will continue for the farmers. Only those farmers in areas with favourable growing conditions and appropriate varieties in respect of water supply and vegetation days will have a chance to survive with the prices the industry is ready to offer.

Fresh / ware potatoes

The third market segment which has to be considered is the traditional segment of ware potatoes. This market has experienced very heavy changes over the last decades in all respects. Not only that the consumption of fresh potatoes has steadily declined over the years (see Figure 2) due to competition with noodles, pizza and rice - also the demand of the consumers has changed. Traditionally, potatoes have been a basic nutrition source for a large majority of the European population and have been considered to be a commodity coming from the region being present on every meal plate. Ware potatoes have mostly been on the market in large volumes. But the quality has varied strongly throughout the year. When the growing conditions were good and it was not too late in the year, the appearance of the potatoes was still quite nice. The more time after harvest passed by or the more difficult the growing conditions were the more unattractive the potatoes became.

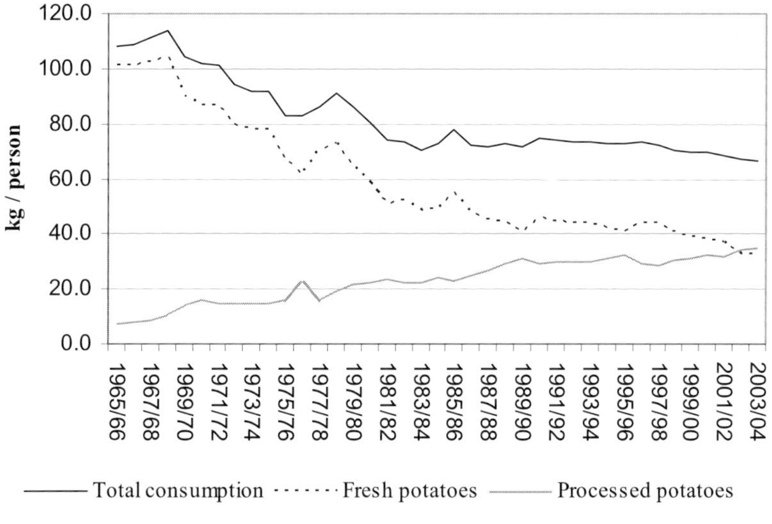

Figure 2. Consumption of potatoes in Germany (German Ministry of Agriculture, ZMP).

About two decades ago the market started to change. New varieties with smoother skin and nicer appearance entered the market. Furthermore, imported potatoes from Mediterranean countries started to gain more and more market share in springtime due to their better look compared to the old crop potatoes. Consequently, step by step the market of fresh potatoes was segmented in two parts: the nice-looking shiny high price top quality potatoes and the standard merchandise at discount price level. There are certainly preferences of variety types like for example red skin varieties in Portugal or Hungary, round white potatoes in the UK, yellow varieties in Central European countries or more pale flesh varieties in the Western European countries. But taste was not really considered, it is still very often today in fact only important to have a perfect optical quality or a cheap discount potato. For some years now additionally to the outer appearance also taste is becoming an important criterion for the consumer's purchase decision.

As a result, the potato is increasingly becoming a delicate crop which needs to have a perfect outer appearance combined with a good intensive taste in an attractive packaging. This type of potatoes also has to have a smooth skin in order to boil well and to eat it even with the skin. It can be considered to be a kind of convenience product. These high-end potatoes are nowadays packed more and more into microwave-ready packs or bags and we see increasingly special varieties being used for such products. This high value segment needs varieties protected by Plant Breeders Rights in order to control the market and to gain extra margins for the costly packing and marketing activities. Having been a niche market for many years it is now starting to develop and gain a bigger market share in the Western European countries. The marketers of the packing stations and the supermarket chains require a steady supply of the same varieties around the year in constant quality. Another trend becomes obvious: The importance of imported potatoes from the Mediterranean countries will decline in the coming years. Many potatoes especially those being cultivated in Northern Africa do mostly not have a sufficient taste for the needs of the described high class products. Secondly, not all varieties which are accepted in premium quality programmes are suitable for a profitable production in warm climates. Furthermore, the capacities of refrigerated storages in Central Europe will be increased, so that the required qualities and varieties are available from local stores also in late spring. With these potatoes the supermarkets get what they need: a reliable merchandise in premium quality that can be sold at stable prices.

The other part of the ware potato business will be the discount. Medium quality potatoes packed in different sizes will be a low cost foodstuff to supply the part of the population with lower income. These potatoes will have to be grown by farmers who are able and determined to produce largest volume. Low cost production will be the keyword for such farmers. The breeders will have to provide varieties with high resistances in order to save costs for chemical treatments and with maximum yields to be as cost efficient as possible. Also off grades of the industry will continue to be used for this segment and for sure the profit margins to be earned in this kind of business will remain very low for all companies involved. This market segment will continue to take large volumes of imported early potatoes only when the Central European potatoes are on a high price level so that the price gap is not too big. The 2004/2005 season has dramatically shown that consumers refuse to buy imported potatoes when the old crop is still acceptable in quality and the prices are 7 to 8 times lower than the prices for imported potatoes. Today traders try to establish new supply channels for potatoes from India and China.

India nowadays already provides large quantities of baby potatoes for the high-end market of the USA. There will be trends of supply in big quantities for the discount segment also. The developments in these two important potato growing countries have to be monitored carefully.

As recapitulation of the before given statements we will have the following general trends in European potato trade:

- Production for the starch or processing industry will remain predominantly a contract production, but with continuing pressure for efficiency. We will see new varieties with higher yield potentials and better processing characteristics combined with higher disease resistances. The seed industry will have to find contract models to supply the large factories steadily with healthy seed potatoes grown near the factories in order to save costs on the transport.
- The ware potato market will increasingly fall apart into two segments:
  - The high-end segment where potatoes will be traded as a high value crop. In this segment varieties are needed that give a special performance for the consumer with regards to taste, cooking quality and outer appearance. We will see more and more specialties in this segment regarding shape and color. This segment will be filled with specially bred monopoly varieties that allow the marketers to control the product or to have exclusivity on it. The seed industry will have to supply the seed potatoes in a contract system as well as the ware potato growers will be linked into hundred percent contracts so that the packer has on the one hand the quality control and on the other hand the quantity control.
  - The low price and discount potato market will be more and more dominated with potatoes from the industry or grown out of very high yielding varieties. Already 41% of ware potatoes in Germany are currently sold by discounters (see Figure 3).

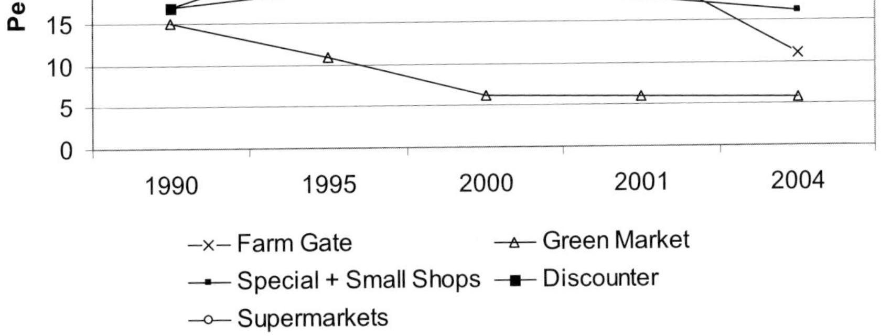

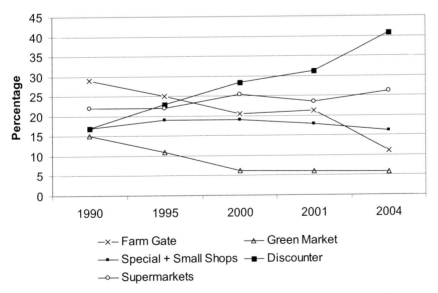

Figure 3. Sales channels for potatoes in Germany (ZMP, GfK Fresh Food Panel).

But this market will not be dominated by a contract system. The only factor which is of importance here will be the lowest possible price which means that the buyers will always chase for every low price offer available and they will not be ready to commit themselves into any contractual liability. For the seed industry the sales will be instable, as farmers will only buy seed potatoes when the season has been acceptable. After a bad season they will use their home-saved "seed" or even remaining ware potatoes.

Curriculum vitae – Jörg Renatus

Since 2002, Mr. Jörg Renatus is working as Managing Director of EUROPLANT Pflanzenzucht GmbH in Lüneburg. He was born in Lübeck in 1962 and holds a Master of Business Administration. His professional experience results from several employments within the potato industry. He started as apprentice for the well-known German potato breeder Kartoffelzucht Böhm. In 1986, he took up a position as Export Manager at SOLANA, the Hamburg-based export company of potato and cereal breeder SAKA-RAGIS. Subsequently he has been appointed as Managing Director of SOLANA and later on of Saka Ragis and other companies of this group of companies. Having gained a vast knowledge in national and international potato trade, he joined the rapidly growing potato breeding company EUROPLANT Pflanzenzucht. Jörg Renatus holds several mandates in German and European associations and working groups such as UNIKA and ESA and is active as RUCIP arbitrator.

Company profile – EUROPLANT Pflanzenzucht - Varieties, Competence & Service

Having been founded as a regional sales company end of the 1980s, EUROPLANT has developed to be one of the leading breeders, seed potato producers and marketers in Europe.

EUROPLANT markets mainly its own varieties as well as the varieties of associated breeding companies based in the main European potato producing countries, but also the trade with varieties from other suppliers is an important part of the business. The highly effective breeding network persistently develops high-performance varieties. Every year about 4 to 5 new varieties with improved characteristics are presented to the market. As a result of 100 years of breeding experience, today EUROPLANT's varieties are at the leading edge of each market segment: the fresh potato market as well as the processing and the starch industry. The actual assortment consists of more than 70 varieties.

EUROPLANT-operated farms and storehouses as well as contract multipliers are forming together an integrated production chain beginning with in-vitro plantlets and ending with the delivery of certified seed potatoes. On top of the strict official German seed certification control EUROPLANT applies its company own Quality Management System to each seed lot offered to the market, which guarantees a maximum quality level for the clients.

EUROPLANT is offering the full range of services. Technical advice to make use of the full potential of the varieties, logistics on demand and punctual service are appreciated by the clients. Through its worldwide contacts within the potato business, EUROPLANT is a reliable trade partner for all kinds of potatoes such as early ware potatoes or raw material for the processing industry. Today, EUROPLANT is present in all major potato producing countries, either through own daughter companies or through representatives.

For more information visitors are invited to contact the EUROPLANT staff in the exhibition or to visit the company's website www.europlant.biz.

Serving the potato market: Danespo-Denmark view

Peter van Eerdt
DANESPO A/S, Denmark

Introduction

Few people know that just after the Second World War - in 1947 - Denmark was the main producer of seed potatoes in Europe with an acreage of almost 40.000 hectares. The production was concentrated around the centre of Jutland. The conditions were optimum for seed production: sandy soils, low virus pressure and a perfect climate for storing potatoes in field stores during winter.

The marketing of the Danish seeds was centralised at the Federation of Danish growers and exporters in the commercial brand of Danpatatas. Competition came in that time mainly from Scotland and the North of Holland. The new Polders, reclaimed from the sea in Holland in the period between 1938 and 1952, offered excellent new possibilities for the Dutch. The virgin soils in areas with few aphids in combination with successful breeding and excellent marketing made serious competition to the Danish seeds. In the beginning of the 1980s Denmark was struck by cases of ring rot. The timing could not be worse, as it was also the time that the seed growing moved "more South", as a result of the introduction of more resistant varieties and new application of mineral oils for reducing virus infections.

Denmark took fierce measures to overcome this unexpected setback. It was decided to clean up all the old stocks by making a compulsory start of multiplication by in vitro culture. Surprisingly, the problem was neutralised quickly and showed to the rest of the world how to handle a sudden outbreak of a bacterial quarantine disease. However, for the potato seed trade's position, the harm had been done, and multiplication acreage dropped to less than 5.000 hectares. The Danpatatas trade organisation got into financial problems. It was not a surprise that in that time the private company DANESPO A/S emerged as an alternative trading partner for the seed growers. DANESPO saw the changing market and started to invest in facilities that could match the increasing requirements for quality, and moved production fields to the clay soil areas of Denmark. The seeds from the sandy soils rapidly lost popularity when more clay grown seeds became available on the market.

Breeding in the Danespo group

LKF Vandel, the only potato research and breeding station of Denmark, has achieved 80% coverage in Denmark with their LKF-varieties. Today's breeding focus is more towards export, starch and the fry industries. The 57 years of research & development have produced exciting developments, which will be mentioned later. Danespo's cooperation with the SARPO KFT research station in Hungary has resulted in *Phytophthora infestans* resistant varieties that are about to be commercialised.

This brings us to the situation of today. Despite the setback in the 1980s, Danish seeds are successfully finding their way into many markets. The parameters for today's and for future success differ a lot from the previous parameters. - A potato world "on the move".

Research

The money for potato research has been reduced a lot as governments have reduced their funding of research programmes to "practical level". International programmes on environmental issues as *Phytophthora infestans* can still find substantial support, but the "making of a variety" is left to the commercial breeders. Driven by the high costs and the need for fast success, breeding has become "very much of the same" in most of the potato breeding stations connected to commercial companies. If you analyse the background of many recent commercialised varieties, the variety x variety cross is the most popular. The very small genetic basis of the new range of varieties "after Bintje" for the French fry industry is an example: At least 8 potential new fry-varieties are deriving of the so-called 'Agria line'.

In Denmark we are about to harvest the advantages of institutional/ fundamental breeding in LKF Vandel for more than 56 years.

An example: The research department of LKF Vandel has looked into the possibility of selecting 2-4 °C storable varieties destined for the fry industry for many years. New parental lines have been detected in LKF, which are now implemented in the commercial breeding programme, and I can predict a major break-through coming up in the field of reducing sugar levels. DNA markers are used as a common tool in the breeding in LKF Vandel. This fingerprinting of the varieties and the clones is used to improve parental selections for crossing, avoid inbreeding and for the screenings of glycoalkaloids.

In the near future potato breeders have an important choice to make: to invest in GM potato techniques or to continue in the costly traditional way. The costs & income ratio of traditional breeding is, in the overall picture, running out of balance when you look at the increasing percentage of free varieties in the market.
The near future will hopefully show that the fear of GMO's has been groundless. I sense a less radical approach as more GM products and applications are presented on the market in a better way than before. In the DANESPO group practically all GM techniques are available for when the time is right.

Trade

The potato market has become much more transparent as communication and travelling are now available to almost everyone. The established seed potato companies have spread out their network into many countries. Importers have become more professionalism and have grown in size. In strategic countries such as Holland and France, DANESPO subsidiaries have been founded to ensure permanent presence in these markets.

The WTO rules and the open EU borders have stimulated the free trade of potato seeds. "Cheaper seeds and easier access to markets for everyone", was the message. You can say that seed potatoes became more a "commodity", whereas previously it was more a privileged and protected product to deal with. Unfortunately the increased transparency, in combination with the ending of protection of key varieties such as Désirée, Nicola and Spunta, have not led to improvement of the financial results for the producers and/or traders. On the contrary, today's market is looking for a new balance that can secure an income.

What are the reasons for today's uncertainty?
- Ending of protection of key varieties.
- The establishment is "out of control".
- Production and demand are not in balance.
- Concentration of buying power.
- Reduction in fresh consumption.
- Increasing transport costs.
- Creation of local seed potato structures.

We are, unfortunately, not talking about a temporary situation of a too large production. Many of these parameters have a structural impact on the market for many years to come. Transport costs, quality- and payment risks have increased and will limit future developments. We do not expect to have an EU-potato organisation that will regulate the market. In EU, the growth of potato consumption, french fries and crisps included, have come to a hold. In a not growing market you have to profile the company on its added values. DANESPO is focusing on adding value to partnership by:
1. Offering guarantee conditions to contracted seed growers.
2. Creating an overflow for fluctuations in yield and demand.
3. Development and marketing of new varieties.
4. Constant support of farmers by our technicians.
5. Own quality control and bonus system.
6. Making structural agreements with market partners whenever possible.
7. Giving transparency and information when needed.
8. Usage of an excellent sales network and organisation.
9. Experienced and skilled personnel.
10. Strong and supportive shareholders.

DANESPO wants to be a key player in the potato world with the profile of a European company with ambition. The brand has to stand for quality seeds, exciting new varieties and partnership.

Curriculum vitae – Peter van Eerdt

Peter van Eerdt has 30 years' of international experience in the potato business. In 1976 after his agronomic studies he joined a potato breeding station as research worker and until 1998 he was engaged in potato breeding as well as staff assignments in quality control, production and sales within different companies in Holland. Peter works now for over 7 years as Commercial Director in DANESPO A/S.

Company profile – DANESPO A/S

DANESPO A/S, founded in 1986, is one of the major players in the international seed potato trade with their head office in Give, Denmark. With subsidiaries in Holland, France, Algeria, Morocco, Hungary and Ukraine as well as worldwide representation of varieties from LKF Vandel breeding station, DANESPO A/S is well equipped to supply the market with "quality from the top of Europe".

Production of potato and seed potato in Russia

Boris V. Anisimov
All-Russian Potato Research Institute (VNIIKH), Lorch Street,23, Lyuberetskiy rayon, Korenevo, 140052, Moscow region, Russia

Potato production, usage and market

The total area of potato fields in Russia during the last 5 years (2000-2004) has been established at a level of 3.1-3.2 million ha for all categories of farms. The average potato yield can be expected to be at a level of 10-11 t/ha and gross potato yield 32-36 mln tons (Table 1).

The structure of the usage of the earned potato yield is given in Table 2.

According to the forecast of the Ministry of Agriculture the total annual:
- demand for food potato in Russia is 17-18 million tonnes (with the average rate of consumption per capita of 120-125 kg);
- demand for seed potato for all categories of farms is 9 millions tonnes;
- demand for potato suitable for processing - 0.5 millions tonnes;
- demand for fodder - 6 millions tonnes. So the total annual demand for the potato will be about 36 millions tonnes.

Table 1. Potato production and consumption in the Russian Federation 2000-2004 (according to the data of the Ministry of Agriculture).

Figures	2000	2001	2002	2003	2004
Arable land, thousand ha	3252	3240	3232	3190	3100
Gross yield, thousand tonnes	33980	34965	32871	36644	35080
Yield, t/ha	10.4	11.0	10.2	11.6	11.0
Production, kg/person	233	241	228	254	236.5
Consumption, kg/person	118	122	115	128.6	119.7
Potato products, thousand t	7.9	10.9	11.9	12.2	11.4

Table 2. Estimation of the structure of the usage of the potato yield earned in 2003.

Usage structure	Millions tonnes (estimation)	%
Food	17	49
Seeds	9	26
Fodder	6.5	18
Processing	0.5	1.5
Losses	2	5.5
Total	36	100

In order to maintain the total demand for potato at the rate of 35-36 million tonnes all facilities of farms, agricultural enterprises and private household allotments will be used.

Different categories of farms have their own peculiarities of areas under crops, gross yield, crop capacity and the technical level and efficiency of potato production.

According to the forms of management within the last 10-12 years agricultural producers can be divided into three main categories:
- agricultural enterprises (AE);
- peasants farms (PF);
- small private farms (SPF).

Agricultural enterprises (AE) include the collective farms (former kolkhozes), the state farms (former sovkhozes), the associations, the joint-stock companies and the agricultural producing cooperative societies.

In 2004 the potato in AE was grown on an area of 170 000 ha, gross production was more than 2 million tonnes or 6.8% of the total volume of potato production in all categories of farms. The average crop yield in AE was 12 tons/ha.

The basic reasons why agricultural enterprises lose their interests in potato growing are high costs of labour and energy (Table 3). It is also necessary to mention the sharp increase in prices of farm machines and equipment, fertilizers, pesticides and services. At the same time the potato producers are not able to increase their prices to compensate the growth of production costs due to the declining life standard of the population of the Russian Federation.

Peasant farms (PF) have a new form of management in Russia. They were formed about 10 years ago. At present more than 250 000 peasant farms are registered and they have about 170 000 ha. In 2004 these farms produced about 640 000 tonnes of potatoes that is 1.8% from the total production volume in all categories of the farms.

Small private farms (SPF) are represented by a great variety of minor forms of management. This category includes country residents' plots, gardens and country plots of city dwellers, and some other. These minor forms comprise 16 million small farms. In 2004 SPF produced 32.4 million tonnes of potato, that is 92.4% of the total volume production in all categories of the farms.

Table 3. Increase in production costs.

Indicators	2000	2001	2002	2003
Production cost, rubles/100 kg	202	230	276	284
Price per 100 kg, rubles	306	298	389	451
Profit, %	44	23	31	44

Thus the analysis of the structure and volume of potato production according to the different categories of the farms shows that in Russia at present the major part of potato production is transferred to the private sector. The share of large agricultural enterprises in the gross potato yield was reduced from 34% in 1990 to 8% in 2004 (Figure 1). Minor areas with small production units of the potato production prevails. Potato production by SPF is to a great extend done manually using primitive technologies and minimal mechanization of labour.

The sector of larger agricultural producers belonging to agricultural farms produced 6.8% and peasants farms 1.8% from the total annually produced potato.

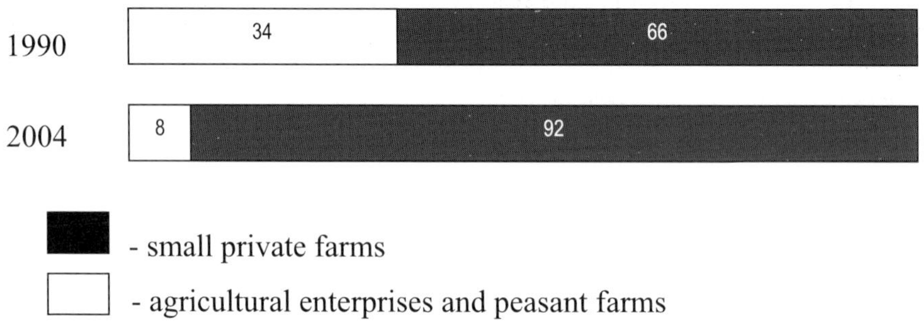

Figure 1. The share of farms of different categories in potato production (%).

The potato market price is formed under the influence of SPF (small private farms). They are dominant suppliers on the potato market and do not depend on the seasonal changes of the sales volume, i.e. they determine the price level during the whole period of sale. SPF sale is 70-75% of potato. But taking into account that a major part of the harvest is used for their own consumption, the level of potato traded is low (about 10%). Any change in the production volume in SPF can influence the demand change on the market both from the population side and the processing industry that increased the potato production traded during the last years.

It is necessary to note one more fact. In spite of the widespread opinion, the potato market does not depend on import. As potato usage from imports is about 2% only, this figure can't influence greatly the potato market.

Cultivar resources, seed quality and prices

There is a total of 200 potato cultivars present in the Russian State Register in 2005 which consists of 121 Russian cultivars (about 60%), 15 Belarusian, 4 Ukrainian, 25 Dutch, 25 German, 8 Britain and 1 Finnish cultivar.

Prevalence of the early and medium late maturing cultivars meets the requirements of the agroclimatic conditions of most country regions. The vegetation period of potato plants is limited by the late cultivars that do not have enough time to reach maturity. Immature tubers can be seriously damaged at harvest and heavy losses may happen at the storage of such tubers.

All foreign cultivars included in the State register are of prime interest mainly because they are resistant to the golden potato nematode and to some viruses and their complexes, and they are suitable for potato processing (for dry potato puree, French fries, crisps, etc).

The analysis shows that the leading potato cultivar in Russia is Nevsky. In the commercial seed production the Nevsky cultivar takes the primary position as to the original, elite and reproduction seed material production. Next according to the upgrading scale are Udacha, Lugovskoy, Romano, Elisaveta, Rosara, Sante, Zhukowskoi early, Petrburgsky, Lukjanovsky, Ilinsky, Snegir, Latona, Bimonda, Alvara, Aroza, Scarlet and other cultivars.

There are many problems concerning the seed potato quality. The yield potential for the most modern potato cultivars is rather high. The breeding development has grown reasonably and attainable yields are now 50-60 t/ha. That is practically similar to the level of western countries with advanced development of potato production. Nevertheless the average yield in some farms is not more than 10-11 t/ha. Thus actual yields are only 25-30% of the yields attained at the official state cultivar tests. According to the test results of the seed potato resources only 60% of the potato meets the requirements of the state quality standards.

Seed potato prices are 1.5 times higher than table potato prices especially when traded via wholesale buyers (intermediaries). In the spring time the prices goes up to 20-35 RU per kg because the cultivar's popularity influence the price more than the seed potato category and class. National seed potato prices are very often higher than foreign prices due to a wide range of intermediaries who first buy them. For example the foreign company seed potato price for the class A potato (which is the 1st generation after elite) comprises 350-380 € per ton (i.e. 13-14 RU per kilo). This is a reasonable price for the Russian potato producers by direct deals on the market because selling seed potato through intermediaries always doubles prices.

Nowadays it is most important to create an infrastructure to achieve a rational seed potato market.

New directions of seed potato production development

The improvement of scientific provision and the development of original, elite and certified seed potato production is planned to be introduced in the following directions:

- Creating an All-Russian Bank of Healthy Potato Varieties (BHPV) in the conditions of Archangelsk region which are most favourable and suitable for this purpose and clean in phyto-sanitary respect. This BHPV will be basic and a new source of original seed potatoes;
- Ensuring a faster adaptation of scientifically grounded schemes and technological requirements for the production of original, elite and reproduction seed potatoes on the basis of BHPV as applicable to various agro-ecological conditions of Russia;
- Adopting a modern certification scheme based on the application of strict requirements, norms and methods of seed control, with regards to their unification and harmonization with international requirements;
- Raising elite and reproduction seed potato production to a new level of quality in the volume necessary with regard to the requirements of farms and processing enterprises as well as peasant farms and private small farms (Figure 2)

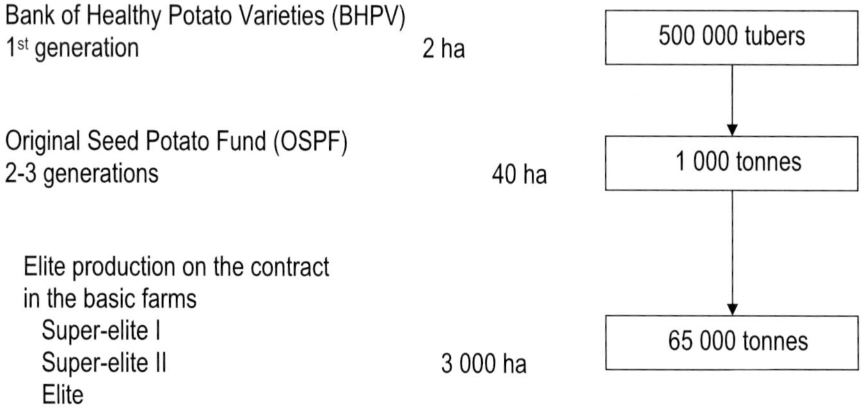

Figure 2. Provisional volumes of production on the basis of BHPV.

All the above will be facilitating the conditions for the creation of commercial seed production of the best domestic and international varieties at federal and regional level, the improvement of integration in the sphere of international seed potato trade between Russia, CIS and EU.

The exceptional value to Russia will be the overall improvement of potato yields on farms to the level of 18-20 tonnes/ha, reduction of losses by 25-30%, and the improvement of economic parameters of production. That will enable us to attract investments from non-sanitary sources (venture capital, entrepreneurs, interested foreign planters and other) into the development of infrastructure, renovation and modernization of the storage base, improvement of technical and technological levels in potato breeding.

The Ministry of Agriculture of the Russian Federation provides the following essential measures in order to improve potato production and marketing:

- to assist and support specialized potato growing farms as the major producers and suppliers of high quality seed and ware potato for large industrial centers, the northern regions and processing enterprises;
- to increase the efficiency of potato production in agricultural enterprises;
- to introduce new technologies;
- to establish special enterprises that will specialize in all-year-round growing of healthy basic material for elite potato;
- to pay more attention to providing private sector, farms and private households with high-quality seed material.

Curriculum vitae – Boris V. Anisimov

Boris V. Anisimov studied at the Agricultural University in Stavropol where he received a Master of Science degree in 1959. In 1966, after a short period of practical work, he was invited to the All-Russian Potato Research Institute, where in 1970 he received a PhD in Biology.

In 1970-1980 Boris Anisimov worked in the Potato Research Institute as a Head of the Department of Seed Potato Research. In 1981-1986 he worked as a Director of Scientific Pathological Laboratory (SPL USSR) of the Ministry of Agriculture, Ambo, Ethiopia. In 1987-1992 he worked as a Head of Breeding and Seed Potato Centre. In 1993-2003 - Head of Potato Section Programme Development at the Ministry of Agriculture of the Russian Federation. Since 2003 until present time - Deputy Director of All-Russian Potato Research Institute.

Boris V. Anisimov was an initiator and coordinator of seed potato research production and seed certification programme development in Russia, the breeding and seed potato programme development at the Potato Research Institute, was a team-leader responsible for the breeding and seed potato programme (1970-2003).

At the present time he is an initiator of the development of a new concept for potato seed production in Russia for the period 2005-2010. Boris Anisimov is an active member of the Russian Academy of Natural Science since 2001. He took part in national and international potato congresses including the Netherlands Congress in 1995, the Congress in Finland in 1999 as a member of the Organizing Committee, and the Netherlands Congress in 2005.

Company profile – All-Russian Potato Research Institute

All-Russian Potato Research Institute (VNIIKH) was established in 1930 in order to improve yield and quality of potatoes in Russia.

The main objectives of All-Russian Potato Research Institute are coordination of the national potato research and development programme; improvement of genetic variability and breeding of new varieties; production of virus-free original material and basic seed potatoes; improvement of potato technology (growing, harvesting and storage).

The Institute includes the following research departments: genetics, breeding, biotechnology, seed potato, plant protection, technology, mechanization, storage and processing, marketing and management.

Researches and specialists of the Institute are working in laboratories, greenhouses, trial plots, basic potato breeding farms, and implementing general projects in cooperation with other farms and organizations. The key activities of the Institute are the following: breeding new potato varieties adopted to the local conditions, resistant to diseases, best use of environmental and technological resources; production and delivery of original seed potatoes of high biological quality; improvement of certification and quality control methods; promotion of modern growing technologies; testing and promotion of pesticides used in potato production; assessment of foreign varieties for processing and their promotion in Russia; marketing studies, quality management studies and research of potato growing location and training of potato growers.

Main results of All-Russian Potato Research Institute include the following facts: The breeders of the Institute created more than 90 potato varieties, amongst which there are new high yielding disease-resistant and very popular varieties, such as Jukovskiy Ranniy, Udacha,

Lukianovskiy and others. The Institute annually produces about 80 000 micro-plants in vitro, over 1.000.000 mini-tubers on which basis the Institute's experimental farms produce about 10.000 tons of original seed material (pre-basic seed material of the varieties recommended for growing in different regions). Up to 2 000 diagnostic kits for testing potatoes for viruses and bacterial pathogens are annually produced for the seed certification programme development. Fifty young specialists and farmers and 120 regional trainers annually attend the practical training course organized by the Institute.

All-Russian Potato Research Institute (VNIIKH), Lorkh Street, Lyuberetskiy rayon, Korenevo, 140052, Moscow region, Russia

Production and marketing of potato in the process of full membership of Turkey to the European Union

Aziz Satana
Trakya University, Havsa Vocational College, 22500, Edirne, Havsa, Turkey

Abstract

Today Turkey is a developing country in agriculture and industry and also a candidate country for the European Union. There are 72 million people in Turkey and the youth constitutes most of the population. People in Turkey have a habit of eating cereal crops mainly. The production and the marketing of potato are important as an alternative to cereals in nutrition in recent years. In the 1950s the planted area was about 100 thousand hectare, the production and the yield of potato have increased gradually and the planted area has increased two fold and the production five fold. The planted area of potato in Turkey is approximately 200 thousand hectare and the production is 5 million t and Turkey ranks tenth of all European Union countries. In recent years in comparison with the planted area, the production has increased considerably and this has occurred as a result of increased potato yield per ha. Although potato yield is 26 t/ha today it isn't at the desired level of the European Union.

Most of the potato harvest is offered to the domestic market and the value is approximately $640 million. Some production is exported to foreign markets such as the Middle-East an Arabian countries. Potato is not imported in Turkey except for suitable seed potatoes that are needed from time to time. Although potato is grown in Turkey, important problems concerning suitable seeds haven't been solved yet. For this reason suitable seed potatoes are imported from European Union countries such as Germany, Austria, the Netherlands and also from the USA and even Canada. As most of potato varieties grown in Turkey are local varieties that are unknown in foreign countries, there are some problems regarding marketing, packing and transport and there is a wide-spread belief in European countries and especially in England that potato grown in Turkey is infected. There are privileges that European Union countries apply exemption from duty among themselves, the export is affected negatively leading to fluctuations in years. Since the planted area of potato is infected with diseases such as *Corynebacterium sepedonicum* and *Pseudomonas solanecearum*, this leads to export problems.

In 2004, potato wart *(Synchytrium endobioticum)* was determined in some places of the Nigde and Nevsehir provinces that are located in the middle of Turkey and where 60-70% of potato production takes place. The area was subjected to quarantine measures. Unfortunately this condition signifies that products will not be made in the coming 20-30 years and there will be problems in marketing.

Keywords: Turkey, EU, production, marketing

Introduction

Although potato species are found in America and many countries in the world as wild ones, cultured crops that produce tubers are encountered in South America only. The gene centre of potato has not been defined certainly up to now. But as a result of recent genetic studies, the gene centre of the potato has been accepted as the area from Chile to Mexico, Peruvian coasts and the Andes. Produced potato and eaten potato were first seen in 1537 by Spanish sailors in the Andes region in South America.

The potato was first brought to Europe in 1570 by Spanish sailors and in 1590 by English sailors from America. In 1745, they entered Germany and Austria from Italy and then its culture began. Afterwards it spread to the Netherlands and Belgium. Potato production has speedily increased in France since 1773. In the middle of the 18th century it entered Northern countries such as Norway, Sweden and Denmark from Scotland and its agriculture began (Hawkes, 1956).

The potato brought from South America in 1570 by Spanish sailors was first grown in Southern Spain and then taken to Italy and Portugal. The potato brought to England was first grown in Ireland and then they spread to Scotland and Wales (Langer and Hill, 1982; Simmonds, 1976; Incekara, 1978). After the potato had been recognized in Europe, the planted area of the potato expanded in the early 18th century and the consumption of the potato increased in Ireland. The records show that potato consumption of the Irish farmers exceeded 3 kg per person each day at the end of the 18th century. For this reason in some literatures the potato is called the "Irish Potato". In 1845, the consumption of potato rose to 4-6 kg in Ireland. But the potato mildew which grew into an epidemic caused the eradication of the potato, famine arose, and also led to the death of one million people, the unemployment of three million people and the immigration of one million people from Ireland to North America (Esendal, 1990; Arıoglu, 1997).

There are many points of view about the entree of the potato into Turkey. Turks began to grow potato in the Erzurum plain at the end of the 19th century. The people in Eastern Anatolia call the potato "Kartol" just as people in Russia say and this suggests that the potato entered to Turkey, from Russia and Caucasia to Eastern Anatolia and the Black Sea region (Ilisulu, 1986). According to some points of view, the first places where potato was grown are Sakarya and Adapazarı. Wherever the potato was first grown, it has a history in Anatolia for about 120-130 years.

The potato has had great importance as human food for years. When nutritional habits are investigated it is possible to define people eating wheat (Turkey etc.), those that eat rice (Far-East countries such as China and Japan) and those eating potato. European Union countries are included in the third group (Germany and the Netherlands etc.).

The reason why potato is consumed in human nutrition is that it has not only high nutritional quality but also has different ways of consumption. Carbohydrates formed in starches constitute the main part of the dry matter (10.96-22.13%). When compared with some vegetables, the percentage of dry matter, calorie value, starch, carbohydrate and protein levels are lower. Although the potato has a low percentage of protein, it is of very good quality.

Since it has the lowest calorie level in daily consumed nutrients and the lowest potassium concentration (507 mg/100 g in bread, 3 mg/100 g in potato) the potato is a good part of the diet.

The aim of this paper is to compare the potato production in Turkey and marketing with European Union countries and to benchmark Turkey's situation to potato production in the world.

Potato production in Turkey

As is shown in Table 1; the production of the potato in the world was 328 million t in 2004, planted area was 19 million ha and the yield was 17 t/ha. When we look at 5 years data, there were no important fluctuations both in planted area and in the yield. The significant countries where the potato is grown are respectively China (75 million t), Russian (37 million t), India (25 million t), USA (20 million t), Ukraine (19 million t), Germany (13 million t), Belarus (8 million t) and the Netherlands (7 million t) (FAO, 2005).

As is shown in Table 2, in 2004, the production of the potato of the European Union was 67 million t, planted area was 2 million ha, and yield was 28 t/ha. Five countries that have the largest planted area are those countries that produce the most potato. But this ranking changes in potato yield: Belgium replaces Poland that produces the most potatoes in the European Union (FAO, 2005).

The population of Turkey is about 72 million. According to the last five-year data, the average area planted with wheat is 9.5 million hectare. The production is based upon Turkish nutrition habits. Almost all Turks eat wheat or wheat products (primarily bread etc.); in daily nutrition bread is eaten during every meal. Nevertheless, the area planted with potato, the production and yield have increased in recent years.

Table 1. The potato production in the world in 2004.

Countries	Production (Million t)
Belarus	8.5
Canada	5
China	75
France	6.9
Germany	13
India	25
The Netherlands	7.5
Poland	15
Russian Federation	37
Ukraine	19.5
United Kingdom	6
USA	20.5

Table 2. The planted area and production of potato in significant EU countries in 2004.

Countries	Planted Area (thousand ha)	Production (million t)
Austria	23	685
Belgium	65	3,000
Czech Republic	43	1,000
Denmark	41	1,400
Finland	30	658
France	160	6,900
Germany	295	13,000
Greece	47	850
Hungary	31	650
Ireland	14	500
Italy	81	2,000
Latvia	54	739
Lithuania	110	1,600
The Netherlands	161	7,500
Poland	800	15,000
Portugal	80	1,250
Spain	99	2,800
Sweden	31	857
United Kingdom	140	6,000

As shown in Table 3, from 1950s to 2004 the planted area increased two fold, the production 5 fold and yield 2.8 fold.

In production Turkey places in the 13th rank in the world, and in the 6th rank in Asia. If Turkey had been a member of the European Union, it would have placed in the 6th rank in production and in the 3rd rank in planted area. But the yield of Turkey is below the average of the European Union. Turkey has to work hard to increase the potato yield; this work should be based upon science and technology in agriculture.

Although the planted area in 2003 was the same as in 2004, in 2004 the production decreased a little. The cause of this is that in Nigde and Nevsehir, which are the most significant potato

Table 3. The planted area, production and yield of potato in Turkey.

Years	Planted Area (thousand ha)	Production (million t)	Yield (t/ha)
1952-56	106	1	0.98
2000	205	5.3	26
2001	200	5	25
2002	198	5.2	26
2003	200	5.3	26.5
2004	200	4.8	24

production cities, potato ward (*Synchytrium endobioticum)* was observed. For this reason the Ministry of Agriculture had to apply quarantine to the planted area in some villages.

In Turkey, Middle-South Region takes in the first rank among agriculture regions for potato farming. The Black Sea and Middle-North follows this region in production proportion. Southeast region has the least potato production. In this region, the climate is too hot for potato production so it covers 1% of total production. Sanlıurfa, Siirt, Sırnak and Batman are the cities where potato is never grown and these cities are in this region.

On the other hand, potato farming is limited in the Mediterranean and Middle-East Regions. Because of hard and long winter and low annual precipitation in the Middle-East Region, potato needs irrigation. Otherwise it is impossible to obtain high yield but in these conditions potato farming in Sivas and Tokat is very important in the Mediterranean Region. Middle-South Region has the highest yield (33.6 t/ha). When South-East and Mediterranean Regions are excepted owing to disadvantages of the climates, yield in the Black Sea Region can be considered low. Because of poor seed and some problems with keeping seed, water shortage and infestations of *Erwinia* and nematodes, yield is very low in this region.

As shown in Table 4, Nigde has the highest yield among the areas, in the Mediterranean Region. The yield of Icel is 17.9 t/ha, in the Black Sea Region the yield of Ordu and Kastamonu is 12.5 t/ha the lowest yield (DIE, 2001).

According to the last five-year data there are not large fluctuations in the potato production for seed in Turkey. In 2000, 400 thousand t, in 2001, 396 thousand t, in 2002, 2003 and 2004 400 thousand t of potato for seed were produced (FAO, 2005). Since Bolu is located at high altitude, Izmir (Odemis), Erzurum and Ordu have levels of virus and other diseases; they are therefore important centres of seed production.

Table 4. The planted area and production of potato in the cities in Turkey.

Cities	Planted Area (thousand ha)	Production (thousand t)
Nigde	32.7	1.000
Nevsehir	24.6	784
Izmir	12.5	409
Bolu	10.8	279
Afyon	9.5	278
Ordu	9.5	115
Trabzon	8	145
Erzurum	6	122
Konya	5.6	108

Potato marketing in Turkey

Although potato amount that this exported and imported in the world fluctuates, it tends to increase. This appears because of general economic situation of the world and system changes in agriculture of some countries. In 2003, both potato import and export had about the same volume of approx. 9 million t and a value of approx. $ 2 billion.

Europe ranks first among the continents both for import (6.9 million ton) and export (6.8 million t). In export Asia (1 million t), North-Central America (780 thousand t), Africa (381 thousand t) and South America (29 thousand t) follow Europe. In import Asia (929 thousand t), North-Central America (834 thousand t), Africa (408 thousand t) and South America (74 thousand t) follow respectively Europe (FAO, 2005).

The most important party of world import is the European Union; these countries also have 80% of world export so European Union is the centre of world trade. In 2003, the export of European Union was 6.5 million tons and the value was $1.4 billion. The import of European Union was 5.8 million t and the value was $1.2 billion. The countries that have the most import in European Union are the Netherlands, France, Germany and Belgium, the export portion is respectively 1.7 million t, 1 million t, 680 thousand t, 585 thousand t and 560 thousand t (FAO, 2005).

Since in the European Union countries duty is 0% the countries which aren't members of European Union have no chance to export potato to European Union, countries but in the frame of customs Union arranged between Turkey and European Union, potato trade will have a new dimension with the disappearance of duty funds in crops in 2007. On the other hand potato varieties are well-known in EU countries and in quality and product standard people are accustomed to these varieties. Because they believe that these varieties are clean and reliable in diseases and insects.

It can't be said that Turkey has an important place in import and export. There have been significant fluctuations in export and import in the last five years as is shown in Table 5 (FAO, 2005). Political and economical instability in Turkey in 1999-2002 affected potato trade very much; especially in 2001 economical crisis and devaluation led to falls in import and soon after that in 2002 it led to falls in export.

Table 5. The export and import of potato in Turkey.

	Exports		Imports	
Years	Quantity (thousand t)	Value (million $)	Quantity (thousand t)	Value (million $)
1999	65	15	24	11
2000	140	25	11	5
2001	106	13	2	0.8
2002	25	2	12	6
2003	172	16	9	5

Potato export is affected negatively and there were fluctuations in years because of potato varieties in Turkey that are native and unknown in other countries. Problems concern standardization, packing and transport to England and other European countries, the widespread thought that Turkish potato was infected and the exemption from duty within EU countries. In production areas in Turkey there are some diseases such as Bacterial Ring Rot (*Corynebacterium sepedonicum*) and Erwinia (*Pseudomonas solanecearum*) in addition to these diseases new diseases sometimes appear in epidemic form; potato wart (*Synchytrium endobioticum*) seen in Nigde and Nevsehir led to important harms. In the following 20-30 years the infected area will be under the risk in potato production. Even if the potato is produced in this region it will be infected and it won't be able to be exported.

The most important part of potato production in Turkey is sold in the field or as soon as it is harvested. For this reason it is stored by the merchants. But the producer stores the seed for himself. As storing conditions aren't modern seed losses are 15-20%. In addition to this, 1% of the potato seed for the need is sold with a certificate (Atakisi, 1992).

The major part of potato in Turkey is consumed in domestic market, a small amount of it is exported to the foreign markets such as Middle East and Arabian countries. In the years when the production is low almost all of it is bought by the merchants so prices increase. In the years when the production is large the control is out of the merchants and prices decrease. In the last years potato has been consumed directly and also it has been processed as chips, puree and starch. Puree is a new variety in these processed products for Turkey and its production and consumption are limited (2 thousand t/year). The consumption of other products is increasing. For the spread of these products, nutrient habit, cheap packing and the supply of vegetable oil are import.

In the European Union countries potato tubers to be sold are transported in deep freeze train and refrigerated trucks. These vehicles arrive directly at selling points incurring 4-6% of losses.

Conclusion

If Turkey had been full member of the European Union, it would have been in the sixth rank with 5 million tons potato production. As the production potential is high, the Turks will produce more potatoes in the future. Then the problems of the potato production and marketing will increase. Now the population of Turkey is 72 million and it is estimated that the population of Turkey will be 78.4 million by 2010. But Turkey will plan its potato production and marketing. Thus, it will be able to solve the problem they encounter.

References

Arıoglu, H. (1997). Nisasta ve Seker Bitkileri (Starch and Sugar Crops). Çukurova Universitesi Ziraat Fakultesi Genel Yayın No: 188, Ders Kitapları Yayın No: 57. 5-11pp. Samsun, Turkiye.

Atakisi, I. (1992). Nisasta ve Seker Bitkileri Yetistirme ve Islahı (Growing and Breeding of Starch and Sugar Crops). Trakya Universitesi Tekirdag Ziraat Fakultesi, Yayın No: 93, Ders Kitabı No: 16, 60pp, Tekirdag, Turkiye.

DIE, (2001). Devlet Istatistik Enstitusu, Tarımsal Yapı (Uretim, Fiyat, Deger) 2001 (The State Institute of Statistics, Agricultural Structure (Production, Price, Value). ISSN 1300-963X, ISBN 975-19-3332-3, Yayın No: 2758, 215-219pp. Ankara.
Esendal, E. (1990). Nisasta Seker Bitkileri ve Islahı (Starch Sugar Crops and Breeding). Ondokuz Mayıs Universitesi Yayınları No. 49. ISBN 975-7636-06-1, Samsun, 17-26 pp.
FAO, (2005). Food and Agriculture Organization of The United Nations, FAO Statistical Databases. (FAO web pages).
Hawkes, J.G. (1956). Taxonomic Studies on the Tuber-Bearing Solanumus: I. Solanum Tuberosum and the Tetraploid Speceis Complex. Proc. Linn. Soc. London, 166, 97-144.
Ilisulu, K. (1986). Nisasta ve Seker Bitkileri ve Islahı (Starch Sugar Crops and Breeding). Ankara Universitesi Ziraat Fakultesi Yayınları: 960, Ders Kitabı: 279, Ankara.
Incekara, F. (1973). Endustri Bitkileri ve Islahı (Industry Crops and Breeding). Cilt 3. Nişasta-Şeker Bitkileri ve Islahı (2. Baskı), Ege Universitesi Ziraat Fakultesi, Yayın No: 101, Ege Universitesi Matbaası, Bornova, Izmir, Turkiye.
Langer, R.H.M. and G.D. Hill (1982). Agricultural Plants. Cambridge University Press., England.
Simmonds, N.W. (1976). Potatoes. Evolation of Crop Plants (Ed. N.W. Simmonds). Longman Inc., New York, USA. ISBN: 0-582-46678-4. 279-283 pp.